9th International Conference on Environmental Science and Development (ICESD 2018)

IOP Conference Series: Earth and Environmental Science
Volume 151

Paris, France
7 – 9 February 2018

ISBN: 978-1-5108-6394-1
ISSN: 1755-1307

Printed from e-media with permission by:

Curran Associates, Inc.
57 Morehouse Lane
Red Hook, NY 12571

Some format issues inherent in the e-media version may also appear in this print version.

Printed by Curran Associates, Inc. (2018)

For permission requests, please contact the Institute of Physics
at the address below.

Institute of Physics
Dirac House, Temple Back
Bristol BS1 6BE UK

Phone: 44 1 17 929 7481
Fax: 44 1 17 920 0979

techtracking@iop.org

Additional copies of this publication are available from:

Curran Associates, Inc.
57 Morehouse Lane
Red Hook, NY 12571 USA
Phone: 845-758-0400
Fax: 845-758-2633
Email: curran@proceedings.com
Web: www.proceedings.com

Table of contents

Volume 151

9th International Conference on Environmental Science and Development-ICESD 2018

7–9 February 2018, Paris, France

Accepted papers received: 2 May 2018
Published online: 4 June 2018

Preface

Papers

Emission Reduction and Climate Management

Resource Management and Sustainable Development

Environmental Biology and Biotechnology

IOP Conf. Series: Earth and Environmental Science **151** (2018) 011001 doi:10.1088/1755-1315/151/1/011001

PREFACE

It is my great pleasure to welcome you to the 9th International Conference on Environmental Science and Development (ICESD 2018) which was held in HOTEL NOVOTEL PARIS CRETEIL LE LAC, Paris, France during 7-9 February, 2018. ICESD 2018 was dedicated to issues related to Environmental Science and Development.

I, Prof. Miklas Scholz, from Faculty of Engineering, Lund University, Sweden, held the keynote speech with title "Wetland Systems Used for Flood and Diffuse Pollution Control: Integration of Hydro-environment, Energy and Decision Making". Prof. Khalil Hanna from Graduate Engineering School of Chemistry of Rennes, France had a keynote speech with title "Implications of Fe chemistry in environmental processes". Prof. Slav Hermanowicz from University of California at Berkeley, USA, held the keynote speech with title "Will Water Be the 'Oil of the 21st Century'? A quest for sustainable water management?". Prof. Susan Richardson from University of South Carolina, USA performed the keynote speech with title "Emerging Environmental Contaminants in Water: State of the Science".

There were eight presentation sessions in which Alvey Hau, Hameed Sulaiman, S M Lam, Parveen Fatemeh Rupani, Dhrumin Patel, Monica Dantas, Ankur Mehta and Ekaterina Golovina have received the best presentations awards. Many researchers, engineers, academics as well as industrial professionals from all over the world have presented their research results and development activities to end the whole conference with a great success.

The major goal and feature of the conference was to bring academic scientists, engineers and industry researchers together to exchange and share their experiences and research results, and discuss the practical challenges encountered and the solutions adopted. Professors from Sweden, France and USA were invited to deliver keynote speeches regarding latest information in their respective expertise. It has been a golden opportunity for students, researchers and engineers to interact with the experts and specialists to obtain their advice on technical matters as well as sales and marketing strategies.

The proceedings present a selection from papers submitted to the conference from universities, research institutes and industries. All papers were subjected to peer-review by conference committee members and international reviewers. The papers selected based on their quality and relevancy to the conference.

This volume presents to the readers the recent advances in the field of Environmental Science and Development as well as various related areas such as Environmental Science and Technology, Environmental Dynamics, Meteorology, Hydrology, Geophysics, Atmospheric Physics, Physical Oceanography, Global Environmental Changes and Ecosystems Management, Climate Changes, Global warming, etc..

We would like to thank all authors who have contributed to this volume and also the organizing committee, reviewers, speakers, chairpersons, sponsors and all the conference participants for their support to ICESD 2018.

Prof. Miklas Scholz
Faculty of Engineering, Lund University, Sweden
April 3, 2018

Conference Committee

Conference General Co-Chairs

Prof. Miklas Scholz, Faculty of Engineering, Lund University, Sweden
Prof. Slav Hermanowicz, University of California at Berkeley, USA

Program Co-Chairs

Prof. Khalil HANNA, Ecole Nationale Supérieure de Chimie de Rennes, France
Prof. Susan D. Richardson, University of South Carolina,USA
Assoc. Prof. Vasileios Drosos, Democritus Univeristy of Thrace, Greece

Contact Chair

Ms Sophia Du, Hong Kong Chemical, Biological & Environmental Engineering Society, Hong Kong

Technical Committee

Prof. Gabriel Lazar, Vasile Alecsandri University of Bacau, Romania
Prof. Hans-Uwe DAHMS, Kaohsiung Medical University, Taiwan
Prof. Bakhshal Khan Lashari, Mehran University of Engineering and Technology, Pakistan
Prof. CHARU LATA SINGH, AFFILIATED TO GGSIP UNIVERSITY, DELHI, INDIA
Prof. SURESH PRASAD SINGH, JHARKHAND RAI UNIVERSITY, RANCHI, INDIA
Prof. Judilynn N. Solidum, Unviersity of the Philippines, Manila, Philippines
Prof. J.Rajaraman, AMET University, Chennai Tamilnadu, INDIA
Prof. Tank Shantilal K, V.N.South Gujarat University,India
Prof. Gilmore G. Solidum, Pamantasan ng Lungsod ng, Manila, Philippines
Prof. Mona Gamal, Egypt-Japan University of Science and Technology, E-JUST, Alexandria,Egypt
Prof. Rogério de Mesquita Teles, Federal Institute of Education, Brazil
Prof. K. D. Ahire, Chhatrapati Shahu Institute of Business Education and Research, India
Assoc. Prof. MOHD RAMZI MOHD USSAIN, International Islamic University Malaysia, Malaysia
Associate professor Defri Yona, University of Brawijaya, Indonesia
Assoc. Prof. NORHA ABU HANIFAH, UNIVERSITI TEKNOLOGI MARA, Malaysia
Assoc. Prof. Leila Vafajoo, Islamic Azad University, Iran
Assoc. Prof. Behzad Kaviani, Islamic Azad University, Iran
Assoc. Prof. Nilofar Asim, National University of Malaysia
Assoc. Prof. Mohammad Shakir Nasif, Universiti Teknologi PETRONAS, Malaysia
Assoc. Prof. Abdullah Trad Sh. Al-fawwaz, Al al-Bayt University Mafraq, Jordan
Assoc. Prof. Christia Meidiana, Brawijaya University, Malang, Indonesia
Assoc. Prof. Plamena Zlateva, IR - Bulgarian Academy of Sciences, Bulgaria
Dr. Ranjan Sarukkalige, Curtin University, Australia
Dr. Kinga Szopińska, UTP University of Science and Technology, Poland
Dr. Serpil SAVCI, Bozok University, Turkey
Dr. Simon Chen, The University of Melbourne, Australia
Dr. Wael Mahmod, Egypt-Japan University of Science and Technology, Egypt
Dr. Sivakumar Nallusamy, Sultan Qaboos University, Oman
Dr. Hameed Sulaiman, Sultan Qaboos University, Oman

IOP Conf. Series: Earth and Environmental Science **151** (2018) 011002 doi:10.1088/1755-1315/151/1/011002

IOP Conf. Series: Earth and Environmental Science **151** (2018) 011003 doi:10.1088/1755-1315/151/1/011003

Peer review statement

All papers published in this volume of *IOP Conference Series: Earth and Environmental Science* have been peer reviewed through processes administered by the proceedings Editors. Reviews were conducted by expert referees to the professional and scientific standards expected of a proceedings journal published by IOP Publishing.

Carbon Footprints in Ecuador: Case of Riobamba city´s Bus Stations

M Córdova[1], D Cordova[2], F C Alvarez[1], M T Chaglla[1], P E Pico[1], L V Pérez[1]

1 G+ BioFood & Engineering Group, Department of Food Science and Engineering, Technical University of Ambato. Av. Los Chasquis & Río Payamino, Z.C. 180150, Ambato
2 SMP Schlumberger, Av. 12 de Octubre, Z.C. 170517, Quito
E-mail: lvperez@uta.edu.ec

Abstract. Analysis of carbon footprint and its environmental impact remains one of the most critical topics in air pollution research. Likewise, the effect of the transportation sector to emissions of air pollutants and greenhouse gases is a growing concern in developing countries, but, it is more complicated in undeveloped countries because of the lousy quality of fuels, exhaust technology and non-regulations of government control agencies. In the city of Riobamba, Ecuador, the growing population associated with the massive use of public transportation have all resulted in significant air pollution and greenhouse gas emissions. In this study, emissions were obtained with the Bacharach ECA 450 analyzer and used to calculate the emission factor of carbon dioxide and nitrogen oxide in kg/TJ. It was found that value of the emission factor of *CO2* was 57690, 13 kg CO_2/TJ. All Riobamba city bus stations contribute significantly to the carbon footprint. The factors that influenced the increase in the generation of greenhouse gases were: technology, the operation of buses and characteristics of fuel (diesel).

1. Introduction

Global climate change has become a significant problem to the environment of the world [1]. To address the problem, environmental factors must be considered in some different types of decisions made by international, national and regional policies on many issues [2]-[4]. The air pollution is a severe problem that affects all people around the world [5]-[7]. The World Health Organization (WHO) estimated that it is possible to assign more than two million premature deaths annually due to the combustion of fossil fuels [8]-[10]. According to the Ministry of the Environment of Ecuador (MAE), between the years 1990 and 2016, Ecuador has experienced an increase of 78, 70% of emissions of carbon dioxide (CO_2), principally generated by public transport [11]. Carbon footprint helps in emission management and evaluation of mitigation measures as a quantitative expression of GHG emissions [12]-[14]. Having quantified the emissions, the most important sources of emissions can be identified, and areas of emission reductions and increasing efficiencies can be prioritized [15], [16]. Globally, one of the most important pollutants in the air is transportation sector, dates shown that this area contributes about 25% of carbon dioxide (CO_2) emissions, with the road transport sector responsible for 80% [17]-[19]. Emissions of gases as traditional air pollutants make the road transport sector a suitable choice for investigating and propose procedures to reduce the air pollution and establish a climate change mitigation procedure [20]. Emissions to the atmosphere are determined by the quantity of fossil fuel consumed by the vehicle, vehicle technology, fuel quality and transportation land-use planning [21], [22].This research aimed to establish the carbon footprint at the Riobamba city bus stations under considerations of the ISO 14064.

2. Materials and Methods

2.1. Identification of Measurement Places and Emission Factors

Riobamba has different areas where buses have places for taking passengers. The areas considered for the investigation were: Santa Faz, San Miguel de Tapi, El Prado, La Dolorosa and Dávalos Market. GHG emissions were quantified and documented by ISO 14064 part 1. In the same way, considering sources of emissions, different scopes were determined. A) Direct GHG emissions (Scope I) were evaluated, this scope considers all cars that belonged to the bus station and transfer areas. B) indirect emissions of GHG (Scope II) consider all consumption of electricity that was generated in the building of the bus station and transfer areas. C) Other indirect GHG emissions (Scope III) take an account buses classified in exhaust technology such as Euro I, Euro II and Euro III which provide their services to the bus stations and areas of transfer of the organization.

2.2. Technology to Quantify GHG Emissions

A Bacharach ECA 450 analyzer, designed for analysis of combustion efficiency, was used to determine and quantify the exhaust gas composition in selected buses in Riobamba bus station, the measurements were developed according to the process described by Boruta, Imiołek [23] and ISO 14064-1. The analyzer measures the percentage contents of oxygen (O_2) and fuel gases (hydrocarbons [HC], the so-called unburned fraction) in the tested gas and the levels of carbon monoxide (CO), nitric oxide and nitrogen dioxide (NO and NO_2) and sulphur dioxide (SO_2) molecules per one million gas molecules and, on this basis, computes for the defined fuels (e.g. NG or LGP) the value of the air excess factor, the percentage carbon dioxide (CO_2) content and the number of nitrogen oxide (NO_x) molecules per one million gas molecules and the amounts of NO, NO_2, NO_x, CO and SO_2 molecules per one million O_2 molecules.

2.3. Sample Size

The sample for the research was calculated with a confidence level of 95% from a population of 709 buses. Likewise, a stratified sampling with proportional affixation was carried out according to the Euro technology to have different strata in which there are homogeneity and a minimum variance.

2.4. Calculation of GEI Emission

The factors corresponding to indirect GHG emissions (Scope II) by energy were obtained using the information provided by the Energy Minister of Ecuador according to the types of fuel used in the generation of electricity. For the calculation of indirect emissions, the factor 0.5062 tCO_2/MWh was used. The average of electrical energy consumption was transformed into emissions of carbon dioxide (t CO_2-e). The emission of carbon monoxide (CO *mg/m3*) in the Scope III with Bacharach ECA 450 analyzer were used for the calculation of the Kilograms of carbon dioxide (CO_2 *Kg*). Through stoichiometric estimates based on the chemical reaction (1) presented by Lipman and Delucchi [24] the nitrogen monoxide (NO mg/m^3) concentrations were transformed into Kilograms and finally the emission factor of CO_2 and N_2O were calculated.

$$2\mathrm{NO} + \mathrm{CO} \rightarrow \mathrm{N2O} + \mathrm{CO2} \quad (1)$$

For the calculation of direct emissions and other indirect GHG, the equation (2) was applied.

$$\mathrm{Emission} = \Sigma[\mathrm{DAa} * \mathrm{EFa}] \quad (2)$$

Where:
Emission = emissions of GHG (Kg)
DAa = fuel sold (TJ)
EFa = emission factor (Kg/TJ).
a = fuel type

IOP Conf. Series: Earth and Environmental Science **151** (2018) 012001 doi:10.1088/1755-1315/151/1/012001

2.5. *Calculation of the Carbon Footprint*

Once quantified the GHG emissions, there were transformed from kilograms of CO_2 to a ton of CO_2. At the end equation (3) was used to convert CO_2 ton to tCO_2-e. According to ISO 14064 part 1, a value of 1 t CO_2 –e for the 100-year global warming potential was used.

$$Emissions\ (tCO_2\text{-e}) = emission * 100\text{-}year\ global\ warming\ potential \quad (3)$$

Once the unit value of the emissions of scope I, scope II and scope III in t CO_2 -e were obtained, all the emissions of the same category were added together to obtain the total emissions of GHG global.

3. Results and Discussion

3.1. *Global Emissions of Greenhouse Gases (GHG)*

Results of GHG emissions were quantified and classified in scope I (direct emissions) and scope II and III (indirect emissions) considering the protocol reported in the standard ISO 14064:1. The most significant contribution of GHG emissions comes from scope III emissions. There were produced by the buses which were at the bus stations and areas of transfer. Results showed that emissions at scope I and scope II were not significant. It was attributable because of in scope I there was not an own vehicle to quantify the emissions. In scope II the consumption of electricity from the bus station building or management office were not possible because there was not building for this purpose.

3.2. *Emission Factors*

Emission factor was obtained from the data of activity and greenhouse gases (GHG) in kilograms, in this study an amount of 57690, 13 CO_2 kg/TJ was established, is it the most significant rate of emission factor obtained. The values of Nitrogen dioxide (N_2O) emission factor shows a value of 0, 08 kg/TJ, methane (CH_4) emission factor was 3, 90 kg/TJ; both were got from the bibliography. However, according to ISO 14064-1, sources of quantification such as equipment or concentration of pollutants are not enough to quantify and could be excluded. On the other hand, CO_2 emission factor average was 57690,13 kg/TJ; it was lower than the emission factor of the diesel which is 74100 kg/TJ [10]. In the same way, CO_2 emission factor generated to take into account the conditions of the technology of buses, geography location, and fuel quality (diesel) used by vehicles to transport people. Also, a GHG average that Riobamba bus station is affected by external situations; for instance, reduced technical support by governmental agencies, defective policy to control GHG´s emissions are reasons, too.

3.3. *Global Emission of Greenhouse Gases*

In Figure 1 present values of CO_2 emissions in function of bus technology shown a decrease. Firstly, Euro I has the maximum emissions because of it has an old technology. Secondly, Euro II technology demonstrated a considerable reduction of emissions which are attributable to changes in systems introduced in the exhaust pipe by manufacturers, to reduce a 60% of CO_2 emissions [11]. However, in Euro III there were little increase emissions despite are new technology, this may be due to the engine power system which is not according to rules for operation in Ecuador.

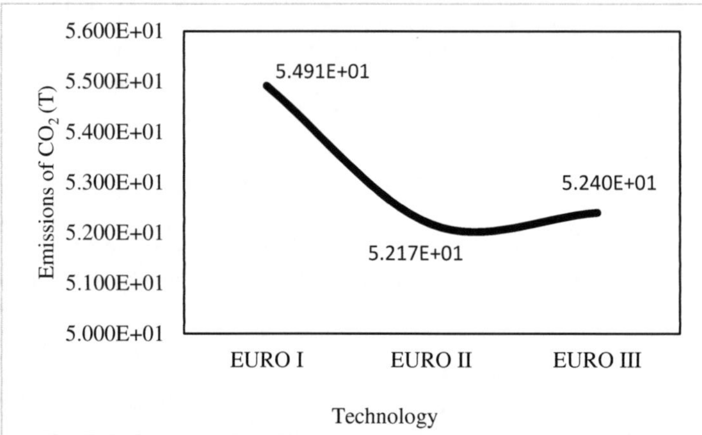

Figure 1. Average emissions of CO_2 by technology

Nitrous oxides are a part of most polluting GHG that exist in current days. As shown in Figure 2 the emissions of N_2O in tonnes increased to Euro II about Euro I, this could happen due to mechanical failure in one of the vehicles, while the Euro III technology decreased, since that this kind of vehicles had the modern and ecological technology to reduce emissions of N_2O.

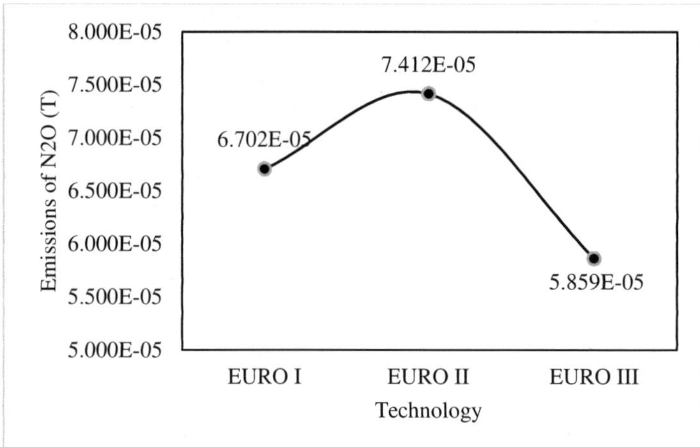

Figure 2. Average emissions of N_2O by technology

In Figure 3 was observed an increase of CH_4 according to the technology, which showed that the Euro III technology had not reduced the production of this gas, on the contrary, emissions have instead increased. This type GHG contributed with a 20% of the greenhouse effect [25]. The calculation of this emission was the theoretical emission factor obtained by the Intergovernmental Panel on climate change IPCC according to the type of fuel.

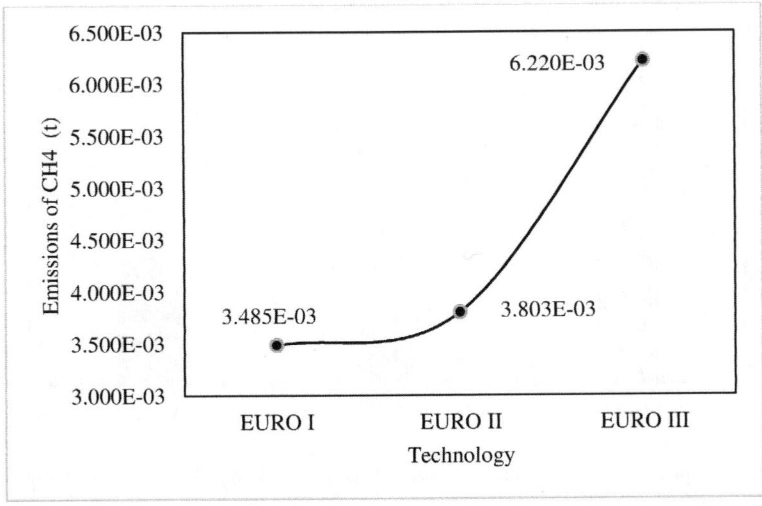

Figure 3. Average emissions of CH_4 calculated with the theoretical emission factor

Average emissions of SO_2 are shown in Figure 4. It shows a drastic decrease of this compound which varied according to the improvement of technology. Sulphur Dioxide would be lower if in Ecuador there would produce fuel with a few sulfur pollutants.

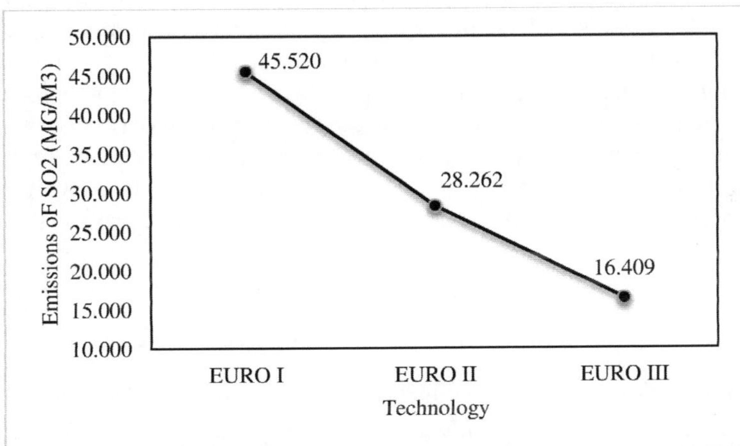

Figure 4. Average for emissions of SO_2 by technology.

3.4. *Assessment of the Carbon Footprint*

The emissions of scope III were those who presented the most significant contribution the global carbon footprint. Scope III was integrated by buses which operate in bus stations such as Santa Faz, San Miguel de Tapi, El Prado, La Dolorosa and Mercado Davalos, being these the primary sources of activity inside and outside of the bus station and affect all Riobamba city by a large number of network routes.

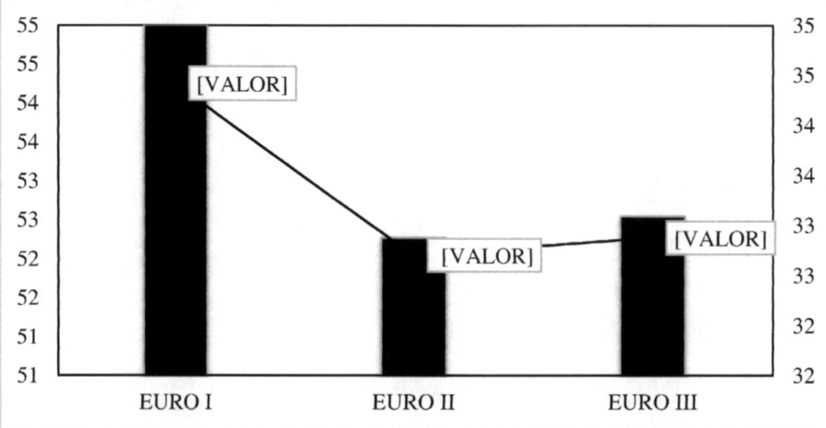

Figure 5. Average and percentage of the calculation of the carbon footprint obtained by the sum of the emissions of CO_2, N_2O, CH_4.

Figure 5 is shown the results of the Global carbon footprint. The carbon footprint emissions were decreasing as the technology improved. In EURO II there was a small decrease in comparison to EURO III. This decline was due to some EURO II buses have proper maintenance and regulation of the injection pump, while some EURO III vehicles were newly purchased and had not passed the review by National Transit Agency (NTA), by which their Jet pump could have been unregulated.

The most significant emissions contribute to the Global carbon footprint was given by buses with Euro II technology. Euro I represented 15% of total emissions, this was lower than other technologies despite that generated the highest emissions, this was due to that the 709 entire buses in Riobamba bus station, only 99 belong to Euro I technology. Table 1 displays the carbon footprint by scopes. The scope I and II did not exceed the emissions of the carbon footprint reported by the organization that controls the operation of the buses in this city (Interurbana S.A). However, the scope III corresponding to the vehicle fleet has the most significant quantity of emissions; this is attributable to the more considerable amount of buses which has Euro III technology and operate in Riobamba bus stations.

Table 1. Carbon footprint in tCO_2–e-relationship between the terminals of Riobamba and Interurbana S.A

Categorization	Riobamba bus station	Interurbana S.A.
Scope I	0,00	84, 07
Scope II	0,55	125, 29
Scope III	37388,23	18264, 40

4. Conclusion

The results in scope III shown emissions which contribute 99.99% in the global carbon footprint of the total GHG generated in the terminals. Similar effects of GHG were found in the construction sector in China, according to the study 72,1 % GHG were emitted [1]. According to ISO 14064, some GHG emission sources can be excluded because total emissions are not significantly, Scope I is an example. In the same way, some emission sources cannot be measured because instruments or supplies do not exist. Therefore, Scope II could be excluded. CO_2 emission factor value did not exceed to that reported by IPCC; it was due to fuel, not only quality but also geographic situation could change it, according to a new investigation transport of fuel can increase the GHG[26]. Likewise, some modern factors can increase this emission factor; for instance, vehicles or buses which are old, or it not are working according to international rules, altitude can be some other factor that together increase CO_2-e. Therefore, carbon fingerprinting is a powerful procedure that has been developed as a tool in the environmental assessment to fuel pollutants (fuel contaminants), and find the influence in GHG production [27]. It established that the carbon footprint of bus stations: Santa Faz, San Miguel de Tapi, El Prado, La Dolorosa and Mercado Dávalos, reported 37388.23 t CO_2–e value. This value was higher than the reported carbon footprint by Interurbana S.A. bus station in 2012, which was of 18264.40 tCO_2–e for the transportation sector (Scope 3). Finally, It investigation show us that the carbon footprint at bus station Riobamba city´s 37388. 23 tCO_2-e, where the 99.99 % concern to bus station terminals, it is similar to GHG´s emission generated by the electric energy consumption of 42000 single Ecuadorian homes per year.

5. References

[1] Chen W, Wu F, Geng W, Yu G. Carbon emissions in China's industrial sectors. Resources, Conservation and Recycling. 2017;117(Part B):264-73.

[2] Obrecht M, Knez M. Carbon and resource savings of different cargo container designs. Journal of Cleaner Production. 2017;155(Part 1):151-6.

[3] Yang C, Lambert P, Zhang G, Yang Z, Landriault M, Hollebone B, et al. Characterization of chemical fingerprints of unconventional Bakken crude oil. Environmental Pollution. 2017;230(Supplement C):609-20.

[4] Finnveden G, Hauschild MZ, Ekvall T, Guinée J, Heijungs R, Hellweg S, et al. Recent developments in Life Cycle Assessment. Journal of Environmental Management. 2009;91(1):1-21.

[5] Kishimoto PN, Karplus VJ, Zhong M, Saikawa E, Zhang X, Zhang X. The impact of coordinated policies on air pollution emissions from road transportation in China. Transportation Research Part D: Transport and Environment. 2017;54(Supplement C):30-49.

[6] Crippa M, Janssens-Maenhout G, Dentener F, Guizzardi D, Sindelarova K, Muntean M, et al. Forty years of improvements in European air quality: regional policy-industry interactions with global impacts. Atmos Chem Phys. 2016;16(6):3825-41.

[7] Shrivastava R, Neeta S, Geeta G. Air pollution due to road transportation in India: A review on assessment and reduction strategies. Journal of environmental research and development. 2013;8(1):69.

[8] Rohde RA, Muller RA. Air Pollution in China: Mapping of Concentrations and Sources. PLOS ONE. 2015;10(8):e0135749.

[9] Cohen AJ, Ross Anderson H, Ostro B, Pandey KD, Krzyzanowski M, Künzli N, et al. The Global Burden of Disease Due to Outdoor Air Pollution. Journal of Toxicology and Environmental Health, Part A. 2005;68(13-14):1301-7.

[10] Zhang J, Smith KR. Household Air Pollution from Coal and Biomass Fuels in China: Measurements, Health Impacts, and Interventions. Environmental Health Perspectives. 2007;115(6):848-55.

[11] Vega Y, Bravo D. Environmental index of the autonomous decentralized provincial governments of Ecuador. Ministry of the Environment of Ecuador. 2015:38.

[12] Barrett J, Scott A. Local Environment. The Application of the Ecological Footprint: A case of passenger transport in Merseyside. 2015;8:16.

[13] Stout SA, Wang Z. 3 - Chemical fingerprinting methods and factors affecting petroleum fingerprints in the environment. Standard Handbook Oil Spill Environmental Forensics (Second Edition). Boston: Academic Press; 2016. p. 61-129.

[14] Pandey D, Agrawal M, Pandey JS. Carbon footprint: current methods of estimation. Environmental Monitoring and Assessment. 2011;178(1):135-60.

[15] Benjaafar S, Li Y, Daskin M. Carbon Footprint and the Management of Supply Chains: Insights From Simple Models. IEEE Transactions on Automation Science and Engineering. 2013;10(1):99-116.

[16] Fang K, Heijungs R, de Snoo GR. Theoretical exploration for the combination of the ecological, energy, carbon, and water footprints: Overview of a footprint family. Ecological Indicators. 2014;36(Supplement C):508-18.

[17] Geng Y, Ma Z, Xue B, Ren W, Liu Z, Fujita T. Co-benefit evaluation for urban public transportation sector – a case of Shenyang, China. Journal of Cleaner Production. 2013;58(Supplement C):82-91.
[18] Chavez-Baeza C, Sheinbaum-Pardo C. Sustainable passenger road transport scenarios to reduce fuel consumption, air pollutants and GHG (greenhouse gas) emissions in the Mexico City Metropolitan Area. Energy. 2014;66(Supplement C):624-34.
[19] CAIT U. Climate Analysis Indicators Tool, Version 2.0. World Resources Institute, Washington, DC, USA. 2007.
[20] Jakhrani AQ, Rigit ARH, Othman AK, Samo SR, Kamboh SA, editors. Estimation of carbon footprints from diesel generator emissions. 2012 International Conference on Green and Ubiquitous Technology; 2012 7-8 July 2012.
[21] Colvile RN, Hutchinson EJ, Mindell JS, Warren RF. The transport sector as a source of air pollution. Atmospheric Environment. 2001;35:1537-8.
[22] Mayrhofer JP, Gupta J. The science and politics of co-benefits in climate policy. Environmental Science & Policy. 2016;57(Supplement C):22-30.
[23] Boruta G, Imiołek M, Piętak A. Test stand for a combined heat and power unit fed with alternative gas fuels. Journal of KONES. 2013;Vol. 20, No. 4:39-45.
[24] Lipman TE, Delucchi MA. Emissions of nitrous oxide and methane from conventional and alternative fuel motor vehicles. 2002:481.
[25] Christensen JH, Tomasi G. 16 - A multivariate approach to oil hydrocarbon fingerprinting and spill source identification A2 - Stout, Scott A. In: Wang Z, editor. Standard Handbook Oil Spill Environmental Forensics (Second Edition). Boston: Academic Press; 2016. p. 747-88.
[26] Hammond GP, Seth SM. Carbon and environmental footprinting of global biofuel production. Applied Energy. 2013;112(Supplement C):547-59.
[27] Alimi H, Ertel T, Schug B. Fingerprinting of Hydrocarbon Fuel Contaminants: Literature Review. Environmental Forensics. 2003;4(1):25-38.

IOP Conf. Series: Earth and Environmental Science **151** (2018) 012002 doi:10.1088/1755-1315/151/1/012002

Methane Emissions from Ecuadorian Hydropower Dams

M A Paucar[1], P I Amancha[2], T D San Antonio[3], L P Acurio[4], A F Valencia[4], C Galarza[4].

1 Department of Design, Architecture and Arts, Technical University of Ambato, Av. Los Chasquis & Río Payamino, Z.C. 180150, Ambato
2 Department of Industrial Design, Pontifical Catholic University of Ecuador-Ambato. Av. Manuela Sáenz, Z.C. 180207, Ambato
3 Department of Civil and Mechanical Engineering, Technical University of Ambato, Av. Los Chasquis & Río Payamino, Z.C. 180150, Ambato
4 G+ BioFood & Engineering, Department of Food Science and Engineering, Technical University of Ambato, Av. Los Chasquis & Río Payamino, Z.C. 180150, Ambato
Corresponding author: pamancha@pucesa.edu.ec

Abstract. Climate change is one of the most critical environmental problems nowadays. This change is caused by greenhouse effect, which is related to anthropogenic activities as gases production by fossil fuels and industrial production. Methane emission was evaluated in five Ecuadorian hydropower dams with the Intergovernmental Panel on Climate Change methodology. The average emission was 168,741 tCO_2eq per year in the five stations. The energy density of the Ecuadorian hydropower dams is higher than Brazilian with similar capacity; this result is due to the Ecuadorian's geography. Despite benefits of the hydropower, to mitigate methane emission is recommended to minimize the environmental impact.

1. Introduction

Energy generation is the main cause of greenhouse gas (GHG) emissions into the atmosphere, with fossil fuel combustion being the major source of these emissions. The most important activities are the use fossil fuels in transport, energy production, and industrial sectors [1]. Since the beginning of the industrial age (1750) until 2013, dioxide carbon (CO_2) atmospheric global average concentrations have increased 41%. The Same tendency is showed in methane (CH_4) concentration (160% higher) and nitrous oxide (N_2O) concentration (20% higher) [2].

CH_4 is a long-lived GHG because it persists in the atmosphere 21 times more than CO_2. About 40% of CH_4 global emission comes from natural sources and 60% of activities as rice cultivation, fossil fuels exploitation, landfills, biomass combustion and hydropower stations. Mainly of these activities are located in the tropical and middle latitudes of the northern hemisphere [2], [3]. CH_4 production in reservoirs has two origins: the decomposition of vegetation, which is accumulated at the bottom of the reservoirs (especially in tropical areas) and the high temperatures recorded in the regions where it belongs [4] [5]. Also, CH_4 emissions can be generated in aquatic ecosystems in three different ways: diffuse emission, bubble emission and degassing emission [6]. A part of the CH_4 is oxidized and converted to CO_2 when it rises to the surface of the reservoir mainly by bubbling. In tropical reservoirs, with little amount of water, the bubbles have less time to oxidize, that means they tend to produce higher CH_4 emission [7], [8]. These emissions occur exponentially in the early stages and cold water environments [9].

According to Organization for Economic Cooperation and Development (OECD), trophic status indicates a nutrient water enrichment, and this generally leads to symptomatic changes as increased aquatic plants production and water quality deterioration [10], [11]. Often, large and deep tropical reservoirs are thermally stratified. This thermal gradient is approximately 10 meters below the surface, which prevents diffusion of deep water and shallow waters [12]. This situation favours a CH_4 concentration profile that rapidly increases with depth until the local saturation level is reached. This profile follows a pattern that may differ between reservoirs or even in the same reservoir, and this depends on the amount of flooded organic matter, nonlocal inputs and water redox condition. These emissions account for about 1.6% of GHG worldwide emissions [6]. Also, CH_4 concentrations fluctuate over time with climatic variations as temperature and rainfall [11]. The northern reservoirs contribute CO_2 to global warming while tropical reservoirs add CH_4.

The IPCC has estimated that reservoirs and lakes are responsible for 22% of the total CH_4 worldwide emission [6]. The most significant CH_4 emission from reservoirs have been measured in warmer latitudes, and this is an argument against the use of hydropower. However, greenhouse gas emissions have been measured in only 18 of the 741 largest dams in the tropics.

Due to a large number of hydrographic basins in South America, hydropower stations have become the main source of electricity generation with a capacity of 161,793 MW [13]. In addition, 16% of the electricity consumed worldwide comes from hydropower [14]. Between 2005 and 2015 the total installed capacity has grown by 39%. This increase was primarily true in Latin American, Asian and African countries [15], and this trend will persist globally [16].

According to Department of Electricity and Renewable Energy (MEER) Ecuador had an installed hydraulic power (reservoir and run-of-river stations) of 2,219 MW by 2013. The most important project is Coca Codo Sinclair (Sucumbíos and Napo with 1,500 MW), which contributed to the National Interconnected System (SNI, acronym in Spanish) with a net energy of 3,605.41 GWh from April 2016 until January 2017. In Ecuador, 62.4% of the produced electricity comes from hydropower stations, followed by the 35.6% which is produced by thermal power stations [17], [18].

This study estimates the reservoirs impact of the hydropower dams in Ecuador have on the environment. CH_4 production was evaluated considering geography, weather, and water trophic status.

2. Methodology

To estimate CH_4 emissions, produced by Ecuadorian hydropower dams, IPCC methodology was used. This protocol classifies according to the types of emissions and depends on geography, climate, water quality and flooded characteristics areas [6]. The parameters required for the calculation are presented in Table 1, which includes daily average of diffuse emissions from the air-water interface (depending on weather of the area where the reservoir is located), and characteristics of CO_2 emissions in the environment. The calculation of these estimates depends on other factors that can alter the results, but these are not considered in this study.

Table 1. Reservoirs information.

Reservoir	Paute - Molino	Pisayambo – Pucará	Daule - Peripa	Amaluza	Compensado r
Province	Azuay	Tungurahua	Los Ríos, Manabí, Santo Domingo, Guayas	Cañar, Chimborazo, Azuay, Morona Santiago	Napo
Flooded Area (km^2)	9	5	340	16	3
Year of commissioning	2010	1977	1987	1983	2016
Dammed volume ($\times 10^6$ m^3)	410	90	6000	120	0,8
Climate (kg CH_4/ha×day)	Warm temperate wet (0.150)	Temperate cold wet (0.061)	Tropical dry (0.295)	Warm temperate wet (0.150)	Tropical very humid (0.630)
Water temperature (ºC)	12.9	13.4	26	15.4	20

Annual generation (GWh)	816	230	980	4,700	8,743
Trophic state	Eutrophic	Eutrophic	Eutrophic	Eutrophic	Oligotrophic

The main sources of GHG emissions in hydropower stations are installations and decomposing biomass in reservoir floods [19]; the latter being the main source of uncertainty. Also, their rate of decomposition is highly dependent, not only in the climatic zone, also in specific aspects of flooded biome. During the first year, one tonne of CH_4 emission into the atmosphere has 72 times more effect in global warming potential against one tonne of CO_2 emission at the same time, this is because CH_4 cycles in the atmosphere are faster than CO_2 cycles over a period of 500 years [6]. The comparison between CO_2 and CH_4 emissions is made with the equivalent of CO_2 (CO_2eq), and it is a measure used to compare greenhouse gas emissions based on their Global Warming Potential (GWP).

2.1. *Diffuse Emission*

This emission is emitted by molecular diffusion through air-water interface and it is estimated by every m^2 (equation 1). This emission is called Level 1.

$$CH_4\ emission = (P_f)\left(E_{(CH_4)diffuse}\right)(reservoir\ surface \times 10^{-6}) = \left(\frac{Gg}{year}\right) \tag{1}$$

In the equation, P_f stands for ice-free period (year); $E_{(CH_4)diffuse}$ stands for daily period of diffuse emissions.

2.2. *Bubble Emission*

This gas emission comes from the sediment-water interface and sediment through the bubble water column. This emission pathway is essential, particularly in temperate and tropical regions, and in reservoirs that have been flooded for a short time (equation 2). This emission is called Level 2.

$$CH_{4\ emissions} = (P_f \times E_{f\,(CH_4)diffuse} \times A) + \left(P_f \times E_{f\,(CH_4)bubble} \times A\right) + (P_i \times E_{i(CH_4)diffuse} + E_{i(CH_4)bubble}) \times A) = \frac{kg}{year} \tag{2}$$

In the equation, P_f stands for ice-free period (year); Pi stands for period covered with ice (year); A stands for flooded area. Equations 3, 4 and 5 should be used to calculate the other parameters.

$$E_{i(CH_4)diffuse} = daily\ average\ of\ diffuse\ emissions\ from\ the\ air - water\ interface\ during\ the\ ice - free\ period \tag{3}$$

$$E_{f\,(CH_4)bubble} = daily\ average\ emissions\ per\ bubble\ from\ air - water\ interface\ during\ the\ period \tag{4}$$

$$E_{i(CH_4)bubble} = daily\ emissions\ per\ bubble\ related\ to\ the\ period\ covered\ by\ ice \tag{5}$$

For this calculation, a table of values for the average daily diffuse emissions from air-water interface is performed according to the weather in the area where the reservoirs are located (Table 2). The results of these estimates should be taken with caution as entirely dependent on other factors that can alter the results [6].

Table 2. Average daily diffuse emission from air-water interface

Weather	Diffuse emissions [free period of ice (kg CH_4/ha×day)]		
	Half	**Minimum**	**Maximum**
Polar/boreal very wet	0.086	0.011	0.300
Temperate cold wet	0.061	0.001	0.200
Temperate warm wet	0.150	0.050	1.100
Temperate warm dry	0.044	0.032	0.090
Tropical very wet	0.630	0.067	1.300
Tropical dry	0.295	0.070	1.100

Comparing both methods, Level 2 is more accurate; however, it needs to carry a study with certain variables as sediments of reservoirs. To calculate gas flows from the sediments Fick diffusion formula can be used (equation 6). This is estimated with data obtained from reservoir sediments or lakes monitoring.

$$J = -\varphi(D_o\, \theta^{-2})\left(\frac{dc}{dz}\right) \tag{6}$$

In the equation, J stands for diffuse gas flow from the sediment-water interface; φ stands for porosity sediment; θ stands for tortuosity; D_0 stands for coefficient of gas diffusion in water.

2.3. Degassing Emission

This emission from reservoirs are due to a change in hydrostatic pressure, and the increasing variation of the air/water exchange surface after water passes through turbines or drain [4]. This is a fundamental channel of CH_4 emission from young tropical reservoirs, reaching 40% in a new-age reservoir [20].This emission is called Level 3. To know the emissions of this level it is necessary to make a reservoirs classification, where can be known if their emissions are important or insignificant through certain characteristics. In addition, some conditions should be given as reservoir shape, circulating way and water movement inside, which directly influences GHG emissions. The age of the reservoir is also preponderant; for example, if reservoirs are newly flooded the carbon present in leaves and trash, can be decomposed quickly, whereas the decomposition of the trunks and little biodegradable organic matter is realized later and slowness. The entrance of organic matter and nutrients to the reservoirs accelerates the proliferation of aquatic plants.

3. Results and Discussion

CH_4 contributes more than CO_2 to global warming because it has higher thermal potential. The CO_2 equivalents normalized emission used as a global warming factor is 100-year 23 kg_{CO_2}/kg_{CH_4}.

3.1. Diffuse CH_4 Emission in Ecuadorian Reservoirs

Table 3 shows the results obtained by calculating the diffuse emissions of CH_4 from the reservoirs. The values obtained for the Daule-Peripa reservoir were significantly higher, compared to those obtained for the rest of the analysed reservoirs. It has significant diffuse emissions because it is located in a tropical humidity zone, and temperature plays an important role in the process of eutrophication of organic matter entering.

Table 3. Diffuse emissions of CH_4 in Ecuadorian reservoirs.

Reservoirs	Gg CH_4/year	t CH_4/ year	t CO_2 eq
Paute-Molino	0.0876	87.6	1,839.6
Mazar	0.0493	49.275	1,034.78
Pisayambo	0.0111	11.133	233.78
Daule- Peripa	7.8183	7,818.3	164,184.3
Compensador	0.0690	68.985	1,448.69

3.2. Comparison of Diffuse CH_4 Emissions from Ecuadorian Reservoirs with Brazilian Reservoirs

Brazil has several investigations that provide relevant information about CH_4 emissions in hydropower dams [21]-[24]. For the estimation of diffuse emissions, between Brazilian and Ecuadorian reservoirs, the tropical zone, MW installed, and flood area (energy density) were considered.

A comparison was made with some Brazilian reservoirs (Tucuri, Samuel, Barra Bonita, Tres Marias, and Balbina), which are part of the First Brazilian Inventory of Anthropogenic Greenhouse Gas Emissions [25]. Table 4 shows that hydropower stations with a higher energy density have smaller flooded areas. This analysis indicated Ecuadorian reservoirs have higher energy density and fewer flooded areas compared to Brazilian with similar capacity.

The CH_4 total amount estimated emitted by reservoirs is 8.035 tCH_4/year, which is equivalent to 168,741 tCO_2eq (table 4). According to the Inter-American Development Bank (IDB), in 2011 the recorded CH_4 emissions of the energy sector in Ecuador added up to a total of 3.433 tCO_2eq; this leads to the conclusion that the analysed reservoirs in this study account 5% of the total emissions made by this sector. In this study, the CH_4 emissions calculated using the IPCC methodology for diffuse emissions provide a general idea about the analysed reservoirs. Also, this reinforces the assertion that the analysed hydropower dams emit GHGs, especially CH_4, because of damming water in their reservoirs. Publications regarding CH_4 emissions in reservoirs have increased the interest in this subject and validate the obtained results [26].

Table 4. Energy density (ED) values for the Ecuadorian (e) and Brazilians (b) reservoirs.

Reservoirs	MW	km^2	Reservoir age	ED MW/km^2
Paute-Molino (e)	1075	16	31	67.2
Compensador (e)	1500	3	0	500
Pisayambo (e)	74	8	37	9.3
Daule-Peripa (e)	213	340	27	0.6
Mazar (e)	160	9	4	17.8
Tucuri (b)	8370	2875	17	2.9
Samuel (b)	216	560	14	0.4
Barra Bonita (b)	140.76	334.3	38	0.4
Tres Marías (b)	396	1155	39	0.3
Balbina (b)	250	2360	19	0.1

In the medium term, it is expected that Ecuador can take measurements to reveal the real state of the existing reservoirs. Therefore, the GHGs that may be generated should not be dismissed, because they also contribute to global warming. The results obtained from this study show big differences between the energy density of Ecuadorian and Brazilian reservoirs. Even though, both countries are in the same tropical zone, their geography causes different flooded areas. Another important factor is the height of waterfall since they show whether the water requirements of hydropower projects are low or high.

4. Conclusion

According to the latest 2013 IPCC report, Latin America expects to experience significant impacts about climate change; for that reason, is essential that mitigation measures be implemented to reduce GHGs emissions. Based on hydropower, Ecuador has set policies for reducing these gases, which are aimed at using the existing water potential in the country to meet the projected medium-term demand. Today, hydropower stations are conceived as a clean source of electricity generation and are primarily implemented in emerging countries like Ecuador. Considering the existing research, this theory has weakened, since hydropower emits significant amounts of gases that can contribute to global warming. CH_4 emission from the reservoirs of the hydropower stations contribute significantly to global warming and this depends of station characteristics as geographic location, reservoir age, carbon and nutrients amount, water flow, rotation time, area, depth, water level fluctuations, turbines, and spillways location. For instance, in tropical areas, the emission is more than in temperate and boreal areas. Considering the Ecuadorian context, the creation of a model that allows estimating the emissions of any reservoir based on the station characteristics is recommended.

5. References

[1] Höök M and Tang X 2013 Depletion of fossil fuels and anthropogenic climate change—A review *Energy Policy* 52 797-809

[2] Team C W, Pachauri R and Meyer L 2014 IPCC, 2014: Climate Change 2014: Synthesis Report. Contribution of Working Groups I *II and III to the Fifth Assessment Report of the intergovernmental panel on Climate Change. IPCC, Geneva, Switzerland* 151

[3] Giles J 2006 Methane quashes green credentials of hydropower. Nature Publishing Group)

[4] Bastien J, Demarty M and Tremblay A 2011 CO_2 and CH_4 diffusive and degassing emissions from 2003 to 2009 at Eastmain hydroelectric reservoir, Québec, Canada *Inland Waters* 1 113-23

[5] Watanabe F S, Alcantara E, Rodrigues T W, Imai N N, Barbosa C C and Rotta L H 2015 Estimation of Chlorophyll-a Concentration and the Trophic State of the Barra Bonita Hydroelectric Reservoir Using OLI/Landsat-8 Images *International journal of environmental research and public health* 12 10391-417

[6] Change-IPCC I P o C 2006 Guidelines for National Greenhouse Gas Inventories: agriculture, forestry and other land use *Japan: Institute for Global Environmental Strategies (IGES)* 4

[7] Fearnside P M 2004 Greenhouse gas emissions from hydroelectric dams: controversies provide a springboard for rethinking a supposedly 'clean'energy source. An editorial comment *Climatic Change* 66 1-8

[8] Xiao S, Wang Y, Liu D, Yang Z, Lei D and Zhang C 2013 Diel and seasonal variation of methane and carbon dioxide fluxes at Site Guojiaba, the Three Gorges Reservoir *Journal of Environmental Sciences* 25 2065-71

[9] Lima I B, Ramos F M, Bambace L A and Rosa R R 2008 Methane emissions from large dams as renewable energy resources: a developing nation perspective *Mitigation and Adaptation Strategies for Global Change* 13 193-206

[10] Vargas J, De La Fuente L and Arumí J L 2012 Balance hídrico mensual de una cuenca Patagónica de Chile: Aplicación de un modelo parsimonioso *Obras y proyectos* 32-41

[11] Gunkel G 2009 Hydropower–A green energy? Tropical reservoirs and greenhouse gas emissions *CLEAN–Soil, Air, Water* 37 726-34

[12] Ramos F, Bambace L, Lima I, Rosa R, Mazzi E and Fearnside P 2009 Methane stocks in tropical hydropower reservoirs as a potential energy source *Climatic Change* 93 1

[13] IRENA2017 2017 Renewable Energy Statistics (Abu Dhabi: The International Renewable Energy Agency)

[14] REN21 2016 Renewables 2016 Global Status Report. (Paris: REN21 Secretariat)

[15] WEC 2016 World Energy Resources (London: World Energy Council)

[16] IEA-WEO 2016 World Energy Outlook, 2016. (Paris: International Energy Agency)

[17] Ministerio de Electricidad y Energías Renovables 2012 Plan Maestro de Electrificación 2012-2021. (Quito: Ministerio de Electricidad y Energñias Renovables)

[18] Ministerio de Electricidad y Energías Renovables 2012 Visión del Sector Eléctrico Ecuatoriano. (Quito: Ministerio de Electricidad y Energñias Renovables)

[19] de Lucena A F P, Szklo A S, Schaeffer R, de Souza R R, Borba B S M C, da Costa I V L, Júnior A O P and da Cunha S H F 2009 The vulnerability of renewable energy to climate change in Brazil *Energy Policy* 37 879-89

[20] Delmas R, Galy-Lacaux C and Richard S 2001 Emissions of greenhouse gases from the tropical hydroelectric reservoir of Petit Saut (French Guiana) compared with emissions from thermal alternatives *Global Biogeochemical Cycles* 15 993-1003

[21] Fearnside P M 1997 Greenhouse-gas emissions from Amazonian hydroelectric reservoirs: The example of Brazil's Tucuruí Dam as compared to fossil fuel alternatives *Environmental conservation* 24 64-75

[22] Demarty M and Bastien J 2011 GHG emissions from hydroelectric reservoirs in tropical and equatorial regions: Review of 20 years of CH_4 emission measurements *Energy Policy* 39 4197-206

[23] Ometto J P, Cimbleris A C, dos Santos M A, Rosa L P, Abe D, Tundisi J G, Stech J L, Barros N and Roland F 2013 Carbon emission as a function of energy generation in hydroelectric reservoirs in Brazilian dry tropical biome *Energy policy* 58 109-16

[24] Rosa L P, Dos Santos M A, Matvienko B, Dos Santos E O and Sikar E 2004 Greenhouse gas emissions from hydroelectric reservoirs in tropical regions *Climatic Change* 66 9-21

[25] Fearnside P M and Pueyo S 2012 Greenhouse-gas emissions from tropical dams *Nature Climate Change* 2 382

[26] Barros N, Cole J J, Tranvik L J, Prairie Y T, Bastviken D, Huszar V L, Del Giorgio P and Roland F 2011 Carbon emission from hydroelectric reservoirs linked to reservoir age and latitude *Nature Geoscience* 4 593

IOP Conf. Series: Earth and Environmental Science **151** (2018) 012003 doi:10.1088/1755-1315/151/1/012003

Carbon Sink Potential of *Avicennia marina* in the Al-Qurm Nature Reserve, Muscat, Oman

Abdullah Al-Nadabi[1] and Hameed Sulaiman[1]

1 P.O.Box 36,PC 123,Biology Department, Sultan Qaboos University, Muscat, Sultanate of Oman
E-mail: hameed@squ.edu.om

Abstract. Climate change is a global issue, caused by increased levels of carbon dioxide in the atmosphere resulting in many adverse effects. One of the many responses to the climate change impact is to sequester carbon dioxide from the atmosphere. Al-Qurm Nature Reserve in Muscat city of Oman is a mangrove vegetation protected by law as conservation area. Carbon sequestration was estimated in three distinct zones of this natural mangrove vegetation with *Avicennia marina*. The mean carbon stock in the middle zone was 7.7±0.3 kg C/m^2, and for the seaward zone was 5.3±0.7 kg C/m^2, and for landward 18.8±0.1 kg C/m^2, which is 3.5 times higher than seaward zone and slightly more than 2 times than middle zone. The 0.82 km^2 of Al-Qurm Nature Reserve occupied purely by *Avicennia marina* was estimated to sink about 8692 t C (0.0087 Mt C) equivalent to about 0.032 Mt of $CO_{2e.}$ These estimates suggest a high carbon storage and carbon sequestration potential of *Avicennia marina* in Al-Qurm Nature Reserve despite their relatively small area.

1. Introduction

Mangroves are unique ecosystems that are widely distributed along the coasts of tropical and subtropical regions [1], also survive in a wide range of salinity levels [2]. They provide several benefits to the environment and human beings [3]. Recently mangroves are recognized for their carbon storage and sequestration of carbon dioxide (CO_2) from the atmosphere [2], [4]. The amount of carbon that can be stored by mangrove forest is two to three times higher than that of terrestrial forests [5]. Large amounts of carbon can be stored by mangroves per unit area compared to many terrestrial systems [6]. So that, mangrove forests play a major role in the global carbon budget [7]. There are many species of mangroves, the specie *Avicennia marina* is widely distributed in many regions of the globe [8].

1.1. Mangroves in the Sultanate of Oman

Oman has a coast line of about 3165 km, connecting three seas that are the Arabian Gulf, the Sea of Oman and the Arabian Sea. Mangrove forests are located in thirty three sites along the coast of Oman with a total area of about 1,100 ha (11 km^2), or 0.0007 % of the total area of mangrove in the world. One of these sites is Al-Qurm Nature Reserve (QNR) which occupies 171 ha (1.7 km^2) of land within the Muscat capital area which contains about 82 ha (0.82 km^2) of mangrove trees. The Government of Oman designated this reserve area for conservation under Royal Degree No (38/1975). Also this reserve was registered as the first Ramsar site in Oman, and the accession to the Ramsar Convention entered into force on 19/8/2013.There is only one species of mangrove, *Avicennia marina* in Oman. This species can grow successfully in arid and harsh environment with high salinity [9].

1.2. Estimation of Carbon Sequestration

Estimations of forest carbon stocks are dependent on forest biomass [10]. Biomass is defined as the total of all parts of a tree consisting of aboveground biomass and belowground biomass including roots, trunk, leaves, branches, flowers and fruits. Actually, carbon is stored in five pools, called aboveground biomass (AGB), belowground biomass (BGB), dead wood, leaf litter and soil carbon stock in forest ecosystems [11]. The focus of this study was only on the first two pools namely above and below ground biomass. During photosynthesis, vegetation sequesters carbon dioxide (CO_2) from atmosphere and stores it as biomass in their tissues. Carbon stored as biomass is known as carbon stock. Biomass estimation of the forest ecosystems allows us to estimate the amount of carbon dioxide that can be sequestered from the atmosphere by the forest [10]. In order to estimate carbon sequestration we have to calculate the total biomass per hectare or other units of area and then convert this biomass to carbon equivalents using an appropriate conversion factors [12]. Such estimates are significant in modelling the potential consequences of climate change, and also for carbon accounting and monitoring requirements [13].Quantifying the mangrove biomass and structure are very important in order to address climate change adaptation and mitigation [14].The carbon content is calculated from biomass by multiplying by a factor of 0.48 (48 %) for aboveground and 0.39 (39 %) for belowground biomass [15]. Eventually, the amount of carbon dioxide (CO_2) that will be sequestered is estimated by multiplying carbon content of the mangrove ecosystem by 3.6663 [14].

2. Methodology

Studying carbon sequestration requires fieldwork for both aboveground biomass and belowground biomass estimation. Due to surrounding environments, working in mangrove forests is almost inaccessible making fieldwork and sampling a tedious task [14].During the past years, forest ecologists have developed different methods to estimate the biomass of forests [16]. One of these methods is to employ of allometric equations. This method are widely used by researchers as a non-destructive method to estimate mangrove biomass [16],[17]. Also, this estimation depends on measuring trees characteristics such as trunk diameter [18].

2.1. Use of Allometric Equations

The basic theory of allometric relationship is that in many organisms, the growth rate of one part is proportional to that of another. For example, the trunk diameter of a plant is highly correlated with trunk mass. Allometric equations for mangroves have been established for several decades to estimate biomass and subsequent growth [16].They are developed by creating relations between some easily measured individual plant parameter(s) and some variable which is much harder to measure [19]. Diameter at breast height (DBH) (i.e. 1.3 meter above the ground) is commonly and frequently used as predictive variable [13],[17],[20]. These variables are used in models (equations) and input are gathered through forest inventories [6].

Relationships have been developed by many researchers using diameter at breast height (DBH) to predict aboveground biomass of mangroves in many areas in the world [20]-[22]. In the case of multi-stemmed (or branched) mangroves such as *Avicennia marina*, every stem is treated as separate tree [14],[20].The logarithmic model (equation) gives the best explanation of the relationships between biomass (B) and diameter (D) i.e. $B= a \times D^b$ where a and b are constants compared to different forms of regression [17]. Allometric equations are also used to estimate belowground biomass for mangrove trees [14]. Most allometric equations focus on aboveground biomass, while a few belowground biomass functions exist for mangroves due to difficulties extracting mangrove roots and the labor-intensive nature of sampling belowground biomass in intertidal habitats [6],[23].

The study area was stratified in to three zones (i.e. landward- middle-seaward). A non –destructive method (i.e. allometric equations) based on the DBH of stems was used to estimate aboveground biomass and belowground biomass and eventually carbon storage. Table 1 shows the equations which are used in this study to estimate biomass.

Table 1. Allometric equations used to estimate biomass for *Avicennia marina* trees, B= biomass (Kg), D= diameter at breast height (cm).

Avicennia marina size class	Estimate	Equation	source
DBH > 4cm	Aboveground biomass	$B= 0.1776 \times D^{2.299}$	[16]
DBH > 4cm	Belowground biomass	$B= 1.28 \times D^{1.17}$	[10]
DBH < 4cm	Aboveground biomass	$B= 200.4 \times D^{2.1} \times 0.001$	[14]
DBH < 4cm	Belowground biomass	$B=0.92307 \times$aboveground biomass	[10]

3. Results and Discussion

3.1. *Estimation of Aboveground Biomass and Aboveground Carbon*

The trunk diameter of the trees were measured and then the equations were used to estimate the aboveground biomass and eventually aboveground carbon .The total aboveground biomass in the study sites was about 218.0 kg/m^2 and the total aboveground carbon was approximately 105.0 kg C/m^2. The average aboveground biomass and aboveground carbon content can be ordered as follows: landward > middle > seaward. The range of overall aboveground biomass was 4.1-28.7 kg/m^2 and aboveground carbon 2.0- 13.8 kg C/m^2. Table 2 shows the summary of estimated aboveground biomass and aboveground carbon of the three zones.

Table 2. Summary of estimated aboveground biomass and aboveground carbon of the three zones.

Zone	Aboveground biomass (kg/m^2)	Aboveground carbon (kg C/m^2)
Average seaward	6.5	3.1
Average middle	9.5	4.6
Average landward	27.6	13.3
Overall average	14.5	7.0

3.2. *Estimation of Belowground Biomass and Belowground Carbon*

Diameter at breast height of the tree trunk was used as variable to predict the belowground biomass by utilizing the allometric equations .The total belowground biomass in the study sites was estimated to be about 155.0 kg/m^2 and total belowground carbon about 60.0 kg C/m^2. The average belowground biomass and belowground carbon content can be ordered as follows: landward > middle > seaward. The range of overall belowground biomass was 4.2-14.7 kg/m^2 and belowground carbon 1.6-5.7 kg C/m^2. Table 3 shows the summary of estimated belowground biomass and belowground carbon of the three zones.

Table 3. Summary of estimated belowground biomass and belowground carbon of the three zones.

Zone	Belowground biomass (kg/m^2)	Belowground carbon (kg C/m^2)
Average seaward	5.7	2.2
Average middle	8.1	3.1
Average landward	14.3	5.5
Overall average	9.4	3.6

3.3. *Total Carbon Stock of Al-Qurm Nature Reserve*

The overall total mean carbon stock, including aboveground biomass and belowground of Al-Qurm Nature Reserve was estimated to be 10.6±1.6 kg C/m^2, which is equivalent to 38.9 kg CO_2/m^2.Although, Oman is located in arid and semi-arid region, mangrove trees have high potential to sequester and store higher amount of carbon dioxide from the atmosphere. The mean carbon stock in landward zone being higher than the seaward zone and middle zone may to be due to accessibility to fresh water facing Wadi(Dry stream) Aday that may supplies nutrients and encourages growth leading to more mature trees that have high circumference. The total area of *Avicennia marina* forest cover in Al-Qurm Nature Reserve is 0.82 km^2. The value of mean stock was extrapolated and estimated that a substantial sink capacity of 8692 t of C (0.0087 Mt), which is equivalent to 0.032 Mt of CO_2. Figure 1 shows carbon stocks of three zones in Al-Qurm Nature Reserve.

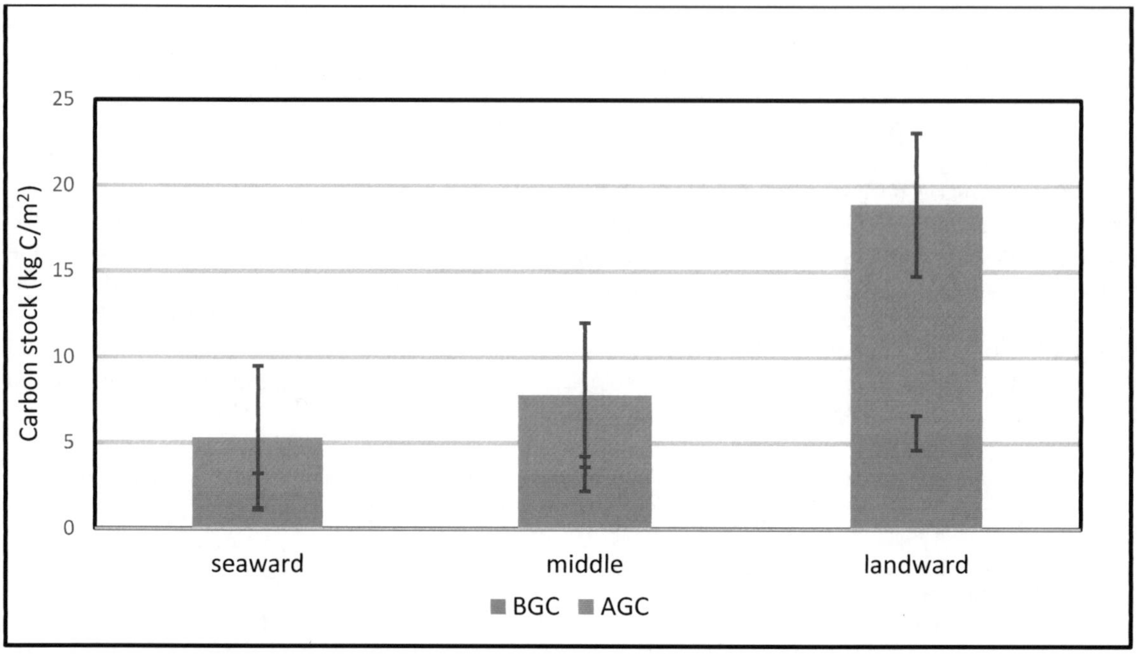

Figure 1. Carbon stocks of three zones in Al-Qurm Nature Reserve.

3.4. Relationship between above and below Ground Biomass

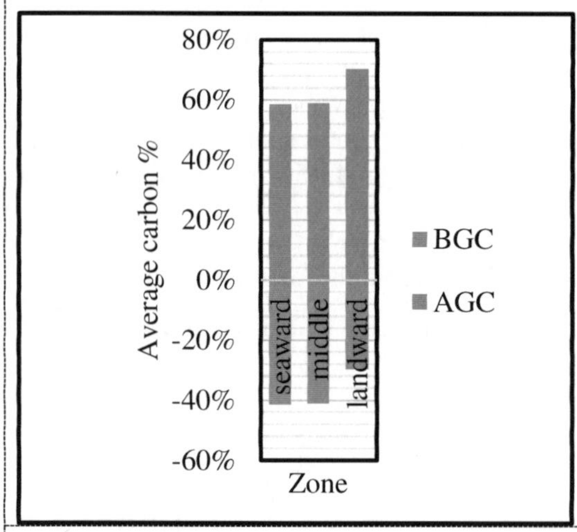

Figure 2 (a). Percentage of aboveground carbon to belowground carbon. (Note: minus sign here indicates belowground).

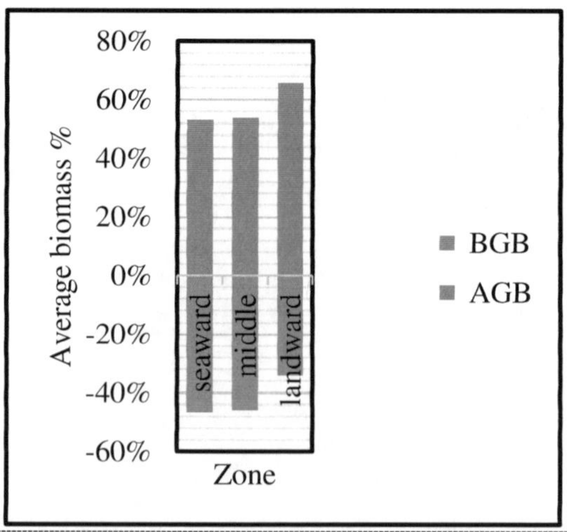

Figure 2 (b). Percentage of aboveground biomass to belowground biomass. (Note: minus sign here indicates belowground.

Figure 2 (a) and figure 2 (b) show the percentage of aboveground carbon to belowground carbon and the percentage of aboveground biomass to belowground biomass respectively. Tree carbon content distributions between aboveground and belowground showed variation between zones. In seaward and middle zones about 40% of the tree carbon and biomass content were in the belowground and about 60% were in the aboveground. However , in the moving to landward zone where larger trees are present with about 70% of tree biomass in the aboveground and only 30% in the belowground. It seems that mangroves in landward side have more matured trees compared to seaward and middle.

4. Conclusion

The three zones showed considerable differences in terms of its productivity to sequester carbon dioxide in their aboveground and belowground biomass. The result from this study concluded that landward zone has the highest value of aboveground carbon and belowground carbon, since more mature trees with high circumference are found in this zone. In contrast, the seaward zone has the lowest value for both aboveground carbon and belowground carbon.

5. References

[1] Santini NS, Reef R, Lockington DA, Lovelock CE. The use of fresh and saline water sources by the mangrove Avicennia marina. *Hydrobiologia*. 2015 Feb 1;745(1):59-68.

[2] Noor N, Shaukat SS, Naseem S. Evaluation of Elemental Composition of Avicennia marina and Associated Soil, Karachi Coast, Pakistan. *Journal of Coastal Research*. 2015 Sep 24;32(5):1142-8

[3] Kauffman JB, Heider C, Cole TG, Dwire KA, Donato DC. Ecosystem Carbon Stocks of Micronesian Mangrove Forests. *Wetlands*. 2011 Apr;31(2):343–52.

[4] Ilman M, Dargusch P, Dart P. A historical analysis of the drivers of loss and degradation of Indonesia's mangroves. *Land Use Policy*. 2016 Jul 31;54:448-59.

[5] Tue NT, Dung LV, Nhuan MT, Omori K. Carbon storage of a tropical mangrove forest in Mui Ca Mau National Park, Vietnam. *CATENA*. 2014 Oct;121:119–26.

[6] Njana MA, Eid T, Zahabu E, Malimbwi R. Procedures for quantification of belowground biomass of three mangrove tree species. *Wetl Ecol Manag*. 2015 Aug;23(4):749–64.

[7] Eid EM, Shaltout KH. Distribution of soil organic carbon in the mangrove Avicennia marina (Forssk.) Vierh. along the Egyptian Red Sea Coast. *Regional Studies in Marine Science*. 2016 Jan 31;3:76-82.

[8] Reef R, Markham HL, Santini NS, Lovelock CE. The response of the mangrove Avicennia marina to heterogeneous salinity measured using a split-root approach. *Plant and soil*. 2015 Aug 1;393(1-2):297-305.

[9] Al Saidi AM. Observational studies of canopy-scale variation in the microclimatic drivers of photosynthesis (Doctoral dissertation, Kagoshima University)2009.

[10] Vashum KT, Jayakumar S. Methods to estimate above-ground biomass and carbon stock in natural forests-A review. *J. Ecosyst. Ecogr*. 2012;2(4):1-7.

[11] Sahu SC, Kumar M, Ravindranath NH. Carbon stocks in natural and planted mangrove forests of Mahanadi Mangrove Wetland, East Coast of India. *Curr Sci*. 2016;110(12):2253.

[12] Kathiresan K, Anburaj R, Gomathi V, Saravanakumar K. Carbon sequestration potential of Rhizophora mucronata and Avicennia marina as influenced by age, season, growth and sediment characteristics in southeast coast of India. *Journal of coastal conservation*. 2013 Sep 1;17(3):397-408.

[13] Comley BW, McGuinness KA. Above-and below-ground biomass, and allometry, of four common northern Australian mangroves. *Australian Journal of Botany*. 2005 Aug 31; 53(5):431-6.

[14] Patil V, Singh A, Naik N, Unnikrishnan S. Estimation of Carbon Stocks in Avicennia marina Stand Using Allometry, CHN Analysis, and GIS Methods. *Wetlands*. 2014 Apr;34(2):379–91.

[15] Adame MF, Santini NS, Tovilla C, Vázquez-Lule A, Castro L, Guevara M. Carbon stocks and soil sequestration rates of tropical riverine wetlands. *Biogeosciences*. 2015 Jun 23;12(12):3805-18.

[16] Komiyama A, Ong JE, Poungparn S. Allometry, biomass, and productivity of mangrove forests: A review. *Aquat Bot*. 2008 Aug;89(2):128–37.

[17] Fromard F, Puig H, Mougin E, Marty G, Betoulle JL, Cadamuro L. Structure, above-ground biomass and dynamics of mangrove ecosystems: new data from French Guiana. *Oecologia.* 1998;115(1):39–53.
[18] Fayad I, Baghdadi N, Guitet S, Bailly JS, Hérault B, Gond V, El Hajj M, Minh DH. Aboveground biomass mapping in French Guiana by combining remote sensing, forest inventories and environmental data. *International Journal of Applied Earth Observation and Geoinformation.* 2016 Oct 31;52:502-14.
[19] Fan KC. Population structure, allometry and above-ground biomass of Avicennia marina forest at the Chishui River Estuary, Tainan County, Taiwan. *Quarterly Journal of forestry research.* 2008 Jun 1; 30(2):1-5.
[20] Clough BF, Dixon P, Dalhaus O. Allometric relationships for estimating biomass in multi-stemmed mangrove trees. *Australian Journal of Botany.* 1997;45 (6):1023-31.
[21] Clough BF, Scott K. Allometric relationships for estimating above-ground biomass in six mangrove species. *Forest Ecology and Management.* 1989 May 1;27(2):117-27.
[22] Saenger P, Snedaker SC. Pantropical trends in mangrove above-ground biomass and annual litterfall. *Oecologia.* 1993 Dec;96 (3):293–299.
[23] Abohassan RA, Okia CA, Agea JG, Kimondo JM, Morag M M. Perennial biomass production in arid mangrove systems on the Red Sea coast of Saudi Arabia. *Environmental Research Journal.*2012; 6 (1), 22-31.

IOP Conf. Series: Earth and Environmental Science **151** (2018) 012004 doi:10.1088/1755-1315/151/1/012004

Building of Platform for Development of Integrated Model to Assess Climate Change Impacts and Vulnerability

S H Chae[1] and M J Lee[1]

1 Korea Environment Institute (KEI), Sejong, Republic of Korea
E-mail: mjlee@kei.re.kr

Abstract. Climate change, by its nature, will affect various realms such as society, economy and the environment. In this sense, it is necessary to develop technologies to assess climate change impacts that can pre-emptively consider individual fields including health, water management, forest, agriculture and ecosystems as well as accidents and disasters, and the interaction among the fields in order to analyse climate change impacts. To provide a platform for the integrated assessment models to develop the climate change integrated impacts and vulnerability assessment models and address political issues, MOTIVE (Model of InTegrated Impacts and Vulnerability Evaluation of climate change) has been conducted as a part of a national R&D project since 2014. Under the project, the study built the common DB with 101 integrated items and 1,355 layers that can be commonly used by each field and the DB utilization system, and also developed a tool that displays the results of the climate change impacts and vulnerability assessment models by field. The tool maximized convenience by providing various analysis functions such as visualization, spatial distribution, bias and schematization of the vulnerability assessment results by field. Using the platform of the DB utilization system and the resulting display tool of MOTIVE, it is possible to compare and analyse the results of the climate change impacts and vulnerability assessment models by field, which is likely to be used as an effective means of decision making in the future. This platform will serve as a foundation for building the basis upon which integrated models can be utilized to assess climate change impacts and vulnerability in the future.

1. Introduction

The fifth IPCC report forecasted that the average temperature of the earth in the late 21st century will rise by 3.7°C and the sea level will increase by 63cm if greenhouse gas emissions remain unchecked [1]. This demonstrates clearly that climate change is one of the most threatening environmental problems facing mankind and the planet and must be addressed immediately [2]. South Korea is not exempt from this phenomenon. Under the Korea Metrological Administration's RCP 8.5 scenario, the temperature of South Korea will go up by approximately 5.7°C compared to the present by the late 21st century and most of the nation is expected to have a subtropical climate. Climate change, which is expected to become severe, will influence and wreak serious damage on the whole of human society including areas such as personal health and industrial profit making. Therefore, it is required to develop national policies to adapt to such changes and reduce emissions by assessing in advance what kinds of harm and impacts climate change will unleash and analyse vulnerable points so that the technical means can be created to support these policies [3],[4]. The concept of vulnerability to climate change has been defined by the IPCC [3], and this definition is most commonly used [5]. Although many researchers around the world have been studying climate change impacts and vulnerability assessment models, most of the studies are about assessment technologies regarding climate change impacts in a single field [6]-[8]. Since climate change, by its nature, will impact on various fields such as society, the economy and the environment, it is necessary to develop technologies in order to assess climate change impacts in a pre-emptive manner so that individual fields and the interaction among such fields can be analysed to accurately evaluate such impacts [9]. For this reason, research on integrated methodology to assess cross-sectoral climate change impacts, vulnerability and adaptation has been carried out globally, and some integrated assessment platforms have been developed. For example, the CLIMSAVE (Climate Change Integrated Assessment Methodology for Cross-Sectoral Adaptation and Vulnerability in Europe) project was implemented during 2010 to 2013 and developed such an integrated assessment platform for Europe [10]. The platform combines ten sectoral models, such as agriculture and forests, biodiversity, water resources, coasts, and urban development [11]-[13] and has been applied widely in climate change impacts, adaption and vulnerability assessment [14]-[15]. In the past, however, there had not been any research or technology development related to the climate integrated assessment platform in Korea. For this reason, a long-term and large-scale project named "MOTIVE (Model of InTegrated Impacts and Vulnerability Evaluation of climate change)" was launched in 2014 and is currently on going. The ultimate goal of this project is to build the Korean integrated assessment platform that can provide reliable scientific information to comprehensively assess vulnerable areas and risks that climate change will produce according to each field. To that end, the study built the impacts and vulnerability assessment models for each field such as health, water management, forest, agriculture and the ecosystem, and these models include accidents and disaster prevention. Currently, the study is seeking to build an integrated assessment platform based on the connection/integration of assessment results of each model. With regard to the first phase of this project, we focused on establishing a DB system to devise an integrated assessment model and model result display tool for project participants (Figure 1).

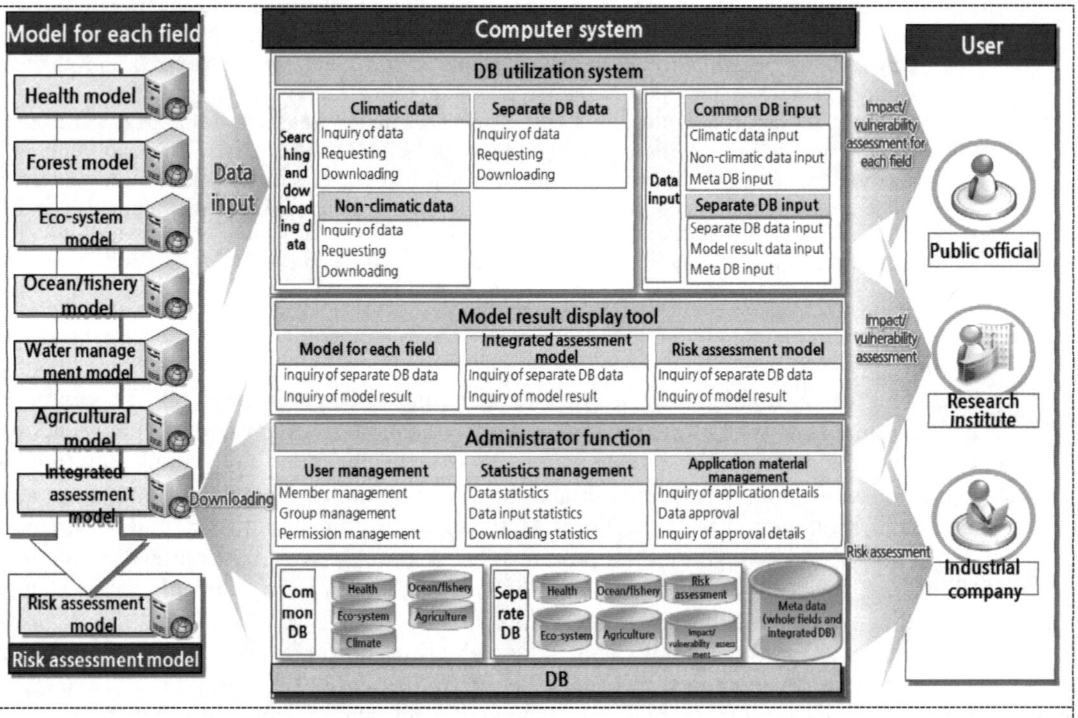

Figure 1. Block diagram of MOTIVE platform.

2. Development

2.1. Building DB system for MOTIVE

To develop a DB system for MOTIVE, a total of 620 literatures were investigated and 20 government organizations were interviewed. Based on the demand survey results, the integrated DB has been

IOP Conf. Series: Earth and Environmental Science **151** (2018) 012004 doi:10.1088/1755-1315/151/1/012004

constructed on 101 integrated items and 1,302 layers related to eight fields including health, agriculture, water management, forest and the ecosystem (Table 1). Also, we were continuing to push forward with standardization for collected DB details and material to make them easier to automate. Developing this system was performed in a web-based manner in order to share climate information from the common DB anywhere, anytime. The system was opened in April, 2016, and is expected to be constantly updated.

Table 1. Building quantity of non-climatic spatio-temporal DB.

Classification	Building quantity		Example
	Number of integrated items	Number of layers	
Health	38	689	Number of emergency medical institutions, people diagnosed with water-borne epidemics, etc
Agriculture	18	96	Apple production, rice production, soil map, etc.
Water management	3	42	Red tide occurrence data, aquatic ecology measurement data, etc.
Forest	27	132	Forest type map, statistics of forest products
Eco-system	2	3	Ecological zoning map, distribution status of endangered species, etc
Combination data	7	33	Land cover map, DEM, etc.
Political map	1	30	Political map
Social economy	5	330	GRDP (Gross Regional Domestic Product), Energy demand, etc.
Total	101	1,355	-

2.2. Building Result Display Tool for MOTIVE

The study developed a tool to display the results of the climate change impacts and vulnerability assessment models by field. The tool was designed to provide the DB and assessment results of the climate change impacts and vulnerability in the form of spatial information on the web. The process of the result display consisted of four stages of 1) data creation, 2) materials upload, 3) service registration and 4) view map (Figure 2). The display tool has three major functions of viewing the display data, uploading the display data and viewing map for each stage. Viewing the display data features the functions of searching data, viewing the list, and viewing and requesting the meta information. Uploading the display data features the functions of registering the meta information, selecting the coordinates, registering the legend information and registering the display data by year. The viewing map features functions including the basic functions of WEB GIS (Zoom In, Out and Pan, etc.) and split viewing the map to compare the vulnerability assessment results.

IOP Conf. Series: Earth and Environmental Science **151** (2018) 012004 doi:10.1088/1755-1315/151/1/012004

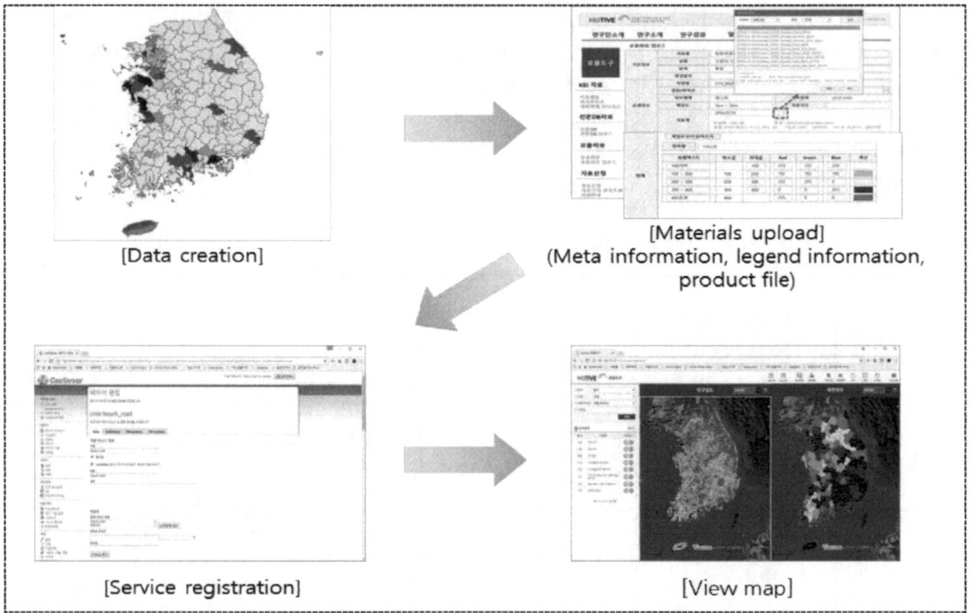

Figure 2. Result Display Process of MOTIVE Platform Using the Display Tool

3. Results and Discussion

The study built the common DB search and download system based on the common DB, which is currently in operation. The common DB search and download system can be used by logging on to the website of MOTIVE (http://motive.kei.re.kr/) and going through the following process: 1) search data, 2) request data, 3) approve data request (administrator) and 4) inquire and download the request result (Figure 3).

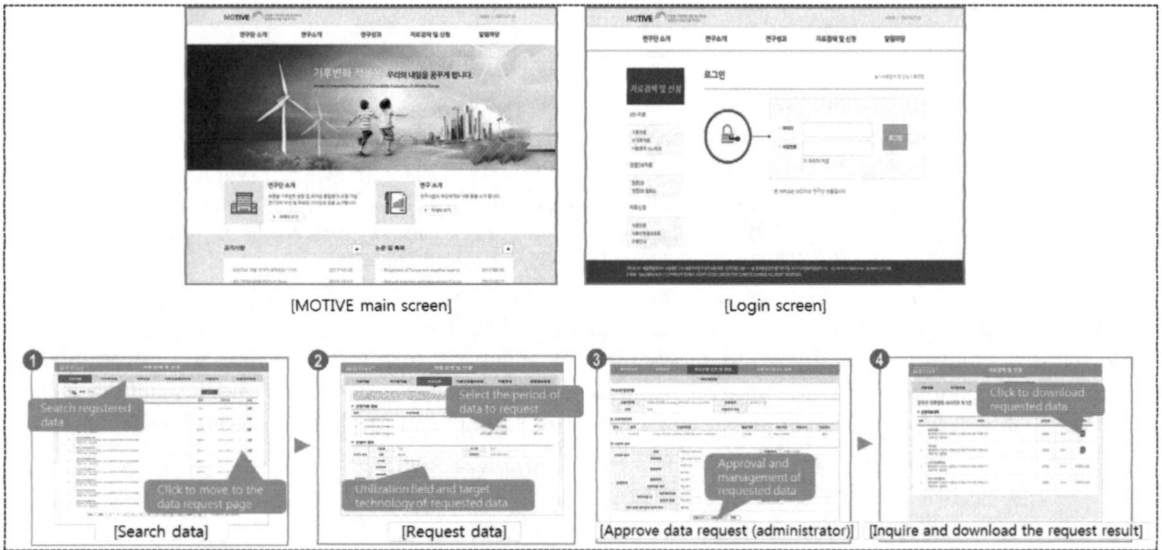

Figure 3. Process of Using the Common DB Search and Download System

As shown in Figure 4, it is possible to view and compare the vulnerability assessment results by field using the functions of viewing the map and split viewing the map to find out the results of the climate change impacts and vulnerability assessment models by field, which were registered by the process of the result display tool. In particular, it is possible to compare the results of the same data by year, different data of the same year and the same points through location-based linkage based on the compared vulnerability assessment results, and additionally view the statistical information. Therefore, it is considered that various decisions can be made using the results.

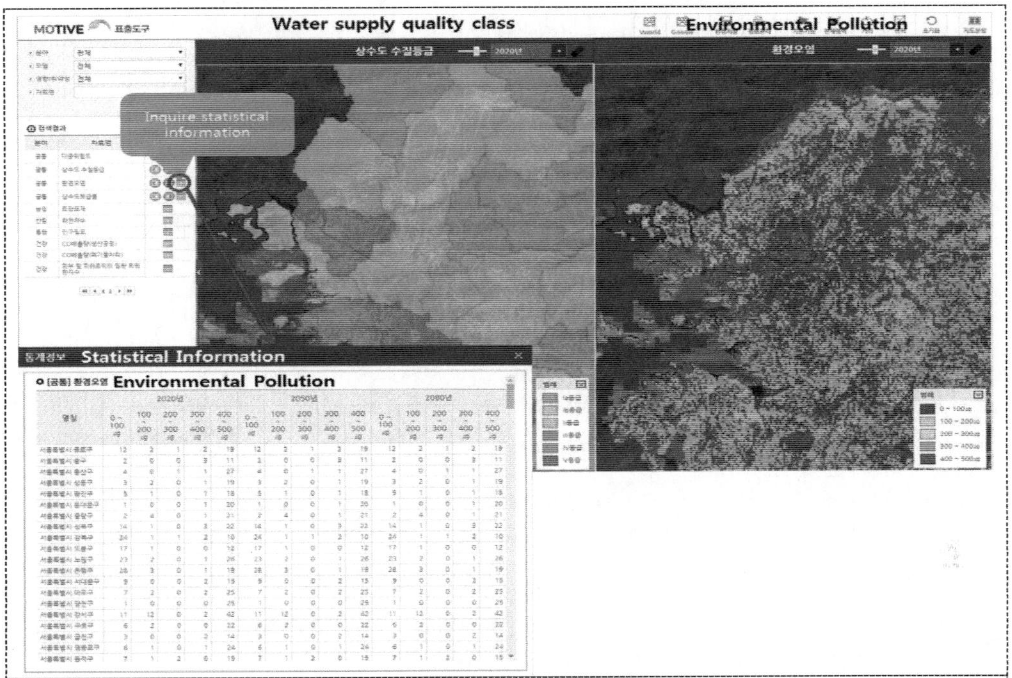

Figure 4. An Example of Comparison of the Impacts and Vulnerability Assessment Results by Field Using the Display Tool

4. Conclusions and Future Work

We have described the platform for the integrated assessment models to develop the climate change integrated impacts and vulnerability assessment models in the MOTIVE project. In order to effectively integrate models for each field, a common DB system with 101 integrated items and 1,355 layers was constructed, and a display tool was developed that can display different results effectively for each field. The expected outcomes and effects of the developed platform are as follows. The expected effects on the policy are development of foundation for efficient national climate change adaptation policy and minimizing mal-adaptation by prioritizing based on scientific analysis. On the industry and economy, this platform supports international support to developing countries via ODA and increasing employment of professional manpower and excavation of new industries for climate change adaptation. Moreover, it can provide scientific grounds for integrated management technology of climate change adaption and national risk management system for climate change. In the future, more meaningful functions will be developed to establish an integrated assessment platform by connecting and integrating the assessment results of each model.

5. Reference

[1] Intergovernmental Panel on Climate Change (IPCC) 2013 *Climate change 2013: the physical science basis: Working Group I contribution to the Fifth assessment report of the Intergovernmental Panel on Climate Change* Cambridge University Press UK pp 1-40

[2] Intergovernmental Panel on Climate Change (IPCC) 2014 *Climate Change 2014: Impacts, Adaptation, and Vulnerability: Regional Aspects* Cambridge University Press UK pp. 1–32

[3] Intergovernmental Panel on Climate Change (IPCC) 2007 *Climate change 2007: impacts, adaptation and vulnerability* Cambridge University Press UK pp. 1–30

[4] Füssel H T 2007 Vulnerability: a generally applicable conceptual framework for climate change research *Global environmental change* 17 2 155-167

[5] Bolin B A 2007 *history of the science and politics of climate change* Cambridge University Press UK pp. 1–42

[6] Moss R H; Brenkert A L; and Malone E L 2001 *Vulnerability to climate change: a quantitative approach.Pacific Northwest National Laboratory operated by Batelle for the US Department of Energy* US pp. 155-167

[7] Thirumalaivasan D, Karmegam M, and Venugopal K 2003 AHP-DRASTIC: software for specific aquifer vulnerability assessment using DRASTIC model and GIS *Environmental Modelling & Software* 18 7 645-656

[8] Krysanova V, Hattermann F, and Wechsung F 2005 Development of the ecohydrological model SWIM for regional impact studies and vulnerability assessment *Hydrological Processes* 19 3 763-783

[9] Harrison P A., Holman, I P, Cojocaru G, Kok K, Kontogianni A, Metzger M J, and Gramberger M 2013 Combining qualitative and quantitative understanding for exploring cross-sectoral climate change impacts, adaptation and vulnerability in Europe *Regional Environment Change* 13 4 761–780

[10] Harrison P A; Holman I P; and Berry P M 2015 Assessing cross-sectoral climate change impacts, vulnerability and adaptation: an introduction to the CLIMSAVE project *Climatic Change* 128 153-167

[11] Wimmer F, Audsley E, Savin C M, Malsy M, Dunford R, Harrison P A, Schaldach R, and Flörke M 2015 Modelling the effects of cross-sectoral water allocation schemes in Europe *Climate Change* 128 229–244

[12] Audsley E, Trnka M, Sabaté S, and Sanchez A 2015 Interactive modelling of land profitability to estimate European agricultural and forest land use under future scenarios of climate, socio-economics and adaptation *Climate Change* 128 215–227

[13] Dunford R, Smith A, Harrison P A, and Hangau D 2015 Ecosystem service provision in a changing Europe: adapting to the impacts of combined climate and socio-economic change *Landscape Ecology* 30 443–461

[14] Harrison P A, Dunford R, Holman I P, and Rounsevell M D A 2016 Climate change impact modelling needs to include cross-sectoral interactions *Nature Climate Change* 6 885–890

[15] Dunford R, Harrison P A, Jäger J, Rounsevell M D A, and Tinch R, 2015 Exploring climate change vulnerability across sectors and scenarios using indicators of impacts and coping capacity *Climate Change* 128 339–354

Acknowledgments
The authors appreciate the support that this study has received from "Development of integrated model for climate change impact and vulnerability assessment and strengthening the framework for model implementation", an initiative of the Korea Environmental & Industry Technology Institute.

Politicization of Global Warming and Energy Restructuring in China

SHI Xiaojin[1],WU Jinxi[2]

1 Assistant Researcher, School of Social Science,Tsinghua University, Beijing, China
2 Associate Professor,School of Social Sciences, Tsinghua University, Beijing, China
E-mail:selina4757@126.com

Abstract. Among the problems concerning climate environment, global warming is one of them drawing the highest attention all over the world. The basic assumption is that the concentration of carbon dioxide in the atmosphere is positively correlated with global temperature. Under the premise of assuming that the concentration of carbon dioxide provided by the nature is constant, carbon dioxide generated from human activities is the crime culprit of global warming. In fact, from the scientific research point of view, a high-degree consensus has not yet been reached in respect to the global warming issue. Since the end of 1980s, it has converted from undetermined statistical prediction in various aspects to certain definitive well-recognized international consensus and become an important factor in the international political game through all efforts of political powers. In recent years, China is devoted to adjust energy structure and playing the role as a big power in terms of emission reduction and limitation. As a result, the leadership in renewable energy sources is quietly transferred from US to China. The move allows China to take the moral high ground and intensify its strength and standing in global governance.

1. Introduction
Among the issues concerning the climate environment, global warming is one of them drawing the highest attention all over the world. The basic assumption is that the concentration of carbon dioxide in the atmosphere is positively correlated with global temperature. Under the premise of assuming that the concentration of carbon dioxide provided by the nature is constant, carbon dioxide generated from human activities is the crime culprit of global warming[1]. The IPCC Climate Change Impact Report in 2007 gives detailed scientific explanation: since 1750, human beings began to use fossil fuels in a large scale way, and the concentration of the greenhouse gas mainly contained in the atmosphere rose from 292ppm prior to the industrial revolution to the 379ppm at present, far more than the record (180 to 330pm) for 650,000 years before Antarctic ice core, resulting in that the average temperature on the earth surface between 1906 and 2005 rose by 0.74°C. Meantime, it also pointed out that, the global temperature currently was relatively warm compared to the past five hundred years, even one thousand years. The warm climate will trigger series of natural hazards resulting in human destruction. This kind of argument with resort to human's fear gradually won great popular support as both media and governments repeatedly propagated and stressed in 1970s~1980s.

2. Global Warming Issue: From Scientific Prediction to Political Game
In 1896, Svante Arrhenius, a Swedish chemist, was the first to put forward the calculation report of "carbon dioxide (CO_2) leading to global warming" and pointed out that the Earth's average temperature will vary with change of content of carbon dioxide in atmosphere[2]. Inspired by him,

David Keeling, an American chemist, concluded through a large number of measurements that "human-caused carbon dioxide emissions are large enough to cause global warming"[3]. By 1970s~1980s, a great number of scientists made demonstration for such conclusion in their books and works, so that the argument of "global warming" gradually won great popular support. Since global warming is involved in humane safety, all kinds of powers started participating in study on global warming and resistance to further deterioration of global warming. In 1988, TPCC was founded, and global warming began to convert from undetermined statistical prediction in various aspects to certain definitive well-recognized international consensus. Through all efforts of political powers, it became an important factor of the international political game.

2.1. Scientific Prediction: Arguments Always Exist

In fact, from the scientific research point of view, a high-degree consensus has not yet been reached in respect to the global warming issue. Especially at the end of 2009, several top-class climatologists' emails and documents were hacked and made public, known as "Climategate" incidence, resulting in heavy discount of the credibility of climate warming. In terms of main arguments for global change theory: Earth temperature is positively related to concentration change of carbon dioxide, and increase of CO2 concentration will cause Earth temperature rise. Some scientists once made the temperature data reconstruction analysis, but it's difficult to judge whether increase of CO2 concentration caused the Earth's temperature rise, or the other way around. Siegfried Fred Singer believed that "However there is no evidence that this high CO2 is making a detectable difference." It should in principle, however the atmosphere is very complicated and one cannot simply argue that just because CO2 is a greenhouse gas it causes warming[4]".Dennis T. Avery believes that the global warming is part of a natural cycle and therefore unstoppable[5]. Water vapor has the action as high as 60% to global greenhouse effect, while the effect by carbon dioxide only accounts for 20%[6]. Neftel A and Oeschger H from Physical Laboratory Institute of Bern University studied the relationship between carbon dioxide concentration and temperature change using Byrd time scale and found that change of carbon dioxide concentration in the atmosphere clearly lagged behind the change of global temperature in the period between 200 and 1200 years[7]. Dominique Raynaud[8], et al from France, Hubertus Fischer[9], et al from Wagner Institute for Polar and Marine Research, and Eric Monnin[10], et al from Switzerland did the corresponding research on change of C02 concentration and temperature change. These research contents showed that carbon dioxide concentration change lagged behind the Earth's temperature change with lag phase from hundreds of years to a thousand years.

In addition to the temperature data reestablishment analysis, the climate change rule and its effect on human life can also be analyzed from archeology point of view. Chu Kochen, a Chinese meteorologist (1890~1974), wrote an article named Initial study on climate change in China for five thousand years in 1972 through study on the climate change using abundant historical materials, which mentioned that China underwent four warm periods and four cold periods in total during five thousand years of history and both cold and warm periods appeared alternately. The following conclusions were made: (1) In the first two-thousand years of the nearly five thousand years, i.e., from Yangshao culture to Anyang Yin Ruins, the annual average temperature in most of time was about 2°Chigher than that at present. The temperature in January was about 3°C～5°C higher than the temperature at present. During the period the temperature went ups and downs, nevertheless, it's impossible to discuss more since the study was only limited to documentation. (2) From then on, the temperature went ups and downs, with the minimum temperature appearing in the years of respectively around BC1000, 400, 1200, 1700, with the fluctuation range of 1°C～2°C. (3) Each period from four hundred to eight hundred years can be divided into small cycles in a unit of fifty to one hundred years per cycle with the temperature range indicating 1°C～0.5°C. The institute of Geographic Sciences and Natural Resources Research of CAS put forward in 2010 that, China had warm period at Western Han and Eastern Han Dynasties (BC210-AD180), warm period at Sui and Tang dynasties (AD570-780), warm period at Song and Yuan dynasties (AD950-1300), warm period at the contemporary age (AD1921-), where the warmth during Sui and Tang dynasties was slightly higher than modern warm period.

More interestingly, the four periods were distributed in the most prosperous duration of Chinese historical culture and China did not suffer any severe catastrophic disaster due to temperature rise.
China Five-thousand-year Civilization left us abundant historical materials, including some records about the emperor's spring ploughing. In ancient China, the emperors' spring ploughing was an important state celebration, so the corresponding record was also prudent and accurate. This could prove that the temperature at Sui and Tang dynasties was slightly warmer than nowadays.
For example, as recorded in Old Book of Tang Biographic Sketches of Emperor Taizong,

> *"In Kuihai of lunar January of the third year of Zhenguan, Emperor Taizong farmed in Xi'an City in person known as JiTian Ceremony".*
>
> *As recorded in New Book of Tang Biographic Sketches of Emperor Xuanzong, "in Yihai of lunar January of the twenty-third year of Kaiyuan, the emperor offered sacrifices to the agricultural ancestor in person".*
>
> *As said in History as a Mirror Tang Record," in Yimao of lunar January of the second year of Ganyuan, the emperor held JiTian Ceremony".*
>
> *The time when the emperors farmed and held JiTian Ceremony as recorded in these historical data was basically between "Beginning of Spring" and the "Rain Water", 10 ~20 days earlier than the days after Insects Awaken at present.*

During the Sui and Tang dynasties (AD570-780), there were a great amount of written record concerning that no ice or snow appeared in winter in Chang'an Area. In New Book of Tang and Old Book of Tang, there were 25 parts with the recording of no appearance of ice or snow for 23 years. The days with no appearance of ice or snow in winter appeared so many times, which could also reflect that the period of Sui and Tang dynasties was indeed an age in relatively high temperature among history of earth. In addition, a great amount of historical data concerning plant growth, farming habit, etc., showed that temperature feature in Sui and Tang dynasties was very similar to that in modern times. It can be seen that the Earth is capable of adapting to the temperature rise and climate warming, and the warming period happens to coincide with the period of economic and cultural prosperity in history, which will not lead to devastating disasters.

2.2. Political Transformation: International Power Game

At the end of 1980s, US broke out the worst high-temperature and drought disaster since 1930, and South Africa also broke out the drought in the same period. The large-scale natural disasters drew people's attention to the living environment, and therefore, the global warming issues previously only discussed in scientific community were begun to be reported massively by media, which gradually attracted political powers in various aspects to participate, resulting in political transformation of global warming issues.
Some scientists began to spontaneously organize to carry out more and more extensive interdisciplinary interactions and exchanges among countries. In Villach Conference held in Austria in 1985, the scientists from all countries proposed that the governments should take measures to intervene the climate warming situation on Earth. At the subsequent Toronto Climate Conference, the scientists from all countries untied together to request the governments from all countries to set up the strict emission reduction targets all over the world and make corresponding policies to control the tendency of global temperature rise.
In 1988, Intergovernmental Panel on Climate Change (IPCC) was founded. It was not a scientific institution or political group in the strict sense, but the mixed body of science and politics, keeping close relationship with the national laboratories, weather bureaus, and scientific institutions in different countries. IPCC is not engaged in any scientific research, but responsible for publishing the climate assessment report every five years by searching the papers concerning climate change released in the influential publications, concluding the existing knowledge regarding climate change, and providing the basis for the political decision-makers in all countries to make related decisions. By far, IPCC has successively completed five assessment reports, with high influential power in the climate change field.

Besides scientific community and Political Science Group, some high-level statesmen also openly support the global warming viewpoints. Margaret Thatcher, Prime Minister of UK, firmly supported the global warming argument and claimed that global warming was a critical political affair worth fund investing for research, which received endorsement and support by other countries in the continent of Europe. Intervention of political power showed that global warming issue had been converted from the undetermined science argument in the laboratory to a chip for negotiation on the conference table among statesmen. Afterwards, aimed at the emission reduction and limitation issue controlling temperature rise, the countries started to carry out long-term political game, global warming gradually became a determined political consensus existing in the field of international political negotiation.

3. Competition of the Political Economic Interests among Countries around the Global Warming Issue

At present, international political negotiation concerning global warming has become an important topic in the process of international political negotiation. One of the theoretical assumptions about global warming is that, increase of concentration of carbon dioxide emitted by human activities will lead to rise of the Earth's temperature, resulting in large-scale natural disaster. Therefore, emission of carbon dioxide among countries becomes a core issue for negotiation. This is also related to the production mode and economic development pattern of the countries all over the world, involved with political and economic interests of political powers among countries, and also concerns redistribution of political powers and economic benefits in international society and establishment of moral banner. Through the international political negotiations over the years, it can be seen that the international society is divided into three camps: European Union, Umbrella Group represented by US, and the developing countries including China, India and Brazil:

3.1. European Union

In terms of global warming, European Union is a primary enabler all the time from official initiation until now. The European Union undergoing industrial age has both capital and technology advantages in respect to new energy and environmental protection industry, with the clean energy accounting for a large proportion in the energy structure. Conclusion of emission reduction and limitation agreements around the world is not only good for European Union's development in energy, industrial and agriculture, and production mode, but also in favor of occupying the leading position in new energy competition for EU, indicating positive influence in determining and solidifying its future international standing. Therefore, EU always regards the global warming issue as the international moral consideration concerning future human social development and spreads all over the world. At the stage when UN Framework Convention on Climate Change and Kyoto Protocol was coming into effect through negotiation, EU played a leading role.

3.2. US and Umbrella Group

US, Japan, Canada, Australia, and New Zealand, and other western developed countries other than EU haven't made any positive response to global warming issues as emission reduction and limitation will cause loss to economic interests of these countries since their manufacturing industries are developed and greenhouse gas emission is enormous in these countries. In the meantime, establishment of any international political structure and dominant right are very important, so holding dominant right means the country is able to obtain the benefit for itself from the international political structure. At present, the core of international cooperation for global warming control is to manage the carbon dioxide emissions, and EU who is actively promoting global emission reduction all the time has already held the initiative in planning and supervision of carbon emission, built the climate political pattern in favor of EU. Therefore, America is always resistant to the existing climate political and political pattern. Japan, Australia, and Canada are also pendulous about the emission reduction issue, who sometimes advocate emission reduction, while drop out the agreed protocol other times. These developed countries with fluctuating attitudes about high carbon emission lead to little effect on global emission reduction. From the carbon emission point of view at present, emission load in developed countries has reached to peak, and these countries have gone on the way of using clean energy or industry transfer, and technological advantage to realize carbon emission reduction. In fact, from actual range point of view, emphasis and stress of emission reduction is placed on developing countries including China, India, and Brazil, etc.

3.3. Developing Countries

Throughout global history, nearly all the developed countries have their process of modernization completed through an enormous amount of energy consumption and have undergone the stage of high emission of carbon dioxide. At present, the developing countries are in the process of the industrialization construction, and most of countries have to undergo the highenergy consumption stage. Reaching the emission reduction target forcibly will take enormous impact on economic development of developing countries, resulting in certain shock to economic prospect of developing countries. Politicization of global warming basically forces developing countries to go on the road of transformation of economic development and adjustment of energy structure. Since China is the world second largest economic entity, also as a country with high carbon emission, therefore, China always bears huge pressure during climate negotiation. As a developing country, China has to not only keep rapid economic development but also take emission reduction responsibilities, which is an enormous economic and social challenge.

4. Adjustment of China's Energy Mode

The conversion of global warming towards politicization makes it important to monitor carbon dioxide emission all over the world. Controlling carbon dioxide emission in their own country becomes a major factor of each country's national obligations（Table1）. As two largest countries concerning carbon dioxide emission in the world, China and US, as believed by international society, of course should take great responsibilities(Figure 1).

On June 3, 2017, US announced to withdraw Paris Agreement in spite of objections from all other countries. China, as the world first carbon emission country, becomes the focus of attention immediately in the world, and therefore bears heavier international responsibilities in terms of emission reduction. Nevertheless, China always keeps positive attitude and gradually takes new clean energy structure targeted at carbon emission reduction.

As predicted in National Energy Development Strategy (2014~2020), the total primary energy consumption in 2020 will be controlled to be 4.8 billion ton of standard coal. As predicted in 13th Five-year Energy Planning Report, the total primary energy consumption in 2020 will be controlled to be 5 billion ton. As described in China Energy Outlook 2030, the growth rate of the future energy demand is expected to slow down and the total primary energy consumption in 2030 will be controlled to be 5.3 billion ton of standard coal. Meantime, the primary energy consumption structure will be continuously optimized, with the coal proportion decreasing dramatically, clean energy consumption increasing continuously. By 2020 and 2030, the coal proportion will be respectively 60% and 49% and non-fossil energy proportion will be respectively 15% and 22%.

In 2006, China issued and implemented the Renewable Energy Law, in order to guarantee development of renewable energy source by law and provide effective fiscal subsidies and tax preference for solar energy, wind energy, and biomass energy. It mainly includes three aspects: the first is to make further detailed plan within the national range, including renewable energy power industry development; the second is to provide guarantee for power enterprises to take over the renewable energy enterprises; the third is to increase the renewable energy fund in order to expand the proportion of sales electric energy. At present, Chinese government is vigorously promoting construction of natural gas facilities of decentralized energy system, in order to centralize the energy intensity by constructing the decentralized energy system including 1000 projects and distribute the renewable energy sources in the rural area. In general, China is pursuing to establish the modern energy industrial system to guarantee stable, economic and practical, and clean energy development (Figure 2).

Table 1. Top 10 Carbon Emission Countries in 2014

Country	1988	2014

CHINA	2,369,502	10,291,927
United States	4,892,526	5,254,279
India	527,564	2,238,377
Russian Federation		1,705,346
Japan	989,082	1,214,048
Germany		719,883
Iran, Islamic Rep.	176,115	649,481
Saudi Arabia	202,554	601,047
Korea, Rep.	221,960	587,156
Brazil	209,364	529,808

Data source: World Bank Website (https://data.worldbank.org)

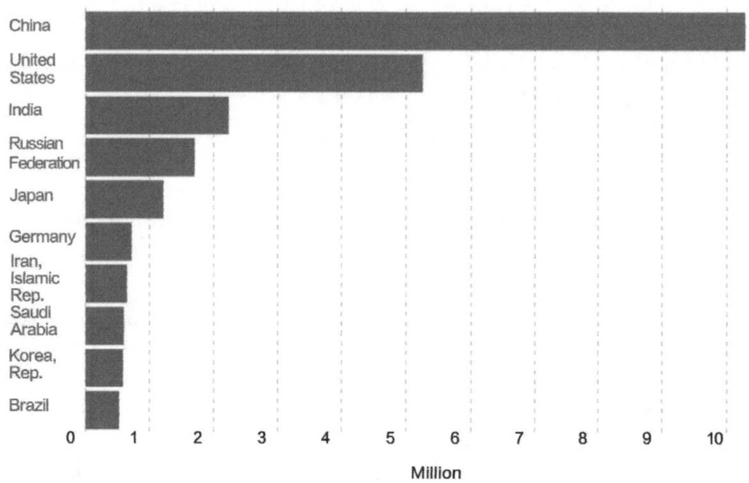

Figure 1. Emission Comparison of Major Carbon Emission Countriesin the World in 2014 (in Kt)

Data source: World Bank Website (https://data.worldbank.org)

4.1. Petroleum

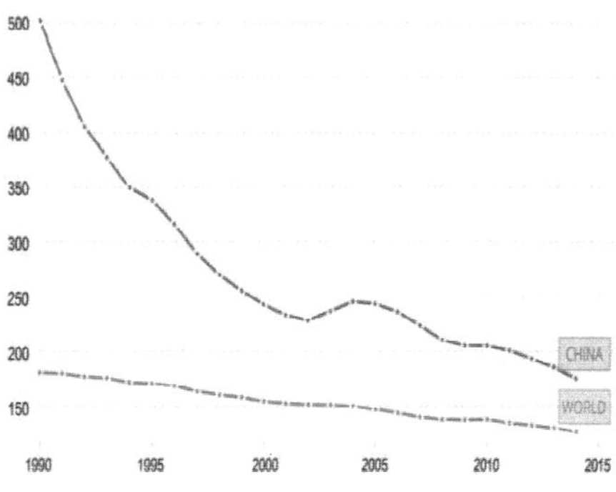

Figure 2. Energy use (kg of oil equivalent) per $1,000 GDP (constant 2011 PPP)

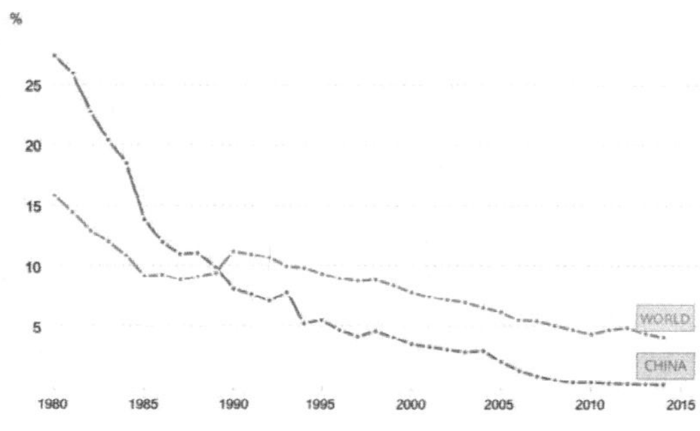

Figure 3. Electricity production from oil sources (% of total)

China has high oil consumption and high oil external dependence. Currently, China's energy policy is aimed at promoting clean development of fossil-fuel energy, developing the petroleum resources by advanced technological means, and eliminating production structure with outdated capacity. Meantime, the exploratory development is intensified in order to improve recovery efficiency of the old oil extraction rate and guarantee steady improvement of the capacity. The alternative clean energy industry shall be developed at the same time. At present, the power generation with petroleum as the energy in our country drops significantly, up to the low level from 8.103% in 1990 to 0.168% in 2014 (Figure 3).

4.2. Coal

China is the largest country with coal production all over the world and also the one with the highest carbon emission due to coal consumption. At present, China uses coal as the main energy for power generation, with its proportion far higher than the world average level (Figure 4). To realize the emission reduction target, it's necessary to promote transformation of related industries to the clean energy industry. China is devoted to transforming the backward production capacity of the coal by means of modern technological means. Meantime, special attention is paid to investment of clean coal technology and creation of green mining area. For example, the liquid perfluorocarbon project is tried out in the Inner Mongolia now, to prevent the coal from burning in atmosphere using carbon capture, fossil fuel manufacturing, and other technologies. The technology is expected to be a key growth area for China energy transformation. At the beginning of 2016, China also issued a serial of reform measures in order to consume the excess capacity in domestic coal industry, hence to improve the productive forces and profit capacity. The focus of the reform was to eliminate the small-scale coal mine with low production efficiency and encourage large-scale merging and reorganization. The effect of reform was quite remarkable. As a result, the domestic coal production decreased sharply and coal price increased rapidly. In 2016, the annual coal production decreased by 7.9% (-140 million oil equivalent), forming the maximum price drop throughout history (Figure 5).

Figure 4. Electricity production from coal sources (% of total)

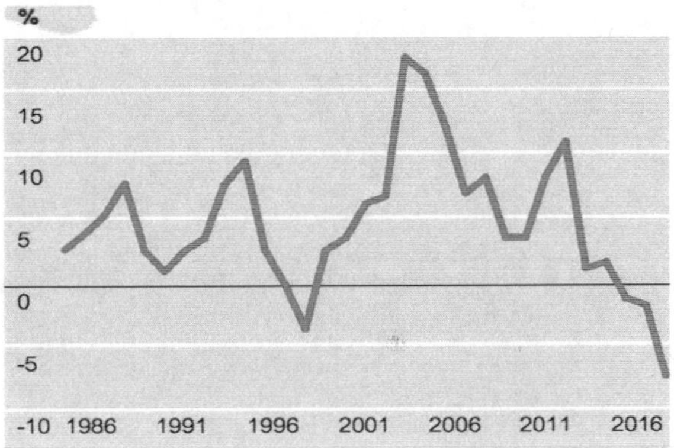

Figure 5. Annual change rate of coal production growth in China (Data source: BP Yearbook of World Energy Statistics, June 2017)

4.3. Natural Gas

The natural gas, as the cleaner energy than other fossil fuels, has an important strategic position in undertaking the international responsibilities of emission reduction and limitation. However, China does not have sufficient natural gas reserve, with the energy characteristics of "rich coal, poor gas, and little oil". Therefore, China puts forward the "equal stress on oil and gas" policy to gradually increase the natural gas production and seek for balanced utilization of natural gas resources. As seen from the reserve capacity of natural gas in China, the natural gas reserve in China is basically distributed in east area and west area, while the important refining and chemical industry base is distributed mainly in China Yangtze Delta, Pearl River Delta, and Circum-Bohai Sea Region. Therefore, China is devoted to creating the trans-regional integrated management mode integrating oil & gas exploitation and oil & gas extraction.

4.4. Nuclear Energy

Nuclear energy, wind energy, and solar energy are used as the green energy. Nuclear energy is used in order to facilitate to optimize natural energy structure and reduce carbon dioxide emission. Therefore, China will always focus on nuclear energy development in recent years. China has built more than 11 nuclear power stations by far and looks forward to promoting the technological innovation with nuclear energy as the core by more and more investment (Figure 6). Since uranium as the main element of nuclear raw material is a comparatively scarce element and the proved reserves in China don't have long service life, China is seeking for uranium resource in other countries of the world, in order to further develop the nuclear energy in the country(Figure 7).. In 2016, China put five nuclear reactors into production, indicating the maximum annual grown in the nuclear energy history of China. In addition, there are additional 20 nuclear reactors under construction.

In 2016, all the net growth of nuclear energy all over the world comes from China. Annual increment of nuclear energy production in China (+9.6mtoe) became the maximum value among annual increments in all the countries since 2004.

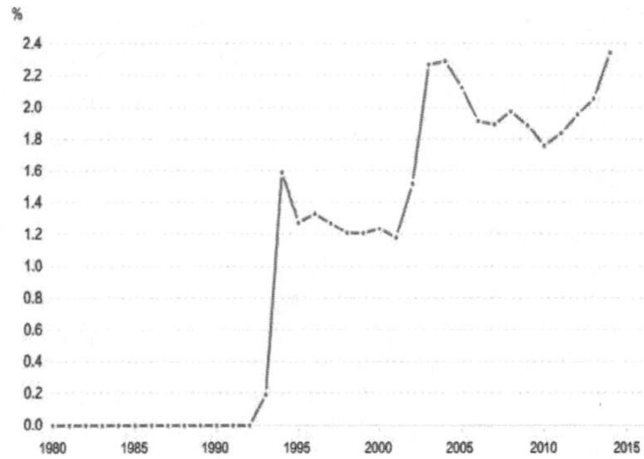

Figure 6. Electricity production from nuclear sources (% of total) (China)

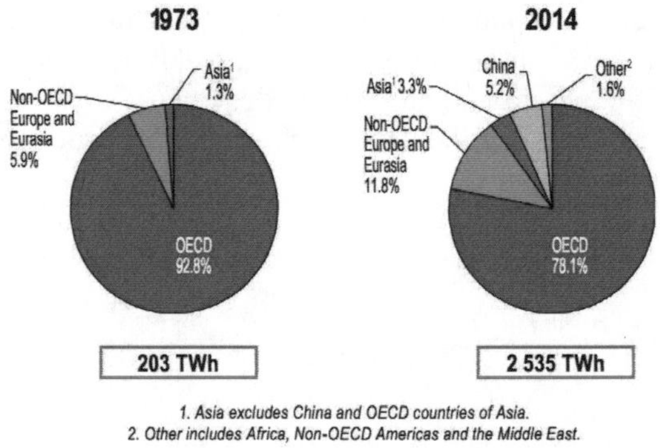

Figure 7.1973 and 2014 regional shares of nuclear production

4.5. Renewable Energy Sources

Through years of development, we have made great achievement in development of renewable energy sources in our country and become the world largest producer and consumer of renewable energy sources at present. In 2016, annual consumption of renewable energy sources in China increased by 33.4%. Over ten years, the proportion of China's renewable energy consumption has risen to 20.5% in 2016 from 2% (Figure 8), where the growth rate of solar energy consumption is the fastest, up to 71.5%. The renewable energy development in our country has come to the stage of the wide-range increment alternative and regional reserve alternative. Meantime, overseas investment for renewable energy sources has accelerated in China and foreign countries. In 2016, the overseas investment is in total of USD 32 billion, increasing by 60% compared to last year. China has taken the leading role in the field of renewable energy resources, owning five of the six largest solar cell manufacturing companies, as well as the world largest wind turbine manufacturer, Li-ion manufacturer, and electric companies. Four out of five energy deal companies in 2016 are Chinese companies.

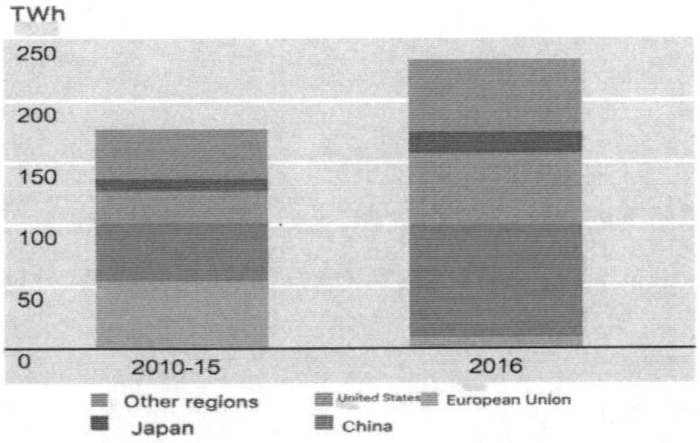

Figure 8. Annual growth and distribution of renewable sources

5. Conclusion

In conclusion, political transformation of the global warming issue makes carbon dioxide emission and clean energy substituting traditional fossil energy become national responsibility and moral principle. As a big power in developing countries, China is always be devoted to emission reduction and limitation and manifests its big power responsibility awareness. As US President Trump made it clear to support fossil fuel and withdraw the climate agreement, the leading role of renewable energy sources will be transformed from US to China. In the future, China will take over USA's climate leadership, which will allow China to occupy the moral high ground and strengthen its power and prestige in global governance.

6. References

[1] IPCC Climate Change 2007: Working Group I: The Physical Science Basis. Cambridge, United Kingdom and New York, NY, USA: Cambridge Press, 2007, 996

[2] Svante A, *On the Influence of Carbonic Acid in the Air upon the Temperature of the Ground*, London, Edinburgh, and Dublin Philosophical Magazine and Journal of Science (fifth series), April 1896. vol 41, pages 237–275.

[3] Baum S, Rudy M. *Future Calculations: The first climate change believer.*Distillations. 2 (2): 38–39. Retrieved 17 February 2017.

[4] GrayL and FredS,*to speak at climate change sceptics conference*, The Daily Telegraph, November 18, 2009.

[5] Dennis A and Fred S,*The Physical Evidence of Earth's Unstoppable 1,500-Year Climate Cycle,*Archived from the original on December 2, 2008. Retrieved 2009-01-16.

[6] FredS, Dennis A. *Unstoppable Global Warming: Every 1,500 Years,*Updated and Expanded Edition, 2007.

[7] NeftelA, Oeschger H, Staffelbach T, et al. *C02 record in the Byrdice core 50000-5000years,* BP. Nature, 1988, 331:609-611.

[8] Raynaud D, Jouzel J, Barnola J M, et al. *The ice record of greenhousegases*. Science, 1993,259:926-934.

[9] FischerH, Wahlem M, Smith J, et al. *Ice core records of atmosphericC02 aroundthelastthreeglacial erminations*. Science, 1999,283:1712-1714.

[10] Monnin E, Indermuhle A, Dallenbach A, et al. *Atmospheric C02concentrationsoverthelastglacial termination*. Science, 2001,291:112-114.

IOP Conf. Series: Earth and Environmental Science **151** (2018) 012006 doi:10.1088/1755-1315/151/1/012006

Preliminary Analysis of Surface Circulation off the Minjiang River Estuary near the Southwestern East China Sea

Hsien-Wen Chen

Department of Maritime Police, Central Police University, No.56, Shujen Rd., Takang Vil., Kueishan District, Taoyuan City 33304, Taiwan, R.O.C.
Email: hsienwenchen@gmail.com

Abstract. Better understanding the surface circulation in the marine environment is pertinent to the preparedness of pollution response and maritime safety. In the spring of 2012, two GPS surface drifters were deployed off the Minjiang River Estuary to study the surface circulation near the southwestern East China Sea. One drifter followed the China Coastal Current and the other one joined the Taiwan Warm Current. Classical harmonic analysis technique was employed to analyse the current velocities derived from the trajectories of these two drifters. M2 tidal component is dominant in the tidal current spectrum. The residual currents obtained from these 2 drifters agree very well with the previous studies.

1. Introduction

Hypoxia in the East China Sea (ECS) has been referred to as the one of the largest low-oxygen areas in the world [1]. The hypoxia is aggravated by the increasing eutrophication, the frequent algal blooms and global warming in coastal and estuarine regions. The frequent hypoxia incidents and the expanding of its distribution have seriously damaged to the ocean organisms, even the whole ecosystem. In addition to the Yangtze River, the coastal waters near the Minjiang River have been faced by huge stress from anthropogenic activities and population growth along the Fujian coast [2]. Although the priority work is to cut off the land-based sources of ocean pollution [3], a better understanding of the surface circulation will be beneficial to proper response to the pollution accidents if happened.

It has long been recognized that the cold China Coastal Current (CCC) or ZheMin Coastal Current encountered the Taiwan Warm Current (TWC) in the western Taiwan Strait [4]-[6]. Several reports involved with the circulation and tidal residual currents near the Zhejiang-Fujian coast in the western ECS have been published [7]-[10]. However, these studies focused mostly upon the seas in the Zhejiang Province, with little emphasis in the vicinity of Fujian, especially near the seas close to the Minjiang River estuary (Figure 1). Since the oceanic region near this estuary is a traditional fishing ground for both Chinese and Taiwanese fishermen, the knowledge of surface circulation is pertinent to the ecosystem and the maritime safety.

The primary objective of this article is to gain some insight into the sea surface circulation near the Minjiang River estuary by employing the current velocities derived from the trajectories of satellite-tracked drifters. The outline of this paper is as follows. Section 2 presents the methodology used in this study. Results and discussion are shown in the section 3, and section 4 gives the conclusion.

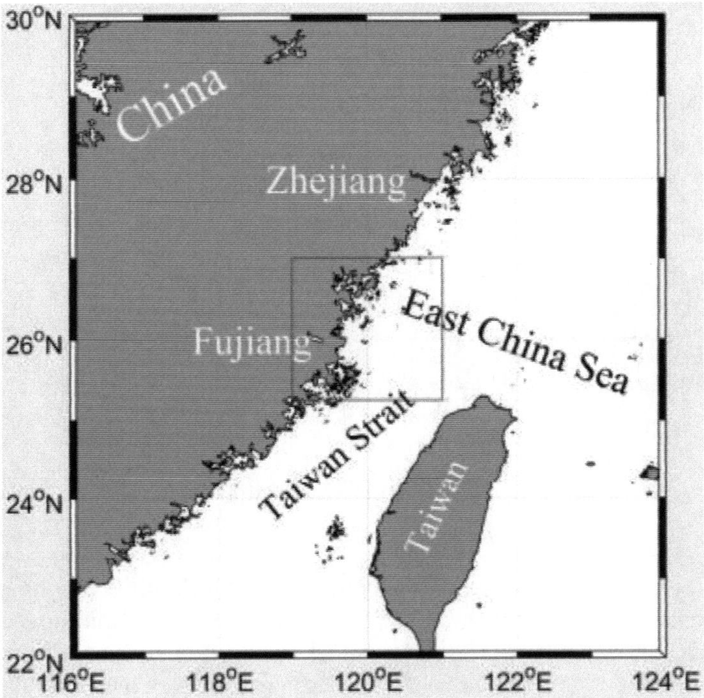

Figure 1. The map indicating the study area (red box) and the surrounding seas. It is located near the northwestern Taiwan Strait and southwestern East China Sea. The detailed topography of the study region will be shown later.

2. Methodology

2.1. GPS Surface Drifter

We need a drifter to be used in the nearshore environment with time scales of several days as suggested by Johnson *et al.* [11]. A compact, low-cost GPS drifter was designed as in Figure 2. It is a plastic tube with 9-cm diameter and 87-cm of length with batteries and one GPS receiver inside, connected with a 15-m wire with 3 holey-socks at 12-m, 13-m, and 14-m respectively, and with a 1.5-kg anchor at the end of the wire. When deployed into the sea, 36-cm of the tube is above the sea surface and 41-cm in the water. The temporal resolution of the received GPS signals is set to be 1-min. As pointed out by Zeng *et al.* [9] that the boundary between the CCC and TWC is located at the 50-m isobath off the China coast. Hence, we deployed 2 drifters in May 24 and June 9 of 2012 respectively in the west and east of the 50-m isobath. The one deployed in May is denoted by 'M' and the other one denoted by 'J'. The duration of drifter M is 05/24-30, and J is 06/6-20. However, from 07:00 of June 11 to 15:00 of June 12, GPS signals of J were not received for unknown reasons. The data before 15:00 of June 12 were not further analysed except showing it trajectory.

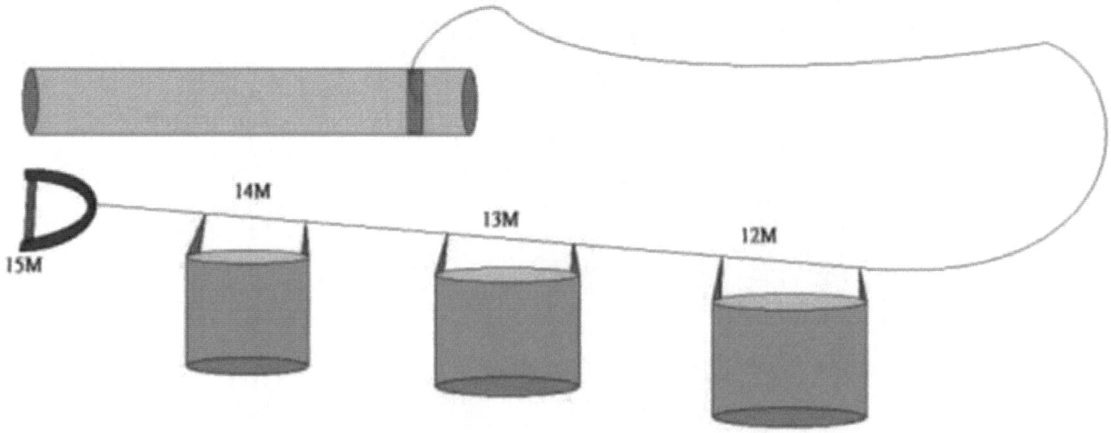

Figure 2. The configuration of the surface drifter employed in this study.

2.2. Method of Analysis

Wang *et al.* [12] analyzed more than 10 cruises of shipboard ADCP observations in the Taiwan Strait and concluded the current field in the strait is dominant by the M2 semidiurnal tidal current. As a result, the first step of analysis here is to assess the tidal contribution to the surface circulation. Poulain [13] recently studied the tidal currents measured by surface drifters in the Adriatic Sea. The harmonic analysis technique was applied to the drifter velocities using the least-squares method and the T_Tide MATLAB package [14] for comparison purpose. The results obtained from the 2 methods agree very well so that we decide to adopt the T_Tide package to analyse our drifter velocities. This software performs classical harmonic analysis using as many as 68 constituents, accounts for unresolved constituents using nodal corrections, and computes the Greenwich phase with nodal corrections applied to amplitude and phase relative to center time. It also estimates confidence intervals for the analysed components.

3. Results and Discussion

Figure 3 illustrates the trajectories of these 2 drifters. M was deployed west of the 50-m isobath, flowing southward with counter-clockwise traveling northward in the beginning and then turning toward southwest. However, J, deployed southeast of M, was moving northward with similar counter-clockwise pattern as M. When it approached the Dongyin island, it was blocked and circulated around the island and then also turned to the southwest.

Figures 4 and 5 present the original, tidal and residual speed of E-W and N-S components of current velocities derived from the trajectories of these 2 drifters respectively. The east-west tidal current is about 0.4 m/s, about twice of the north-south component. After deducting the tidal contributions, the residual currents were very weak and variable for both drifters. In sum, the mean residual current of M drifter is about 0.17 m/s, and that of J drifter is about 0.05 m/s. Compared with the observation work done by Zeng *et al.* [9], the mean current velocity of CCC is 0.24 m/s which is consistent with our measurement because their observations were conducted during winter, the northeast monsoon is stronger then and thus stronger residual current. For TWC, the mean current velocities obtained by Zeng *et al.* [9] are 0.03~0.09 m/s, which also agrees very well with the value of 0.05 m/s derived from the J drifter.

The tidal constituents of current velocity obtained from the J drifter are shown in Figure 6. As indicated in the figure, M2 constituent is dominant for both components, which is consistent with the work by Wang *et al.* [12]. However, the magnitude of tidal current in our study region seems to be smaller than those in the Taiwan Strait. Since the results of M drifter are similar and for the sake of brevity, we do not present the tidal information of M drifter here.

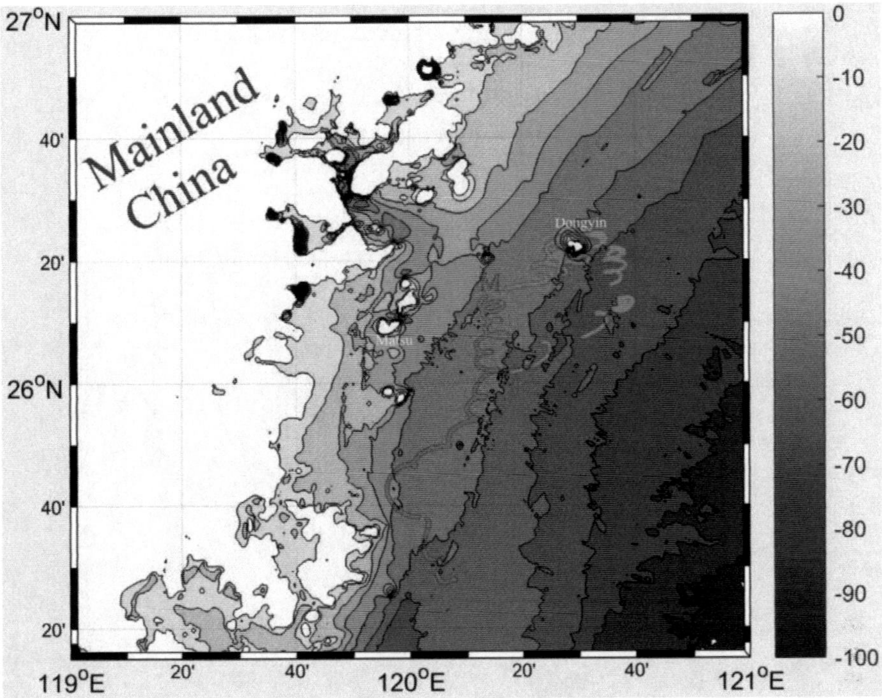

Figure 3. The trajectories of the two drifters. The track of the M is represented by red, and J is denoted by pink. The locations of the letters are the locations of deployment. The color bar at the right indicates the depth (in meters) associated with the colors in the image.

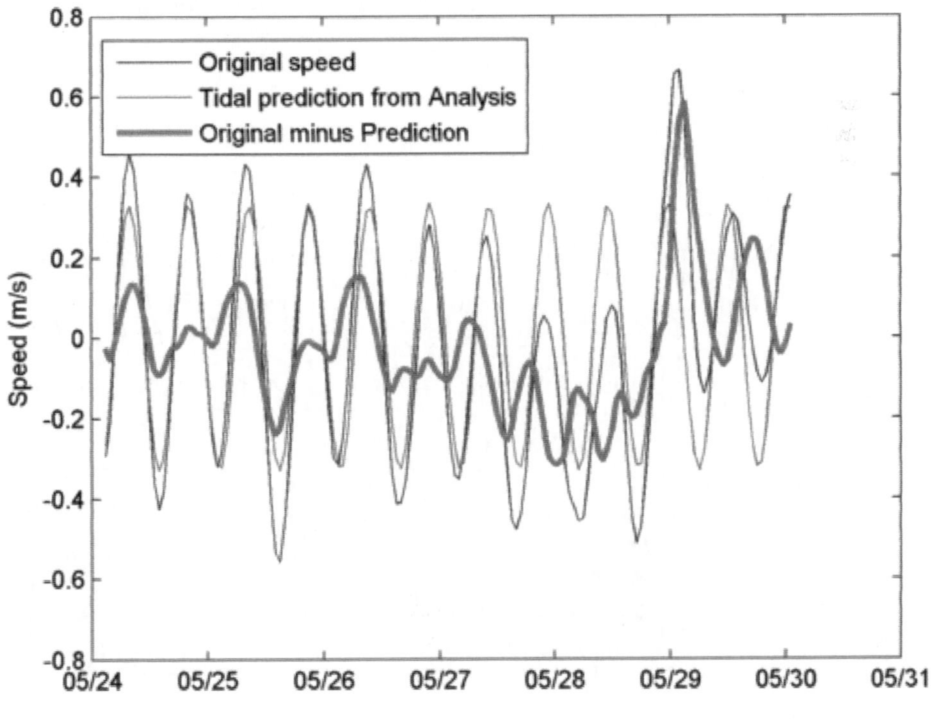

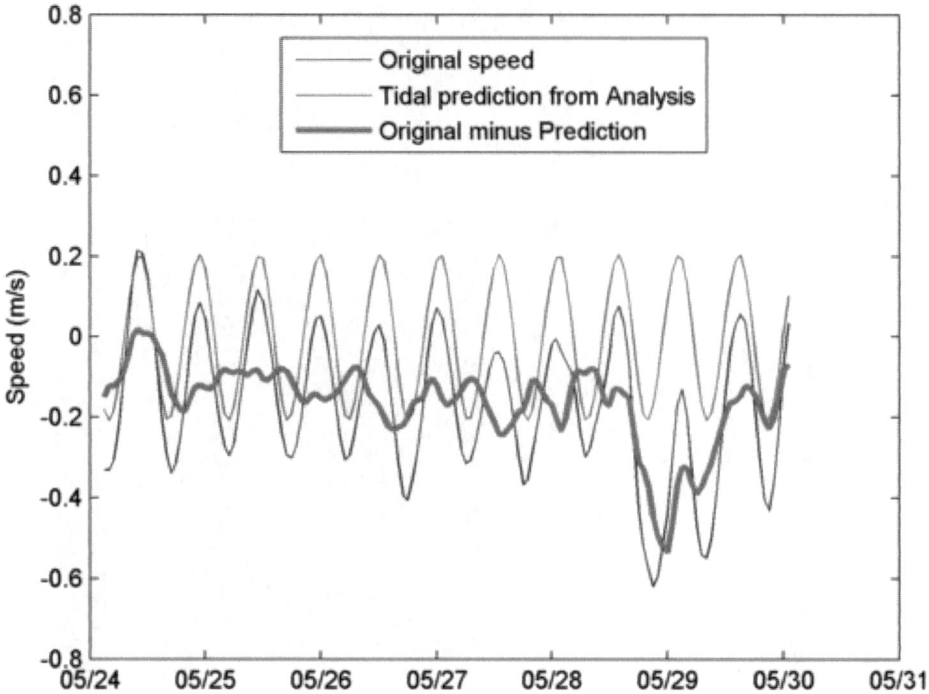

Figure 4. The results of harmonic analysis of current velocities derived from the M drifter. The original velocities are denoted by blue, the tidal currents from harmonic analysis are indicated by green, and the residual currents are red. The E-W component is shown in the upper panel, and the lower panel presents the N-S component.

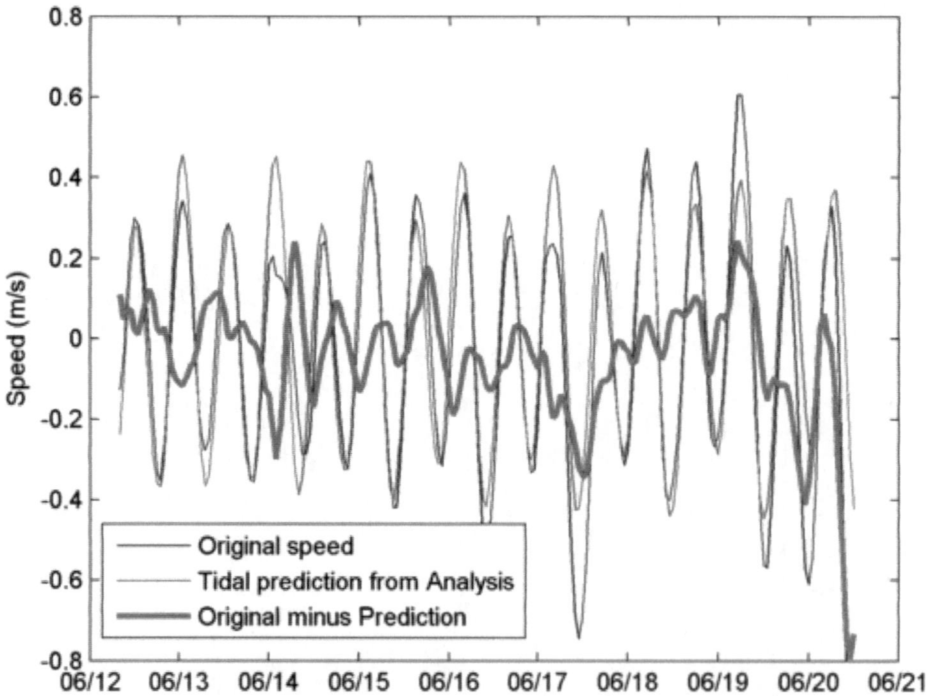

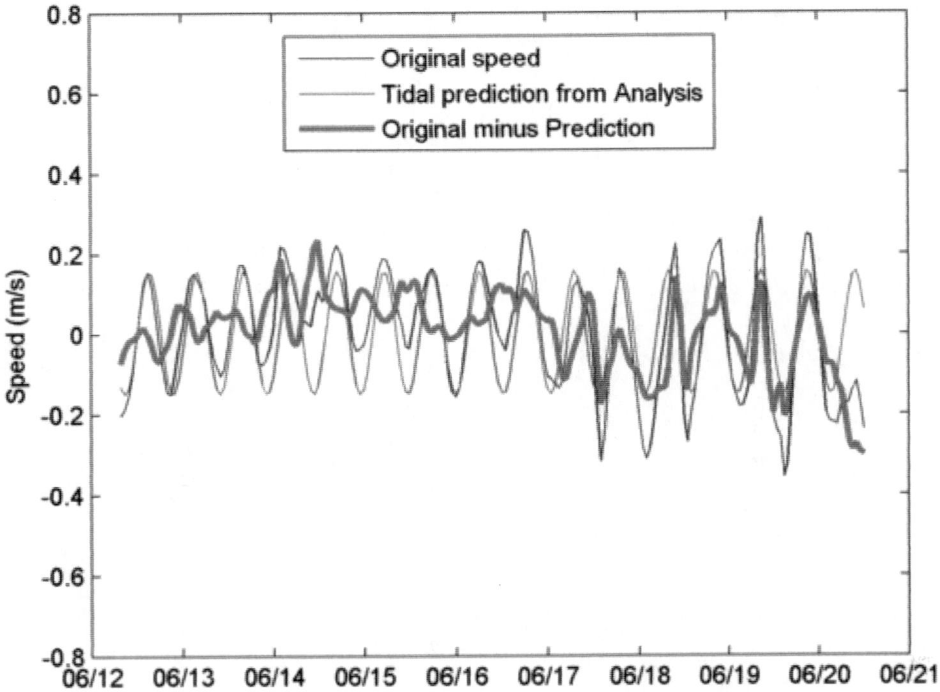

Figure 5. The same as Figure 3, but for the J drifter.

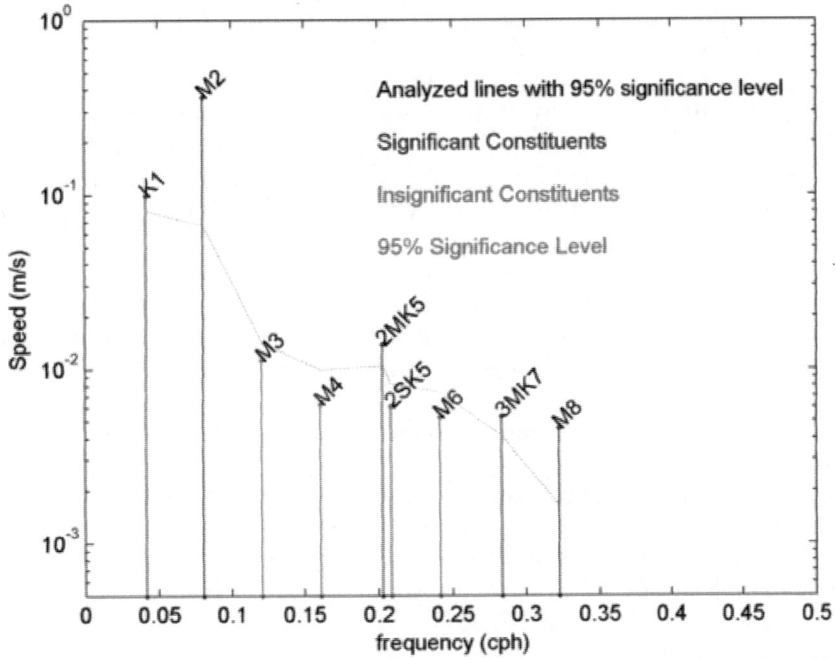

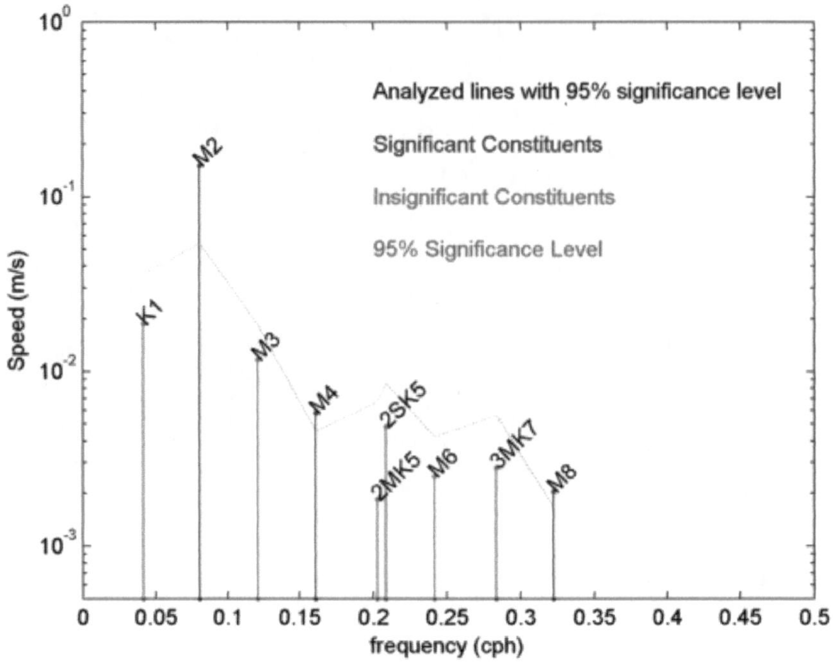

Figure 6. The tidal constituents obtained from harmonic analysis of current velocities derived from the J drifter. The upper panel is for the E-W component of the velocity, and the N-S component is shown in the lower panel.

4. Conclusion

Preliminary analysis of the surface circulation near the MinJiang River estuary in the spring of 2012 confirm the boundary of CCC and TWC to be located near the 50-m isobath. The current velocities derived from the 2 drifters deployed respectively reveal the characteristics of CCC and TWC near this region. The tidal current field is dominant by the M2 tidal constituent. Mean (residual) current of CCC flowed southward and TWC northward with weak and variable features.

5. References

[1] Chen C C, Gong G C and Shiah F K 2007 Hypoxia in the East China Sea: One of the largest coastal low-oxygen areas in the world *Mar. Environ. Res.* 64 399-408

[2] Zhang Z L, Hong H S, Zhou J L, Huang J and Yu G 2003 Fate and assessment of persistent organic pollutants in water and sediment from Minjiang River Estuary, Southeast China *Chemosphere* 52 1423-30

[3] Daoji L and Daler D 2004 Ocean pollution from land-based sources: East China Sea, China *Ambio* 33 107-13

[4] Nitani H 1972 Beginning of the Kuroshio *Kuroshio—Its Physical Aspects* ed. H Stommel and K Yoshida (Seattle: Univ. of Washington Press) pp 129–63

[5] Hu J Y, Fu Z L and Wu L X 1990 Studies on the winter-time current structure and T-S fine-structure in the Taiwan Strait. *Chin. J. Oceanol. Limnol.* 8 319–27

[6] Jan S, Wang J, Chern C S and Chao S Y 2002 Seasonal variation of the circulation in the Taiwan Strait *J. Mar. Syst.* 35 249–68

[7] Zhang C, Shang S, Chen D and Shang S 2005 Short-term variability of the distribution of Zhe-Min coastal water and wind forcing during winter monsoon in the Taiwan Strait (in Chinese with English abstract) *J. Remote Sens.-Beijing* 9 452-8

[8] Zhang H, Du P J and Zheng X Q 2011 Characteristics analysis of residual current of Zhejiang-Fujian water (in Chinese with English abstract). *Mar. Sci. Bull.(China)* 30 151-7

[9] Zeng D, Ni X and Huang D 2012 Temporal and spatial variability of the Zhe-Min coastal current and the Taiwan warm current in winter in the southern Zhejiang coastal sea (in Chinese with English abstract) *Sci. Sin.Terrae.* 42 1123-34

[10] Huang D, Zeng D, Ni X, Zhang T, Xuan J, Zhou F and He S 2016 Alongshore and cross-shore circulations and their response to winter monsoon in the western East China Sea *Deep Sea Res. Part II Top. Stud. Oceanogr.* 124 6-18

[11] Johnson D, Stocker R, Head R, Imberger J and Pattiaratchi C 2003 A compact, low-cost GPS drifter for use in the oceanic nearshore zone, lakes, and estuaries. *J. Atmos.Oceanic Tech.* 20 1880-84

[12] Wang Y H, Jan S and Wang D P 2003 Transports and tidal current estimates in the Taiwan Strait from shipboard ADCP observations (1999–2001) *Estuar. Coast. Shelf Sci.* 57 193-9

[13] Poulain P-M 2013 Tidal currents in the Adriatic as measured by surface drifters *J. Geophys. Res.* 118 1434–44

[14] Pawlowicz R., Beardsley B and Lentz S 2002 Classical tidal harmonic analysis including error estimates in MATLAB using T_TIDE *Comput. Geosci.* 28 929-37

Acknowledgments

This research is funded by the Ministry of Science and Technology of the Republic of China (MOST 106-2621-M-015-001-).

IOP Conf. Series: Earth and Environmental Science **151** (2018) 012007 doi:10.1088/1755-1315/151/1/012007

Problems of Groundwater Extraction from Transboundary Aquifers and Complexes

E I Golovina[1]

1 Department of Economics, Accounting and Finance, Saint-Petersburg Mining University, Saint-Petersburg, Vasilyevsky Island, 21st line 2, 199106, Russia
E-mail: golova2107@mail.ru

Abstract. Water is one of the most important vital types of natural resources used in almost all spheres of life and human activity. It is the availability of water that determines development of industry and agriculture. Its selection and irretrievable losses can have a significant impact on the state of the natural environment. Geology of the earth's surface is arranged in such a way that boundaries of groundwater aquifers and complexes distribution do not coincide with the boundaries of states, and therefore extraction of this priceless resource must be regulated at the international level. The development of a unified universal concept for groundwater production, namely the system of state regulation, monitoring, taxation and planning should be based on the experience of advanced countries such as Russia, Germany, France, etc., where a vast experience of scientific and research activity is accumulated. Many international conflicts can be prevented if one approaches the problems of subsoil use on the basis of a general agreement. The aim of the research is development of an international conceptual model for groundwater production management through regulation of specification systems, monitoring, licensing, control of groundwater extraction. Features of groundwater extraction from transboundary aquifers are considered, international experience of conflict regulation in this sphere is reviewed, recommendations and prerequisites for the development of an international system for the groundwater extraction management are given.

1. Introduction

Groundwater is an important component of the natural water system. Groundwaters are extracted from aquifers and used for drinking and household purposes, in agriculture and production [1]. Groundwaters belong to the category of strategic minerals, along with hydrocarbon resources, so the supply of drinking water will be one of the pressing problems of modern society. According to UN experts, by 2030, about half of the world's population will suffer from a shortage of fresh water, and this can lead to future military actions and conflicts [2].

In this connection, extraction of such valuable minerals as groundwater must be managed rationally. On a combination of a reasonable selection of groundwater with observing the norms of their quality, the concept of "rational use of them" is based. That is, under the rational use of groundwater it is recommended to understand their economically feasible operation, which provides protection from pollution and depletion of operational (mineable) reserves and allows to maintain groundwater resources and environmental conditions at a given level [3].

Over-limited extraction of groundwater and, at the same time, inefficient use cause a rapid depletion of groundwater resources in many countries [4]. In addition, both natural processes and human activities can have a significant impact on groundwater quality and future constraints as a source of water supply. In view of such natural and anthropogenic reasons as population growth, unplanned urbanization, economic development, the amount of groundwater continues to decrease. Therefore, strategies for sustainable management of groundwater production are required. In turn, development of a proper groundwater management strategy requires the study of aquifers characteristics, spatial and temporal groundwater monitoring [5].

Worldwide, about 60% of all freshwater runs within cross-border basins; only an estimated 40% of those basins, however, are governed by some sort of basin agreement [6]. In an increasingly water-stressed world, shared water resources are becoming an instrument of power, fostering competition within and between countries. The struggle for water is heightening political tensions and exacerbating impacts on ecosystems [7], [8]

By its nature, beneficial and rational use of groundwater depends more on socio-economic, institutional, legal, cultural, ethical and political considerations than surface water. Their national development is hampered by weak social and institutional capacity, as well as by weak legal and policy frameworks. In a transboundary context, this can be further strengthened due to contrasting levels of knowledge, opportunities and institutional frameworks on both sides of many international borders [9].

Despite their significance, physical interactions between surface and groundwater have largely been ignored in international water law. While surface water has been given considerable attention as a transboundary natural resource, groundwater has not received the same recognition. International legal doctrines regarding water, such as the 1997 United Nations Convention on the Law of the Non-Navigational Uses of International Watercourses, only recognize done aspect of groundwater, excluding confined aquifers. While the issue of transboundary groundwater in international treaties is becoming increasingly relevant as disputes over groundwater resources come to the fore, it is usually only indirectly mentioned in treaties. Groundwater and surface water should be considered together as part of the hydrological cycle and reflected as such in the legal body. The uncertainty of physical properties is not an excuse for the delay of a concrete framework. An Interactive Coordinated Approach (ICA) is recommended as a guideline for future implementation of transboundary groundwater management [10], [11]. The purpose of this research is to demonstrate the need to develop comprehensive transboundary groundwater management conceptual model.

2. Materials and Methods

2.1. Peculiarities of Transboundary Aquifers

Many of the world's existing groundwater systems (or "aquifers" - so called usable groundwater reservoirs) are transboundary, that is, they lie on the territory of two or more administrative units within the country or two or more countries. Obviously, in the second case, administrative and operational management of groundwater resources is facing additional challenges and requires harmonization of rules and transboundary cooperation between various bodies involved in groundwater mining, based on mutual trust and transparency. There are still not many examples of such cooperation in the world.

The main features of transboundary aquifers are such factors as feeding and discharge areas, extensive distribution, interconnection with other aquifers and open water bodies (Figure 1).

In many cases, aquifer may have a catchment area in the territory of one state, and the water discharge area is located in the territory of other border states. Such schemes are common, as a rule, in submontane and foothill-flat areas. This is typical, for example, for Austria - the Czech Republic, Germany - Poland, Switzerland - France.

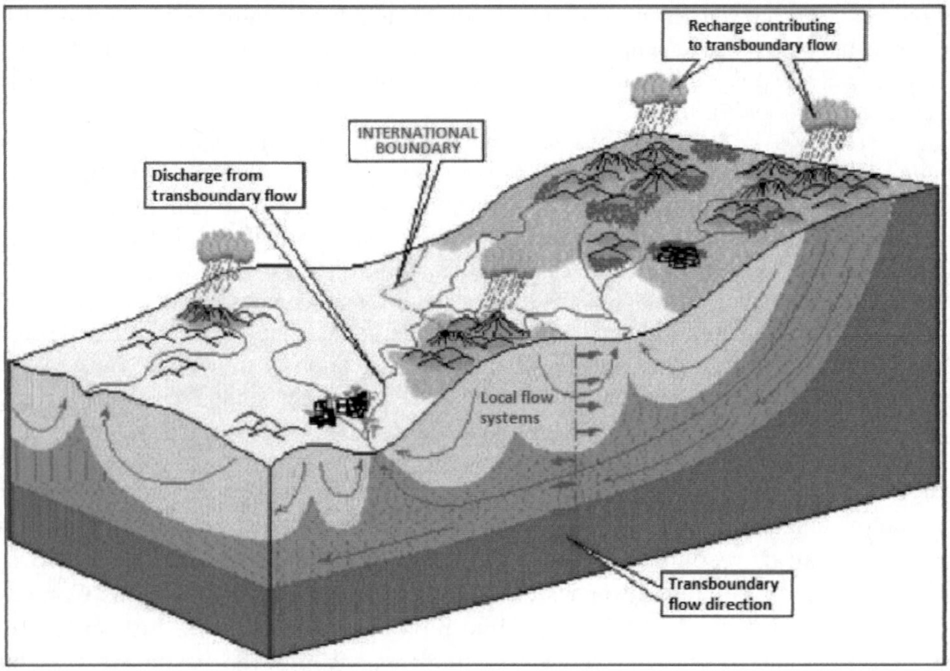

Figure 1. Schematic illustration of a transboundary aquifer [9], [12]

Very few international borders correspond to natural physical features, and water resources can cross them without hindrance. To effectively manage and equitably distribute these resources, scientists assess the resources that cross these boundaries. In hydrogeological terms, these crossing resources can only be estimated by observing and measuring individual hydraulic parameters, similar to the process of assessing other transboundary resources.

There are marked differences in the status of the recognition for transboundary surface water and groundwater. Consequently, transboundary groundwater directives have been omitted from overall water management regulations. The two primary reasons for this absence are also points of contention in transboundary groundwater management. First, groundwater characteristics vary in each aquifer. Groundwater is often deep or unevenly distributed geographically. These uncertainties make groundwater seemingly impossible to regulate, as well as ill defined. The other reason is the transboundary element. Dealing with transboundary issues has been intensively studied in surface water; as a result, the difficulties as well as the necessity for management structures are understood. By contrast, in terms of transboundary groundwater, even the delineations of an aquifer are a challenge. Under the best use of monitoring and modeling techniques to identify groundwater characteristics, the definition of an aquifer cannot provide concrete conclusions about groundwater ownership. Because of these difficulties, in addition to the rapidly increasing population and the rising demand for water, groundwater quality and quantity have become serious environmental, economic, political, and socioeconomic concerns. Therefore, the establishment of an apparent management framework is critical [11].

Even where international borders can follow signs such as rivers and aquifers underlying them, they may not reflect the true transmission of groundwater flows from one side to the other, as it is shown in Figure 2.

Surface waterways, such as rivers and lakes, are regulated by the European Water Framework Directive. All kinds of activities are clearly regulated by international laws. But, extraction and use of groundwater, due to a certain specificity and diverse legislative base, often do not have a single universal management model.

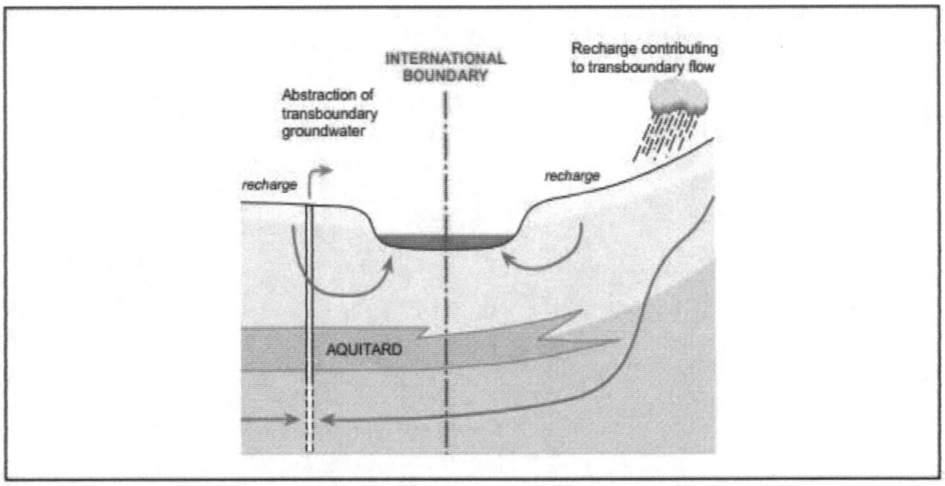

Figure 2. River boundaries over transboundary aquifers [9]

2.2. Transboundary Aquifers Regulation

In any legal agreements that should be drawn up for the management of transboundary resources, the initial stage should be the correct definition of the flow and movement of water with subsequent quantification. Institutional weakness and political pressure may not solve all relevant issues, which can lead to serious consequences for the environment and unsustainable development.

Current growing problems with respect to the quality of groundwater resources and sustainability of groundwater abstraction levels acquire a special political connotation when groundwater flows across the international border of states and, as a result, becomes a "common" resource. Sensitivity to sovereignty, diversity of legal and socio-political systems and various national programs cause certain difficulties. This is exacerbated by the fact that none of the internal "water" laws of neighboring countries can provide acceptable universal management rules. Consequently, such rules must be found elsewhere, that is, in treaties and agreements between interested sovereign states, or in case of non-observance of such treaties and agreements, in the consistent practice of the states themselves [9].

The problem is the following: nowadays there are very few treaties and agreements that provide unified international rules for managing groundwater extraction, as well as protection of common groundwater resources. Particularly acute this problem affects states with a shortage of water resources (countries in Africa, Central Asia, the Middle East, some countries in Europe and North America).

Figure 3 illustrates an overview of global transboundary aquifers made by TWAP (Transboundary water assessment programme) [12].

A clear need has been identified for an international initiative on Internationally Shared/ Transboundary Aquifer Resources Management. The International Association of Hydrogeologists (IAH) established a Commission to investigate the issue in 1999. A meeting of experts was held in parallel with the International Conference on Regional Aquifer Systems in Arid Zones – Managing non-renewable resources organised by UNESCO in Tripoli 20–24 November 1999. The results of the consultation indicated the need to create an international network supported by IAH, UNESCO, FAO and UNECE. Therefore with the support of UNESCO and IAH in co-operation with FAO and UNECE a meeting of experts was held at UNESCO in Paris 27–28 March 2000. As a result of the meeting a programme proposal for an international initiative on internationally shared/transboundary aquifer resources management (ISARM/TARM) has been tabled [9], [12].

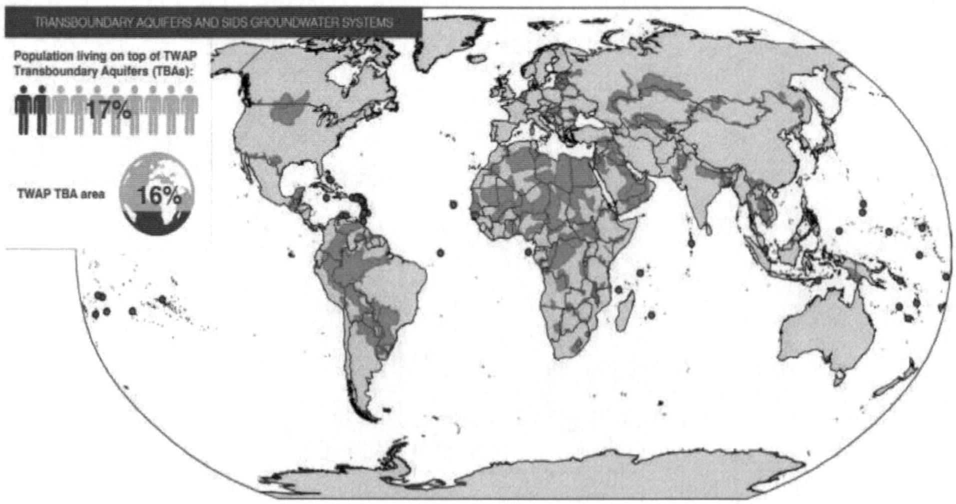

Figure 3. Overview of global transboundary aquifers made by TWAP [12].

Since its start in 2002, ISARM has launched a number of global and regional initiatives. These are designed to delineate and analyze transboundary aquifer systems and to encourage riparian states to work cooperatively toward mutually beneficial and sustainable aquifer development. The ISARM portal is developed and maintained by IGRAC (International Groundwater Resources Assessment Centre) [9], [13].

3. Results and Discussion

Thus, at present legal issues regarding quantitative factors of groundwater extraction from transboundary aquifers and systems are not fully resolved. There is a huge number of countries that have encountered problems regulating the use of transboundary groundwater resources.

For example, the extraction of groundwater from the aquifer of Trifinio - by the name of the region where the borders of the three countries: Guatemala, Honduras and Salvador meet. The most serious problem related to water management in the region is pollution. According to the estimations of the experts of the project GRGTA (Groundwater Resources Governance in Transboundary Aquifers), which is part of the program "Diplomacy and Water Management in Key Problem Transboundary Areas" and funded by the Swiss Agency for Development and Cooperation under the UNESCO International Hydrological Program (IHP UNESCO), in close cooperation with the UNESCO International Center for the Assessment of Underground Resources Water (IGRAC), the International Union for Conservation of Nature (IUCN) and project teams on the ground, in the absence of effective monitoring, the level of groundwater pollution here of time will only increase [14].

The lack of monitoring data (climate, as well as data on groundwater abstraction, water levels and quality) makes it very difficult to conduct systematic diagnostic analysis. Such a problem has arisen in the Stamprietta system of transboundary aquifers, located in a vast arid region in Central Namibia, Western Botswana and the Northern Cape of South Africa [14].

The Pritashkent transboundary aquifer (PTTA), located on the territory of Kazakhstan and Uzbekistan, causes the following problems: inevitable depletion of groundwater resources as a result of their selection. In the long term, groundwater level may drop below a level that is minimally necessary for exploitation from a technical or economic point of view. The decrease in the piezometric level of PTTA will quickly spread in breadth, including beyond the interstate border. Another major problem is potential deterioration of groundwater quality. The gradual lowering of water level in the transboundary aquifer may lead to the infiltration of brackish and saline groundwater from the overlying aquifers. Finally, contamination of the upper aquifers can lead to the abandonment of their use and transition to water intake from PTTA [14].

Of the twelve million people who live within 100 km of the US-Mexico border, 90 percent are clustered in trans boundary sister cities that share common water sources and pollution problems. New institutions created to address environmental concerns over North American Free Trade Agreement (NAFTA) offer the promise of greater financial and technical assistance for water management in border cities. Diversity and geographic dispersion of water conflicts presupposes the possibility of applying an interrelated approach to water negotiations between the United States and Mexico. Back in 1973, Mexico and the United States reached an agreement on specific volume restrictions on the annual production of groundwater in the territories of both countries, within eight kilometers from the international boundary of Arizona-Sonora. In addition, the agreement requires that the two countries consult each other before any new development or substantial modification of both surface and groundwater resources on their territory in the border zone, which could adversely affect another country [15].

The absence of a common reciprocal management system for groundwater extraction from transboundary aquifers and complexes leads to numerous conflicts, including military actions in the territory of Israel, Palestine and Jordan.

For the North-West Federal District of Russia, this problem is relevant for groundwater operation from transboundary aquifers, which is relevant to their operation by neighboring countries - Estonia, Latvia, Lithuania, Poland, Belarus, Finland and Norway. First of all, this is due to the lack of information on the volumes and regime of groundwater production in adjacent territories, as well as the possibility of joint planning for further exploitation of aquifers.

This is not a complete list of examples of the lack of cooperation between states that exploit "common" groundwater resources and resulting disagreements. There are other agreements, in particular in Europe, that cover groundwater resources within a broader framework for cooperation in the management and protection of bordering rivers or in their development. However, groundwater plays an insignificant role in these agreements.

Thus, problems of groundwater production from transboundary aquifers and complexes are:

1. Absence of a unified international system for managing the extraction of groundwater as a particularly valuable type of minerals.
2. Attempt of local, as a rule short-term, laws and by-laws regulating the system of groundwater extraction management from transboundary aquifers. In this case, directives and framework agreements are adopted in each specific situation.
3. Scale of the problem, which lies in the field of its operation. The problem concerns any border subsoil use, in contrast to surface water bodies and streams, which are regulated by the European, Asian and other water directives [16].

During the identification of these problems, it is proposed to use the existing experience of managing groundwater production in Russia, Germany, France and other developed countries to create a unified international concept for managing groundwater production from transboundary aquifers and complexes.

The international management model implies development of a mechanism based on:

- uniformity of the licensing procedure for groundwater extraction (mining permits for individuals and legal entities);
- compatibility of the monitoring system based on uniform standards and reporting on the extraction of groundwater with identification of the main relevant parameters (factors such as dynamic parameters of heads when groundwater reserves are developed, periodic reporting of chemical and bacteriological analyzes);
- access to subsurface resources for all categories of water users with common subsoil use rules for transboundary aquifers;
- unification of the system for planning the development of groundwater deposits, including domestic, drinking, technical, medical and energy purposes based on modern numerical models, allowing complex, real-time analytical calculations (for example, as in Germany);
- system for financing the above systems by optimizing the taxation system (a tax on the extraction of minerals, as an effective way of managing the development of a groundwater deposit) and at the expense of state budget sources;

- planning geological exploration for both existing and poorly studied groundwater deposits, including the creation of a network of observation wells and facilities included in the network of a unified monitoring system using modern remote observation technologies;
- calculation of the limits of influence on transboundary aquifers, the level of critical aquifer operation;
- opportunities for supporting poor countries to provide access for specialists to build observing systems, to identify opportunities for restrictions on groundwater production volume;
- control over the target consumption of groundwater resources, which is especially important for aquifers with difficult water exchange or for unique deposits of medicinal mineral waters.

This model should be built on the basis of interstate agreements based on specific local legislation of individual countries, but within the framework of a single concept.
Advantages of the proposed model:

- simplicity;
- openness;
- understandability;
- universality (using numerical modeling);
- availability (low-cost implementation).

4. Conclusion

The world's groundwater resources have a key importance for sustainable development. Utilizing the full resources of groundwater and effective control of the perennial problems associated with groundwater, such as excessive use and pollution, are very complicated tasks. The exchange of information and experience in the groundwater production management around the world will be necessary and extremely useful in this regard [13].
Aquifer resources management, aiming at the responsible exploitation and adequate protection of the groundwater resources, is therefore of key importance and has to be based on sound hydrological, environmental, economic and social principles. In order to accomplish groundwater resources management goals in the case of transboundary aquifers, a balanced joint strategy is needed [17].
However, practice has shown that management of groundwater resources and their protection is a very difficult task. Groundwaters interact with other components of natural environment and are exposed to them. Often, there is only the most limited knowledge of local groundwater systems and their behavior; In addition, each groundwater system usually unites a multitude of consumers and other parties that often compete or conflict with each other. Adequate administrative measures for information systems, institutions, policies and various forms of support are required for effective measures to manage the extraction of groundwater resources.

5. References

[1] Golovina E I 2014 Economic mechanism of rational groundwater usage. *Proceedings of the Samara Scientific Center of the Russian Academy of Sciences.* 16 № 1 (4) pp 951–955
[2] Antonov A 2010. Water is new oil. URL: http://inosmi.ru/economic/20101024/163785871.html
[3] Yazvin A L 2015 Scientific substantiation of information support system of geological exploration resource potential fresh groundwater. *Abstract for Doctor. geol.-min. science. Thesis*, Moscow p 3
[4] Wada Y and Heinrich L 2013 Assessment of transboundary aquifers of the world-vulnerability arising from human water use *Environmental Research Letters* 8 2 p 17
[5] Sarma P B 2008 Groundwater: Development and Management *Allied Publishers Pvt. Ltd.* 812 p
[6] UNESCO 2010 Sharing water, sharing benefits: working towards effective. Transboundary water resources management *graduate / professional skills-building workbook. Compiled and edited by A.T., Wolf. Published by UNESCO, Oregon State University* p 284.

[7] Rivera A 2015 Transboundary aquifers along the Canada–USA border: Science, policy and social issues Journal of Hydrology: Regional Studies 4 p 623–643
[8] Vörösmarty C, Green J, Salisbury J and Lammers R 2000. Global water resources: vulnerability from climate change and population growth. Science p 289
[9] UNESCO Internationally Shared (Transboundary) Aquifer Resources Management 2001 *A* framework document, Paris 71 p
[10] Eckstein G and Eckstein Y 2003 A Hydrogeological Approach to Transboundary Ground Water Resources and International Law American University International Law Review 19 2 p 201-258
[11] Kyoko M 2002 Transboundary groundwater and international law: past practices and current implications Research paper Oregon State University 67 p
[12] UNEP (United Nations Environment Programme) 2016 Transboundary Aquifers and Groundwater Systems of Small Island Developing States *Transboundary Waters Assessment Programme* 14 p
[13] IGRAC (International Groundwater Resources Assessment Centre https://www.un-igrac.org/ru
[14] UNESCO 2013-2015 GRGTA (Groundwater Resources Governance in Transboundary Aquifers) unesdoc.unesco.org/images/0024/002430/243003r.pdf
[15] Frisvold G and Caswell M 2000 Transboundary water management Game-theoretic lessons for projects on the US-Mexico border *Agricultural Economics* 24 p 101-111
[16] Directive 2000/60/EC of the European parliament and of the council of 23 October 2000, establishing a framework for Community action in the field of water policy
[17] Tujchneider O, Christelis G and Van der Gun J 2013 Towards scientific and methodological innovation in transboundary aquifer resource management *Environmental Development* 7 p 6–16

Acknowledgments
This paper was supported by the grant of the government of Saint-Petersburg (Committee of Science and Higher Education), diploma and reference № 16002. Author also would like to thank Professor Dr. Broder J. Merkel, TU Bergakademie Freiberg (Germany), Professor Dr. Natalia V. Pashkevich, Vice-rector of Saint-Petersburg Mining University, Russia.

Characteristics of Hydrogen and Oxygen Stable Isotopes of Different Water Bodies in Peixian Coal Mining Subsidence Area in Jiangsu Province, China

Mengyu Ge[1], Junfeng Qu[1], Anni Wang[1] and Baozhang Chen[1, 2*]

1 School of Environment Science and Spatial Informatics, China University of Mining and Technology, 221000, China
2 Institute of Geographic Sciences and Natural Resources Research, Chinese Academy of Sciences, Beijing, 110000, China

Abstract. As a result of long-term well mining activities, a large area of coal mining subsidence has formed in the mining area in Peixian County, Jiangsu province, an important energy base in eastern China. This subsidence filled by water made the local water cycle become more complicated and severe geological hazards easy to occur. Oxygen-stable isotope, playing an important role in the study on water-cycle mechanism, is a substantial way to track different sources of water in nature. In this study, we took samples from different water bodies of soil water, the surface waters (river, lake and collapse pit) and the groundwater (well water) seasonally in a year. The analyses of δD and $\delta^{18}O$ were further conducted to trace the transforming relationships between these water bodies. The findings are as follows: (i) precipitation is the main charge source of soil water, along with significant evaporation processes; (ii) with soil depth increasing, δD and $\delta^{18}O$ of soil water showed a trend like that first decreasing to a certain value and then stabilizing and finally increasing to the bottom of soil profile; (iii) when the reclamation soil experienced longer time, the values of soil water δD and $\delta^{18}O$ increased in the whole profile, and this character was more obvious in deeper layers than in shallower and in surface layers; (iv) indicated from the isotopic characters of different water bodies and precipitation that precipitation was the main source of recharge of collapse pit water, and the infiltrated soil water and ground water experienced mixing with river water and precipitation by the water pathways conducted through the entire Cenozoic coal seam by mining.

1. Introduction

Owing to stable water isotopes are very sensitive to water environments, ${}^{1}H_2{}^{18}O$ and ${}^{1}HD^{16}O$, the composition of oxygen and hydrogen elements of water have been approved to be good tracers for water transition processes[1]. Isotopic fractionation occurs during evaporation and condensation processes in water cycle, the characteristics of δD and $\delta^{18}O$ from different water bodies, such as atmospheric water vapor, precipitation, surface waters, soil water and groundwater can be consequently used to trace recharge sources, migration path and retention time of water bodies[2].Research on stable isotope technology research started from 1961, and *the Global Network of Isotopes in Precipitation* (*GNIP*) has been established (currently there are over 800 sampling stations in the world) under the co-sponsorship of *the International Atomic Energy Agency* (*IAEA*) and *the World Meteorological Organization* (*WMO*). Based on these monitoring data sets from the GNIP, Dansgaad[3] analyzed the temporal and spatial characteristics of stable isotopes in precipitation globally, and put forward the concepts of temperature effect, precipitation effect and excessive deuterium. So far in China, a number of researches have been carried out in different areas, such as the Qinghai-Tibet Plateau, the Loess Plateau, the southwestern region, the eastern monsoon region, the Urumqi basin, and the Inner Mongolia's Hetao Plain. However, there are few researches on characteristics of hydrogen and oxygen stable isotopes of different water bodies in coal mining subsidence reclamation area, for example, Lei Zhang et al. (2015)[4] analyzed the source of different water bodies in the subsidence of Huainan coal mining area through hydrogen and oxygen stable isotopes techniques, Junjie Liu et al.(2009)[5] and Tao Ge et al.(2014)[6] made use of the stable hydrothermal isotopes to trace deep groundwater sources in mining areas.

China, as a large coal production country, accounting for half production over the world[7].Coal mining often leads to the surface deformation and forms a large elliptical subsidence, inducing the destruction of original cultivated land, the disturbances of surface water systems, potential geological hazards and so on, which can pose a serious and negative impacts on local people's economics and life. As a large subunit of the China Coal Group, the Datun Coal Electricity Company, located at Weishan Lake at the junction of Jiangsu and Shandong provinces, covering area of 245 km^2, outputs 9,055,900 tons of raw coal and 3,906,700 tons of cleaned coal a year. Long-term coal mining activities have led to severe and contiguous surface subsidence in the large-scale coal mining area, forming a large area of water.

Aiming to make up for the inadequate research on the characteristics of hydrogen and oxygen stable isotopes for the mining areas in China, and to find out the relationships between the different water bodies and the impact of the time of after reclamation activity on the characteristics of hydrogen and oxygen stable isotopes in the mining areas of Peixian county, this study analyzed and tracked the sources of subsidence pit water in the research mining-areas of Longdong Coal Mine, Yaoqiao Coal Mine, Xuzhuang Coal Mine, Kongzhuang Coal Mine in the western part of Weishan Lake in Xuzhou, Jiangsu province. We took samples from the farmland soil water of with varied reclamation years, using a "space replacing time" method, and then analyzed the characteristics of hydrogen and oxygen stable isotopes to detect the impacts of different reclamation years on the characteristics of soil water hydrogen and oxygen stable isotopes. We also measured and compared the isotope characteristics of precipitation, surface waters (river, lake, and subsidence pit water) and groundwater to explore the relationship among these water sources.

2. Research Area Overview

The study area (116.8°E, 34.9°N, Figure 1), located in the western part of Weishan Lake, belongs to the warm temperate zone and semi-humid monsoon continental climate with characteristics of hydrothermal synchronization Its multi-year average annual temperature is 13.7°C, with the minimum and maximum monthly mean temperature of 0.4°C in January and 27.1 °C in July. The muti-years average annual total precipitation is 760 mm, with characteristics of extremely uneven distribution (about 60% in summer and only 5% in winter). The muti-years average wind speed is 2.7 m/s~3.8 m/s, with dominant wind direction from southeast wind and secondary wind direction from northwest in winter. A network of waterways developed well with shallow groundwater and porous alluvial soil type which is suitable for cultivation.

There are several coal mining areas. The Longdong coal mining area located at Longgu Town, which annual production was 1.2 million tons and was put into operation from 1987; the Yaoqiao coal mining area located at Yangtun Town on the western of Weishan Lake was put into operation in 1976 with geological mine reserves of 540 million tons and original designed reserves of 1.2 million tons, and in 1990, it was reconstructed to extend its production capacity of reaching to 360 million tons, which made it to be the largest mining area in Jiangsu Province. The Datun coal mining area, as one of the earliest mining area located at Datun in Peixian County, was put into operation in 1979 with annual designed production of 900,000 tons, actually from 1980 to 2004, the total production of raw coal in this mine was 21.926 million tons. The Kongzhuang coal mining area, located at the southernmost part of

Datun town, covering an area of approximately 570,000 square meters, with annual production capacity of 1.05 million tons.

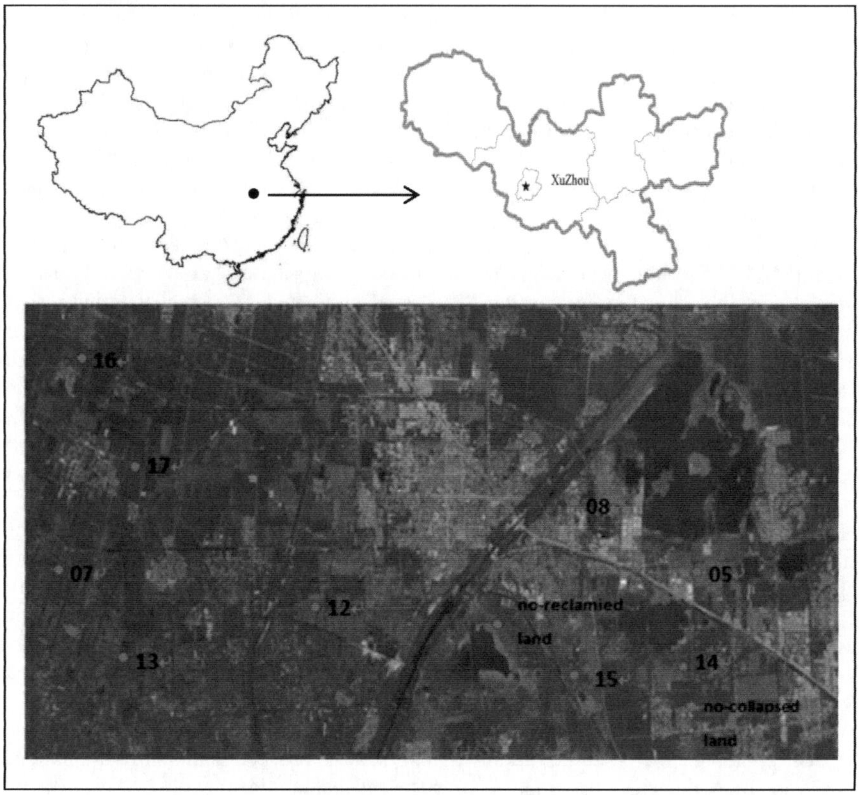

Figure 1. Location of the study area

3. Sampling and Experimental Methods

3.1. Sampling Methods

3.1.1. Sampling of precipitation. During experimental period, precipitation was collected by precipitation collector which was placed on sample plot randomly. Precipitation was collected in a standard rain bucket, and a round funnel was installed above it to collect rain or snow, with a ping pong ball above the funnel mouth to avoid losses of experimental accuracy that evaporation could occur. The collected, precipitation was re-filled to a 30-ml plastic centrifuge tube and sealed with parafilm making sure no air leaching. The tubes were stored in a refrigerator keeping temperature at 4 degrees immediately to prevent moisture from evaporating. By the way, if there were two precipitation samples one day, the averaged value of these two samples was taken as the stable isotope ratio of precipitation on that day. In total, 38 precipitation samples were collected during the study period.

3.1.2. Sampling of soil. According to the method of “space instead of time”, this study conducted sample collection from the reclaimed plots with 0, 1, 2, 3, 4, 5, 9, 10 and 12 years old, as well as one from no-reclaimed plot and one from no-collapsed plot for comparison purpose. The field sizes of these reclamation series were 4.4689, 4.1354, 4.2688, 4.7357, 4.6023, 4.4022, 4.8691, 4.3355, 4.5356, 4.0687 and 4.6690 Mu (Mu = 0.0667 hectares). 3 plots were randomly distributed on each plot, and the soil samples were collected at the depth of 0-100 cm soil profile with 10 cm intervals, which were sealed in bags for a physical and chemical properties analysis and for isotope analysis. Totally, 300 soil samples were collected.

3.1.3. Sampling of the surface waters (river, lake, and subsidence pits) and groundwater (well water). Water samples from river and subsidence pit were taken a few centimeters below the surface to ensure the sampled river water being fully mixing and no isotope fractionation impaction by the possible surface water evaporation. One part of the sampled water was sealed in centrifuge bottles (30 ml) for isotope analysis, and the other part was stored in polyethylene bottles (100 ml) for physico-chemical properties analysis. The sampling method groundwater was same as the surface water. Total 58 water samples were collected.

3.2. Experimental Methods

All samples were processed and analyzed in the Isotope Analysis Laboratory, China University of Mining and Technology.

(1) Soil moisture extraction

Based on the theory of vacuum distillation, the LI-2000 vacuum distillation device for plants and soil moisture was used to extract soil moisture, the extraction time was at least 2 hours long to ensure soil moisture was entirely extracted to avoid experimental errors from incomplete extraction. In addition, the extraction process was conducted once more time in the case of the amount of extracted water was less than 1 ml).

(2) Liquid water analysis

In this study, the LGR908-0008 Liquid Water Isotope Analyzer was used to detect the content of δD and $\delta^{18}O$. As for the precision of this instrument, $^{18}O/^{16}O$ and D/H are above 0.1‰ and above 0.3‰, respectively.

4. Results and Discussion

4.1. Isotopic Composition in Precipitation

As shown in Figures 2 and 3, precipitation was the main source of soil water, surface waters (river, lake and collapse pit), and underground water (well water). Analysis of δD and $\delta^{18}O$ in precipitation is the basis for studying on the interaction between different water bodies. Tab. 1 shows that the value of δD in precipitation ranged from -92.5‰ to -3.5/‰, with average of -50.0‰ and standard deviation of 0.43‰; the $\delta^{18}O$ ranged from -13.1‰ to -2.1‰, with average of -7.96‰ and standard deviation of 0.32‰; the value of d-excess ranged from -2.25‰ to 29.29‰, with average of 13.62‰ and standard deviation of 0.5. Table 1 also shows the seasonal variation of δD、$\delta^{18}O$ that higher isotopic values were found in winter, whereas lower values in summer in general. There are two main factors can be attributed to this trend as follows: (i) On a large scale, isotopic fractionation caused by the source of vapor as well as transportation process while vapor transferring. Our study area is located in the east part of China, which is affected by monsoon severely. During summer monsoon, maritime air mass often condenses into rain times as it goes inland, which depletes the content of heavy isotope. Consequently, the values of δD and $\delta^{18}O$ in precipitation in the research area were lower. By contrast, during winter monsoon, continental air mass carried by westerly zone was dry and cold. Besides, the atmosphere during this season was stable and this makes it hard to form rain which causing isotopic fraction. Therefore, a small amount of heavy isotope depletes in winter; (ii) regional factors were also responsible for this trend, including temperature, rainfall, relative humility. It is found that temperature was low in winter and spring while higher in summer and autumn, and rainfall mainly occurred in summer (from June to August. The values of δD and $\delta^{18}O$ of precipitation in summer were lower than those in winter, indicating the isotope fractionation effects of rainfall and of converse temperature. Similar phenomenon was also seen in other monsoon areas, such as Yunnan[8], Nanjing[9]and Xiamen[10]. Different results were also reported by Vodila[11], Fadong Li[12], Catherine E[13], Congjian Sun[14]because of different atmospheric circulations in their research areas.

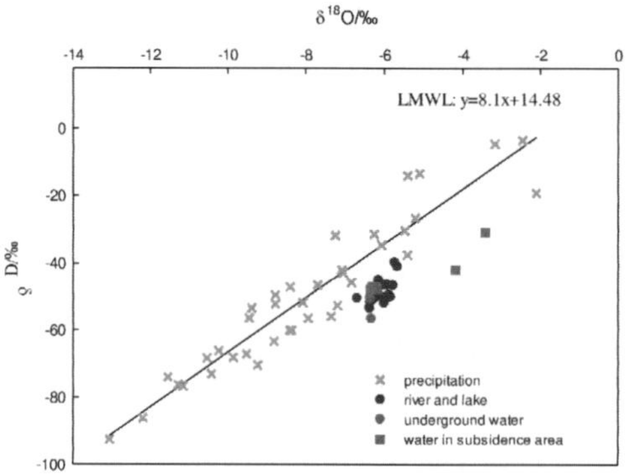

Figure 2. Characteristic distribution of δD and δO[18] of rainwater, mining subsidence cracks and underground water

Table 1. Seasonal variability of δD, $\delta^{18}O$ and d-value in precipitation

Date	δD(‰)	$\delta^{18}O$(‰)	d-excess(‰)	Date	δD(‰)	$\delta^{18}O$(‰)	d-excess(‰)
2016.11.7	-56.46	-9.44	19.08	2017.7.15	-66.19	-10.22	15.61
2016.11.9	-4.58	-3.17	20.79	2017.7.26	-52.65	-7.18	4.83
2016.11.22*	-73.96	-11.54	18.38	2017.8.2	-92.53	-13.05	11.87
2016.12.1	-42.94	-7.05	13.43	2017.8.7	-30.32	-5.49	13.57
2016.12.21	-70.37	-9.22	3.4	2017.8.12	-45.73	-6.83	8.9
2016.12.25	-31.79	-7.25	26.22	2017.8.18	-60.05	-8.36	6.83
2017.1.6	-49.59	-8.77	20.59	2017.8.19	-60.07	-8.41	7.17
2017.1.7	-52.18	-8.77	17.98	2017.8.29	-31.36	-6.26	18.7
2017.1.29	-13.33	-5.1	27.45	2017.8.3	-46.53	-7.69	15
2017.1.31*	-51.79	-8.07	12.74	2017.9.3	-76.22	-11.28	13.99
2017.2.8*	-47.1	-8.39	20.03	2017.9.4	-41.95	-7.09	14.73
2017.4.9	-19.11	-2.11	-2.25	2017.9.6	-86.12	-12.19	11.41
2017.5.3	-3.5	-2.45	16.13	2017.9.25	-68.17	-9.85	10.66
2017.6.5	-13.95	-5.4	29.28	2017.9.26	-63.37	-8.8	7.06
2017.6.1	-26.59	-5.21	15.13	2017.9.3	-76.62	-11.16	12.63
2017.6.23	-55.9	-7.35	2.92	2017.10.4	-37.46	-5.41	5.82
2017.7.6	-68.34	-10.54	15.96	2017.10.5	-34.57	-6.07	13.96
2017.7.7	-56.52	-7.93	6.94	2017.10.1	-53.46	-9.38	21.54
2017.7.13	-73.1	-10.42	10.26	2017.10.11	-67.16	-9.51	8.94

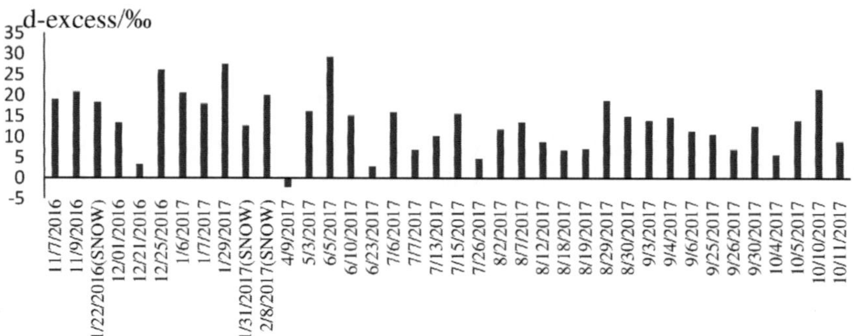

Figure 3. variations in d-value of precipitation

The phenomena of d-excess reflect the thermodynamic condition and vapor equilibrium state when the precipitation vapor forms [15], and also reflect their climatic conditions and geographic characters where precipitation forms. Kinetic fraction intensifies evaporation under arid conditions, this would increase the value of d-excess in precipitation [16]. For example, the d-excess value exceeds 10‰ in Koster (1993) [17], Gat (1994)[18] and Machavarams'[19] studies, and this suggests that evaporation plays an important role in regional water cycle. The distribution of d-excess for a year round sampling in this study can be seen in Figure 3 which presents the values of d-excess vary from -2.25‰ to 29.3‰, with average value of 13.6‰, which is well high compared to the global weighted mean d-excess of +10‰[20]. Among these 38 precipitation samples collected during the study year, 6 and 28 samples had d-excess values above 20‰ and 10‰, respectively, suggesting that evaporation was very important in the study area. There were 26 samples collected during May to October （warm half year）, their average value was 12.15‰, and of which 17 samples' d-excess >10 ‰ and 9 samples' d-excess <10‰.

One thing should not be ignored that the value of d-excess should be less than 10‰ during the warm half year considering the study area is located in the eastern monsoon region which is affected by summer monsoon. However, the higher value of d-excess occurred many times in our precipitation samples, which may be attributed by the strong evaporation fractionation caused by regional evaporation and long-distance transportation from lower latitudes to the study area. The mean value of d-excess is 16.46‰ during the cold half year, especially in January and February when the value reached 20‰. The high value of d-excess may be due to the non-equilibrium evaporation caused by the vapor transport process and secondary evaporation in winter when the weather was rather dry[21]. This research result coincides with theirs', i.e. Huawu Wu[22], Zongxing Li[23], and J Wu's[24].

The relationship between δD and $\delta^{18}O$ is defined as local water meteoric water line (LWML) which is useful for analyzing physical geography, meteorological condition, historical climate, and the sources of vapors[25]. In 1961, Craig[26] proposed a global meteoric water line (GMWL) at the first time when he found the linear relation between δD and $\delta^{18}O$ during reaching precipitation in North America. GMWL can directly reflects the non-equilibrium extend of evaporation and condensation in precipitation, this line has been consequently widely used in analyzing regional geography, climatic conditions and interactions between different water bodies. We can see from Figure 2 that the local LMWL is $\delta D = 8.11\delta^{18}O+14.48$ (R^2=0.9). Obviously, the values of both the slope and intercept of the LMWL were higher than of that of the global GMWL. This is because under the Reyleigh isotope fractionation condition, the fractionation rate of δD is eight times higher than $\delta^{18}O$, and this rate increases with lower air temperature. Owing to the characteristics of moist and rainy in the study area and air temperature is lower when condensation and the secondary evaporation is not weak, leading to higher values of slope and intercept of LWML. LWML in the study area was similar to those areas, such as Tengchong [27], Changsha[28], Yichang29, Guilin[30] because of these areas have similar water sources, climatic characters, and the influence of monsoon. In addition, the LMWL line of cold half year (May to October, $\delta D = 7.34\delta^{18}O+11.61$， R^2=0.85) and of warm half year (November to April in the next year, $\delta D = 8.33\delta^{18}O+15.01$， R^2=0.93) were quite different: the values of the slope and intercept of LWML were higher in the warm half year than those in cold half year, which is attributed from different monsoon in the different seasons (marine vapor is the main source in the warm half year while continental air mass prevails in the cold counterpart).

4.2. Isotopic Composition in Soil Water with Different Reclamation Ages

Soil water links the surface water, groundwater and plant water in the atmosphere-soil-plant system. Table 2 shows the δD and $\delta^{18}O$ characteristics of soil moisture at different soil depths. It shows that the values of δD and $\delta^{18}O$ with different reclamation ages ranged from -92.03 ‰ to -43.26 ‰ and-13.13‰~-5.56‰， with average of -13.13 ‰ and -8.93‰ and standard deviation of 0.16‰ and 0.17‰. The values of mean and standard deviation of δD and $\delta^{18}O$ in soil moisture were much lower than that of δD and $\delta^{18}O$ in atmospheric precipitation of -50.00 ‰ and 0.43, -7.96 ‰ and 0.32‰, respectively. The characteristics of soil moisture with lower negative values of δD and $\delta^{18}O$ and low standard deviation values reflects that the soil moisture was not purely from the precipitation in the current season but affected by mixture of old soil water with previous precipitation. From figure 4, it can be seen that the δD and $\delta^{18}O$ values of soil water were mostly at the right-bottom of the local atmospheric water line, indicating that precipitation was the recharge source of soil water. The linear regression relationship between δD and $\delta^{18}O$ of soil water is $\delta D = 6.34\ \delta^{18}O+8.31$ (R^2=0.95, n=210), and the values of slope and intercept are both lower than that of LMWL, with an average d-excess of 6.53 ‰ which is lower than the global average of 10‰. This suggests that soil water experienced evaporation and fractionation during infiltration process, which was mainly related to the dry season with drought and low air humidity during the study period.

As a result of precipitation infiltration, surface water evaporation, soil texture, soil moisture content and other factors, the values of δD and $\delta^{18}O$ of soil water at different soil depths follow a certain pattern, indicating the degree of soil water affected by infiltration and evaporation[31]. Due to the interactions between environment and the human activities, there were poor regularity and significant fluctuation of water isotopes for the surface soil. There was a decreasing tend in δD and $\delta^{18}O$ from the soil surface to 40cm depth. The effects on δD and $\delta^{18}O$ by evaporation decrease with soil depth following the principle of Rayleigh fractionation principle that heavy isotopes would be enriched in the surface soil water. Below the depth of 40 cm, the values of δD and $\delta^{18}O$ were fluctuated with an increasing trend in general with increasing depth, which is in the line with the findings of Chung[32] and Bristow[33]. This may be mainly because of the shallow groundwater containing high δD and $\delta^{18}O$ content, and steady groundwater environment with weak disturbance and mixing of deep soil water and groundwater. It is noteworthy that the average values of δD and $\delta^{18}O$ are -67.42 ‰ and -9.19 ‰ at the depth of 0-10 cm, and -64.08 ‰ and -8.76 ‰ at the depth of 50-100 cm, which is inconsistent with the results of Zimmermann et al[34] and Barnes et al[35]. This inconsistence may be mainly because their study areas and sampling frequency are different from this research. Their research area is in arid and semi-arid regions which are dry and with small amount of precipitation, precipitation is solely recharge source for soil water but is strongly affected by evaporation, while this study area is located along the Weishan lake with a humid climate. The comparatively low air temperature in the study area led to weak evaporation fractionation of $\delta^{18}O$ at the depth of 50-100 cm, the enrichment of $\delta^{18}O$ in soil water is not obvious compared to the $\delta^{18}O$. The value of $\delta^{18}O$ of subsidence pit water was higher (average value of -5.56 ‰) than that of other surface waters. Located at the subsidence region of western coal mining area in Weishan lake, the study region experienced surface deformation and collapse as a result of long-term mining activities. The subsidence pits were recharged by precipitation, surface runoff and water from river and lake nearby for a long time, water gradually accumulated in the pits. And unlike flowing state of rivers, the subsidence pit is a relatively closed area, leading to weaker evaporation and fractionation[36].The reclamation of soil cover on these subsidence pits made the subsidence pit water as a part of groundwater, therefore the $\delta^{18}O$ value of groundwater increased. The variations in the $\delta^{18}O$ content of soil water of with different reclamation periods ages varies were more obvious in the shallow and deep layers much more than in the middle layers, owing to obvious mixing effect and a combined influence of attributed from infiltration of precipitation, accumulated water in subsidence pit, upward groundwater and water taken by plant root.

As discussed above, soil moisture transportation, affected by many factors, is a very complicated physical-chemical process. And the surface soil is prone to be affected by other factors including human activities. However, below the depth of 60cm, the values of δD and $\delta^{18}O$ of soil water tended to increase with longer reclamation period (Fig. 5). This is because the fractionation effects of evaporation and other processes (i.e. infiltration) became stronger

IOP Conf. Series: Earth and Environmental Science **151** (2018) 012008 doi:10.1088/1755-1315/151/1/012008

as the reclamation time was getting longer. According to the theory of isotope fractionation, lighter isotopes evaporate preferentially from the soil, leaving the heavier isotopes, allowing the fact that the longer the reclamation time, the greater the δD and δ^{18}O content of soil water.

Table 2. Stable isotope values of soil water at different depths

Depth/cm		δD ‰				δO^{18}/‰			
	N	Max	Min	Average	Sd	Max	Min	Average	Sd
0-10	33	-55.23	-92.03	-67.42	0.15	-12.44	-9.19	-9.19	0.18
10-20	33	-48.13	-80.35	-68.11	0.13	-11.14	-9.29	-9.29	0.15
20-30	33	-48.83	-84.44	-68.09	0.13	-13.13	-9.60	-9.60	0.16
30-40	33	-46.30	-74.77	-62.86	0.15	-10.64	-8.78	-8.78	0.17
40-50	33	-43.36	-73.08	-62.59	0.12	-10.53	-8.65	-8.65	0.13
50-60	33	-47.08	-86.37	-63.78	0.17	-11.56	-8.90	-8.90	0.17
60-70	33	-43.91	-77.20	-60.86	0.14	-10.74	-8.43	-8.43	0.15
70-80	33	-46.53	-87.35	-68.58	0.16	-12.25	-9.34	-9.34	0.17
80-90	33	-45.37	-78.71	-63.52	0.18	-11.19	-8.59	-8.59	0.17
90-100	33	-43.26	-82.51	-63.64	0.19	-11.29	-8.55	-8.55	0.18

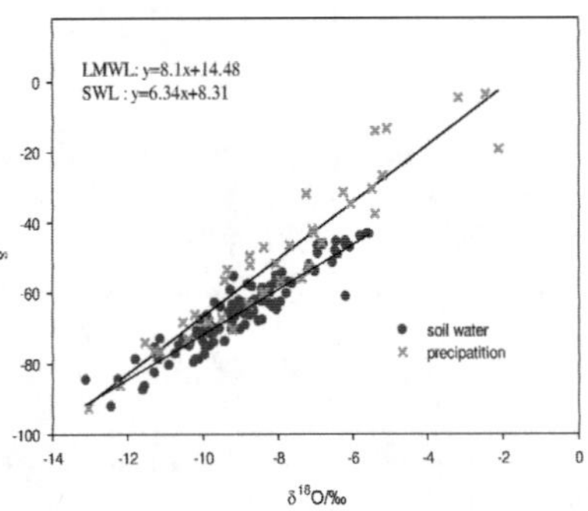

Figure 4. Characteristic distribution of δD and δO^{18} of soil water

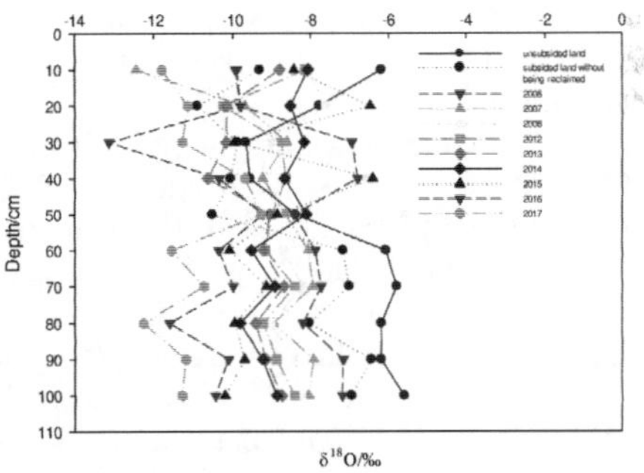

Figure 5. Soil water δ^{18}O profiles with different reclamation ages

4.3. Analyses δ^{18}Oand δD for Surface Waters (River, Lake and Collapse Water) and underground Water (Well Water)

δD in river and lake varied from -53.41‰ to -39.72‰, with mean value of -48.53‰, the values of δ^{18}O varied from -6.69‰ to -5.67‰, with mean value of -6.09‰; the values of δD in collapse water varied from -47.85‰ to -30.79‰, with mean value of -44.23‰, the values of δ^{18}O ranged from -6.35‰ to -7.59‰, with mean value of -5.56‰. The values of δD of underground water (well water) varied from -56.60‰ to -50.58‰, with mean value of -53.59‰, the values of δ^{18}O of underground water varied from -7.88‰ to -7.59‰, with mean value of -7.64‰. All isotopic values of above mentioned water bodies fall on the right-side of the LMWL, indicating that they all received the recharge of rain with isotopic fraction by evaporation which made higher values of δD and δ^{18}O during the recharging processes. The values of δ^{18}O and δD in the collapse water were higher and distributed more distractedly compared with other water bodies. Besides, the values of δ^{18}O and δD of several river and lake water bodies are closely to that of collapse water, suggesting that the river and lake would also be the source for the collapsed pit water. The values of δ^{18}O and δD for most of underground water samples are close to that of precipitation, and this is because that it receives recharge of rainfall. In addition, part river and lake samples have similar δ^{18}O and δD with underground water. Therefore, it is concluded that regional cracks are likely to connect surface water and ground water in the form of infiltration of surface water.

5. Conclusion

Based on the analyses of δD and $\delta^{18}O$ for different water bodies in Xuzhou collapse area, we draw conclusion as follows:

(i) The values of δD and $\delta^{18}O$ of different water bodies had an overall seasonal pattern with higher values in winter while with lower values in summer. The LMWL line for the study area was that, $\delta D = 8.11\delta^{18}O+14.48$ ($R^2 = 0.9$). The values of the slope and intercept of LMWL were bigger than the global GMWL, indicating the local humid weather pattern.

(ii) Almost all δD and $\delta^{18}O$ values fall on the right side of the LMWL, which reflects the effects of evaporation during replenishment course of precipitation. The values of δD and $\delta^{18}O$ of soil water at the upper part of soil profile (0-40 cm) decreased with depth, and then increased with depth below 40cm and remained stable to 100cm depth. The variations in the upper and bottom soil profile were larger than that in the middle. In addition, the overall values of δD and $\delta^{18}O$ of soil water samples show an increased trend with the increase in reclamation age because of the increasing in isotopic fraction effect by evaporation.

(iii) Almost all values of δD and $\delta^{18}O$ of river, lake, collapse water and underground water (well water) also fall on the right side of LMWL, indicating that precipitation was the main source of above water bodies. Some samples from lake and river have similar values of δD and $\delta^{18}O$ to that of collapse water, indicating that the lake and river replenish the collapse counterpart. In addition to this, similar hydrogen and oxygen isotope characteristics between river, lake and underground water suggests that the surface waters were likely to have connection with ground water in the form of surface water infiltration through cracks in the study area between river, lake and underground water.

6. Reference

[1] Gu Weizu and Pang Zhonghe 2011 Isotopic hydrology(Beijing: *Science press*) p105
[2] Hoefs J 1973 Stable Isotope Geochemistry *Springer-Verlag*, p102
[3] Dansgaard W. Stable isotopes in precipitation [J]. *Tellus*, 1964, 16 436-468.
[4] Zhang Lei, Qin Xiaoguang, Liu Jiaqi, Mu Yan and An Shikai 2015 China. *J. Journal of Jilin University*, 45 1502-04
[5] Liu Junjie, Zhao Feng and Wang Daguo 2009 China.*J.Journal of China coal society*, 39-43
[6] Ge Tao, Chu Tingting, Liu Guijian, Fan Xiang and Wu Dun 2014 China. *J. Journal of University of Science and Technology of China*, 44 112-8
[7] Wang Peijun, Hu Zhenqi, Yost R S, and Liu Junting 2016 China. *J. Environmental Earth Sciences,* 75 1-15
[8] Zhang Xinping, Liu Miaojing, Zhongwei Zhengyi, and Zichu Xie 2006 China *.J .Chinese Science* 36 850-9
[9] Song Xianfang, Liu Jianrong, and Sun Xiaoming 2007 China.*J. Advances in Earth Science*, 22 738-47
[10] Chen Yanting, Du Wenjiao, Chen Jinsheng, and Xu Lingling 2016 China.*J. Acta Scientiae Circumstantiae*, 36 (2):667-74
[11] Vodila G, Palcsu L, Futó I, and Z Szántó 2015 France *.J. Journal of Hydrology*, 400 144-53
[12] Li Fadong, Song Xianfang, Tang Changyuan, Changming Liu, Jingjie Yu, and Wanjun Zhang 2007 China *.J. Environmental Geology*, 52 1439
[13] Hughes C E, Crawford J 2013 Australia *.J. Journal of Hydrology*, 489 42-55
[14] Sun Cangjian, Li Xingong, Chen Yaning, RL Stotler.2016 China. *J. Isotopes in Environmental & Health Studies*, 52 1
[15] Jouzel J 1986 American *.J. Terrestrial Environment B*, 61-112
[16] Merlivat L, Jouzel J 1979 American *.J.Journal of Geophysical Research Oceans* , 84 5029-33
[17] Koster R D, De Valpine D P, and Jouzel J 2013 American *.J. Geophysical Research Letters*, 20 2215-18
[18] Gat J R, Bowser C J, and Kendall C 2013 American *.J. Geophysical Research Letters*, 21557-60

[19] Machavaram M V, Krishnamurthy R V 1995America .J.*Geochimica Et Cosmochimica Acta*, 59 4279-83
[20] I Wmo2006 .M. *The GNIP database*
[21] Terzer, Wassenaar, L. I, I L, Araguas-Araguas, J L 2013 France .J. *Hydrology & Earth System Sciences* 17 4713-28
[22] Wu Huawu, Zhang Xinping, Li Xiaoyan, and Jing Li. 2014 China .J.*Scientia Geographica Sinica* 34 4
[23] Li Zongxing, Feng Qi, Yong Song, Jiao Yang, Li Yongge, Li Jiangguo, and Guo Xiaoyan 2016 China .J. *Atmospheric Research* ,182:87-101
[24] Wu Jinkui, Ding Yongjian, and YE Baishengl2012 China .J. *Chinese Geographical Science* 22 531-40
[25] Liu Xin, Song Xianfang, Xia Jun, Yu Jingjie, and Zhang Xuecheng 1961China .J. *Resource Science* 29 59-66
[26] Craig H 1961 America .J. *Science*, 133 1702-03
[27] Li Guang, Zhang Xinping, and Zhang XinZhu 2013 China .J. *Resources and Environment in the Yangtze Basin* , 22 1458-66
[28] Huang Yimin, Zhang Xinping, Sun Xia Yibin Huang, and Naiqiong Wei 2012 China .J.*Scientia Geographica Sinica*,21540-52
[29] Wu Yazun, Wan Junwei , and Yun Lin 2011China.J.*Geological Science and Technology Information*, 30 93-97
[30] Wu Xia, Zhu Xiaoyan, Zhang Meiliang, Xiao Bai, and Biyun Zhang. 2013 China .J. *Resources and environment in the Yangtze Basin*, 22 182-18
[31] Xu Qing, Liu Shirong, Wan Xianchong, and Song Xianfang 2011 China .J. *ESA Convention*. 3800–09
[32] Chung S O and Horton R 1987 America .J. *Water Resources Research*, 23 2175-86
[33] Bristow K L and Horton R 1996 America .J. *Theoretical & Applied Climatology*, 54 85-98
[34] Zimmermann U, Ehhalt D, and Muennich K OJ. 1968 Vienna .J.*International Atomic Energy Agency* , 567-85
[35] Larsbengtsson, Saxena R, and Zegeyedressie 1987 Sweden .J. *International Association of Scientific Hydrology Bulletin*, 32 497-520
[36] Cheng Yuting, Li Peng, Guoce Xu, Zhanbin Li, Tiegang Zhang, and Xiaojun Liu 2014 China.J.*Journal of soil and water conservation*, 28 129-33

Acknowledgements
This work was supported by the research grant (No. 2016067), funded by the Water Resources Department of Jiangsu Province the research grants (No. O88RA901YA and No. O8R8A085YA) by the State Key Laboratory of Resources and Environment Information System and the research grants (No. 41771114 and 41271116) funded by the National Science Foundation of China.

IOP Conf. Series: Earth and Environmental Science **151** (2018) 012009 doi:10.1088/1755-1315/151/1/012009

Species Diversity, Biomass, and Carbon Stock Assessments of Mangrove Forest in Labuhan, Indonesia

M A Asadi[1,2], D Yona[1,3] and S E Saputro[1]

1 Marine Science Department, Brawijaya University, Malang, Indonesia
2 CoReACT, Brawijaya University, Malang, Indonesia
3 MEXMA, Brawijaya University, Malang, Indonesia
E-mail: asadi@ub.ac.id

Abstract. Mangroves provide numerous ecosystem services, including fisheries production, nutrient cycling, soil formation and carbon storage. However, the forests and their habitat have degraded rapidly due to anthropogenic threats. Consequently, the loss of carbon (C) stored in these ecosystem is inevitable. The purpose of this study was to evaluate the species diversity of mangrove forest of Labuhan, Lamongan regency, Indonesia and evaluate its above-ground and below-ground-root biomass and C. Twenty-four plots with a size of 10x10 m were carried out using the quadrat sampling technique to identify, record and measure the DBH of the trees. The Shannon - Wiener's diversity index (H'=1.51) was moderate, having a total of nine true mangrove forests dominated by *Rhizophora mucronata* and *Rhizophora apiculata* with an importance value index of 100.40% and 80.46% respectively. The total C stored in above-ground and below-ground-root was 74.70 ± 15.58 Mg C ha^{-1}, which is equivalent to total CO_2 sequestration of 274.15 ± 57.18 Mg CO_2 ha^{-1}. This study suggests that the Labuhan mangrove forest has a potential to store and sequester a substantial amount of atmospheric C; therefore, it is needed to protect and sustainably manage this important forest.

1. Introduction

Burning of fossil fuels and deforestation are the main sources of greenhouse gases that lead to higher carbon dioxide (CO_2) concentration in the atmosphere. In particular, deforestation and changes in tropical land use account for up to 1.6 ± 1 GtC per year, which is one third of total anthropogenic CO_2 emissions [1]. This has induced many studies of the capacity of forests and wetland ecosystems in sequestrating carbon. Most of the available studies are related to terrestrial forest ecosystem, and there are not enough studies on carbon sequestration potential of mangrove forests. These forests have a high capacity to reduce greenhouse effect via carbon (C) sequestration and storage, and have been reported to sequester up to three times more C per equivalent area than terrestrial forests [2]. However, these forests are vulnerable to climate change impacts, particularly sea level rise, and threatened by rapidly expanding aquaculture and infrastructure development [3].

Globally, up to 35% of mangrove areas has been lost since 1980, and mangrove areas are estimated to disappear as high as 2–8% per year [4]. In Indonesia, mangrove cover has decreased by 40% in the past three decades mainly due to the development of aquaculture, which was the impact of rapid expansion of pond aquaculture between 1997 and 2005 [5]. In order to mitigate the degraded mangrove ecosystem, mangrove restoration and rehabilitation programs have been initiated in many regions in which tourism benefits and shoreline stabilization are added as mangrove conservation goals. Recently, carbon sequestration has also been included as the objective of mangrove restoration for mitigating climate change, in which it is aimed to maintain forest C store through carbon credits financial intensive [6]. This importance has promoted to address mangrove forest in climate change mitigation strategies, and induces the study of mangrove forest capacity in sequestrating atmospheric CO_2.

Furthermore, the total area of mangrove forest in Indonesia is estimated at about 3.2 million hectares, approximately 20% of the world's mangrove forest area. Indonesia also has the richest mangrove diversity in the world by far, in which 45 of 70 true mangrove species have been found in this archipelago [7]. The province of East Java contains 50% of the mangrove forest area in Java Island, 18,253 ha of 34,491 ha, in which some of the forest are naturally regenerated [8]. In Labuhan, Lamongan, East Java, the mangrove forest covers are at least 90 hectares, which is mostly dominated by natural mangrove [9]. Therefore, the aim of this study is to estimate above-ground and below-ground-root biomass and its C-stocks, as well as assessing species diversity of mangrove forest of Labuhan, Lamongan, East Java, Indonesia.

2. Materials and Methods

2.1. Study Area

This study was conducted in the mangrove forest of Labuhan Village, which is located at the northern west of Lamongan Regency, East Java. The area has a precipitation average of 1465 mm, with a temperature average of 27.4 °C, and it has humid tropical savanna climate [9]. The area has diurnal tidal regime, 0.9 m at neap tide and 2.1 m at spring tide [10]. The mangrove forests of the area consist of not only primary forest, but also secondary forest and artificially regenerated forest (mangrove plantation). The forests are fringe mangrove forests that have on average a 50 m wide strip along the coast. On the landward side, the forests are separated by a large pond levee, followed by extensive shrimp ponds. The field study was conducted in March and April 2017 during the rainy season. We conducted survey at 6 stations with 4 sampling plots in each station, totalling 24 sampling plots. Station selection was primarily based on random selection along the mangrove forest of Labuhan Village. The latitude of the stations was between 6°52'23.047"S and 6°52'51.928"S, while the longitude was between 112°12'46.518"E and 112°14'0.189"E. A map of the research stations is presented in Figure 1.

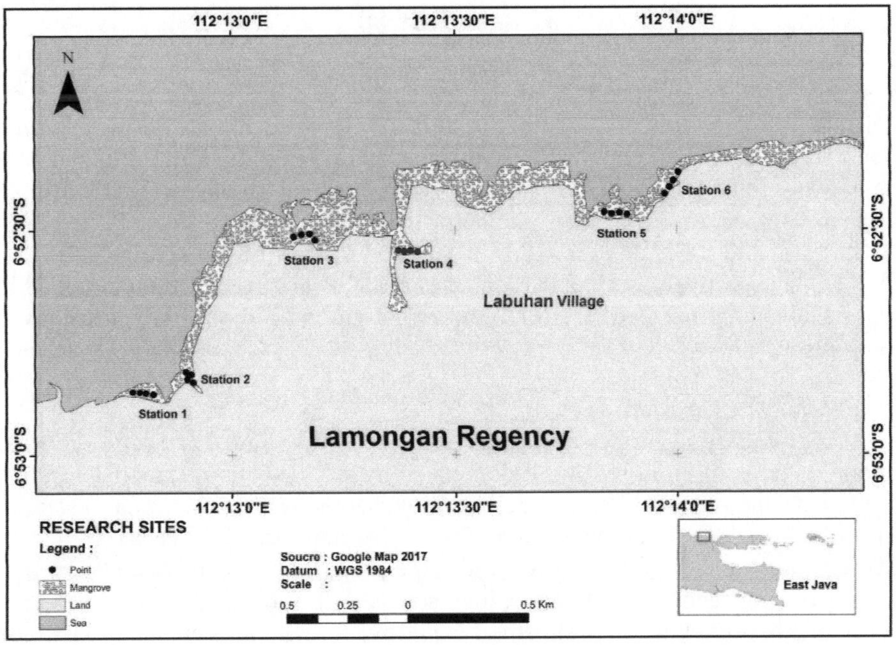

Figure 1. Location map of the study sites in Labuhan Village, Lamongan Regency, East Java, Indonesia.

2.2. Data Collection

Twenty-four plots of size 10 x 10m were laid out to determine the species structure and species composition of the study area. The sampling is a non-destructive quadrat technique with a distance of each plot between 5 and 10 m depending on the forest's characteristics. In each plot, the mature mangroves with diameter of at least 5 cm were identified according to Giesen *et al.* [11], and measured the trunk diameter breast height (DBH). DBH of *Bruguiera* and *Rhizophora* species were determined by measuring the trunk diameter at 30 cm above the buttress and above the highest prop root respectively, whereas the DBH of the rest were measured at 130 cm above ground [12]. Physical and chemical characteristics that were also measured in the research stations were pH, temperature, salinity, phosphate, and nitrate.

2.3. Data Analysis

The structural importance of each species in the community was determined using the importance value index (IVI), which is measured by the sum of relative density (RD), relative frequency (RF), and relative dominance (RDom), where:

$$RD = \frac{Density\ of\ a\ species}{Total\ density\ of\ all\ species} X\ 100$$

$$RF = \frac{Frequency\ of\ a\ species}{Total\ frequency\ of\ all\ species} X\ 100$$

$$RDom = \frac{Dominance\ of\ a\ species}{Total\ dominance\ of\ all\ species} X\ 100$$

Shannon's diversity index was used to characterize the species diversity in the mangrove habitat, where: H' is the diversity index, *Pi* is the proportion of individuals belonging to the ith species, and In is the natural logarithm [12],[13].

$$H' = -\sum Pi\ \ln Pi$$

We estimated standing above-ground biomass (AGB) and below-ground-root biomass (BGB) from three DBH data by using general and species-specific allometric equations. For *Avicennia marina,* the species-specific allometric equations were $0.308 \times D^{2.11}$ (diameter at the DBH, cm) and $1.28 \times D^{1.17}$ for the standing above-ground biomass (AGB) and below-ground root biomass (BGB) respectively [14]. For other mangrove species, the general allometric equations for Indo-West Pacific were used for AGB and BGB estimation, which were $0.251 \times \rho$ (wood density of the species, g cm^{-3}) $\times D^{2.46}$ and $0.199 \times \rho^{0.899} \times D^{2.22}$ respectively [15]. The biomasses of AGB and BGB were multiplied by 0.47 and 0.39 to determine above-ground carbon (AGC) and below-ground carbon (BGC) respectively [16].

3. Results and discussions

3.1. Surface Sediment Characteristics

The variations of grain size in surface sediments of the mangrove forest areas are shown in Figure 2. The distribution patterns of grain size were dominated by silk, except in station 1 where sand was a major contributor to the sediment. In the sediments of the research stations, silk was accounting for 33-69%, while clay and sand were accounting for 6-29% and 8-60% respectively. On average, the proportion of silk was quite high (57±12%), while the proportion of sand and clay particles content were almost similar, but sand relatively had a high standard deviation (22±19% 21±8% respectively). The dominance of silk is common in sediment of mangrove forests and wetland. In Zhangjiang estuary, Cina, silk constituted 61-72% of mangrove forest sediment [17], while in some primary mangrove forests of Sunderbans, India, silk was accounting for 36-62% of the forest sediments [18]. Furthermore, mangrove forests have a significant ecosystem service as sediment traps and play an important role for sinking suspended matter. The complex aerial root structures of mangrove trees catch sediments and function as land builders [19].

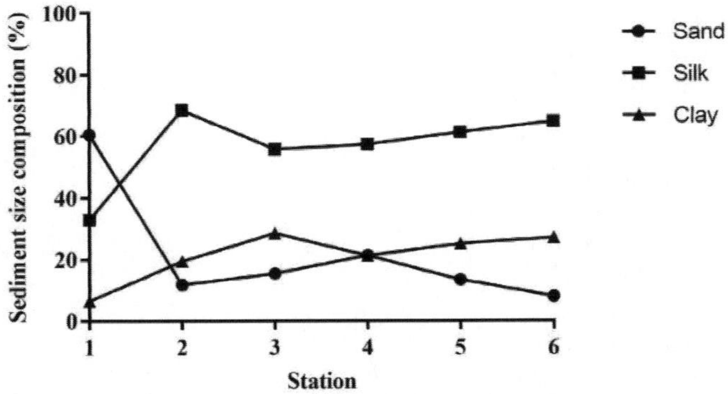

Figure 2. The distribution of grain size in surface sediments in the mangrove forest of Labuhan Village.

3.2. Species Diversity

A total of 236 sample trees representing nine true mangrove species, namely: B. gymnorrhiza, B. parviflora, C. decandra, L. litoria, L. racemose, R. apiculata, R. mucronata, X. moluccensis and X. rumphii, belonging to three families were recorded at the mangrove stand of Labuhan Village (Table 1). Strict or true mangroves species are found exclusively in the mangrove habitat and do not extend into terrestrial plant community and are morphologically, physiologically and reproductively adapted to saline, waterlogged and anaerobic condition [4]. Meanwhile, the DBH of the mangroves varied between 5.1 cm and 34.1 cm, with the average ranging from 9.1 cm to 19.9 cm in each plot.

Mangrove of the genus Rhizopora was found to be dominating the mangrove forest with an IVI of 100.40% and 80.46% for R. mucronata and R. apiculata respectively, which 83 and 49 individuals of these species occurred in 20 and 17 plots respectively (Table 1 and 2). The concept of 'important value index (IVI)' has been developed to measure the dominance and ecological success of a species with a single value [9],[20]. The index was determined based on the total contribution of a species to the community by means of its relative abundance, relative dominance and relative frequency in a study plot or area [12].

R. mucronata and R. apiculata were also among mangrove species with the highest density in the research areas (345.83 ind. ha^{-1} and 204.17 ind. ha^{-1} respectively). Rhizophora is generally the most notable genus in tropical, coastal mangrove ecosystems [21]. The mangrove of Labuhan Village is located along the coastline that is suitable for Rhizophora species, for which their habitat is in intertidal zones that are inundated daily by the ocean. They adapt to this harsh environment using pneutomatophores to respire oxygen even while their lower roots are submerged, and molecular pump mechanism to remove salt from the xylem stream [22]-[24].

The Shannon –Wiener's diversity index (H') of mangrove community of Labuhan Village was 1.51, which was higher than that of the mangrove forest in Palawan, the Philippines (H'= 0.99). This was due to the fact that Labuhan Village possessed a higher number of mangrove species. In contrast to tropical lowland rainforest, the mangroves have very low diversity as few plants have their special adaptations, which are attributed to their unique stands formation and harsh coastal habitat [8]. Furthermore, some species were only represented by one individual mangrove stand (Table 2); therefore, management actions need to be implemented to ensure the long-term persistence of the mangrove biodiversity.

Table 1. Species composition and DBH of the mangrove forest of Labuhan Village, Lamongan Regency, East Java, Indonesia.

Plot	No. of tress	Species	DBH (cm)			
			Mean	Min.	Max.	SD

1	13	Ra, Rm, Bp	11.2	6.7	16.6	3.4
2	34	Ra, Rm, Bg	9.4	5.1	23.6	3.5
3	12	Ra, Rm, Bg	10.1	5.1	14.3	3.1
4	12	Ra, Rm, Bg	10.5	6.4	15.6	3.1
5	7	Ra, Bg	19.9	9.6	28.7	6.5
6	4	Rm, Ra	17.6	13.1	16.9	4.4
7	12	Rm	18	15	27.1	3.3
8	3	Ll, Cd, Rm	15.1	12.7	18.8	3.3
9	12	Ra, Rm, Bg, Bp	13.2	8.3	22.9	5.2
10	10	Ra, Rm, Bg	11.2	5.7	15.6	2.5
11	11	Ra, Lr, Rm, Bp	18.8	7.0	15.9	3.5
12	10	Ra, Rm, Bp, Lr	14.1	8.6	22.6	4.3
13	14	Bg, Lr, Rm	12.4	6.7	22.9	5.5
14	12	Ra, Rm, Lr, Bg	13.6	8.9	34.1	7.3
15	11	Bg, Lr	11	5.7	26.1	5.9
16	5	Bg, Lr	10.7	8.3	11.8	1.5
17	9	Rm, Ra	18.2	8.6	28.7	7.3
18	7	Ra	14.5	12.4	15.9	1.3
19	10	Ra, Rm, Bg	11.2	5.7	13.1	2.6
20	4	Ra, Rm	17.6	13.1	16.9	4.4
21	5	Cm, Rm, Xr, Cd	14.6	5.7	22.9	6.9
22	6	Ra, Rm, Cd	10	8.0	15.9	3
23	7	Ra, Rm, Bp	14.3	11.1	19.7	3
24	6	Rm, Bg	10.5	8.0	12.7	1.6

Note: Bg = *Bruguiera gymnorrhiza,* Bp = *Bruguiera parviflora,* Cd = *Ceriops decandra,* Ll = *Lumnitzera litoria,* Lr = *Lumnitzera racemosa,* Ra = *Rhizopora apiculata,* Rm = *Rhizopora mucronata,* Xm = *Xylocarpus moluccensis,* Xr = *Xylocarpus rumphii.*

Table 2 Analysis of importance value index (IVI) of the mangrove forest of Labuhan Village, Lamongan Regency, East Java, Indonesia.

Family	Species	No. of Individuals	RF (%)	Rdom (%)	RD (%)	IVI (%)	*H′*	Density (Ind ha^{-1})
Rhizophoraceae	*Rhizopora apiculata*	49	25.76	31.15	23.56	80.46	0.33	204.17
	Rhizopora mucronata	83	30.30	33.08	37.09	100.40	0.37	345.83
	Bruguiera gymnorrhiza	71	18.18	16.53	29.80	64.51	0.36	295.83
	Bruguiera parviflora	13	7.57	4.54	6.25	18.34	0.16	54.17
	Ceriops Decandra	3	4.54	1.81	1.44	7.80	0.06	12.50
Combretaceae	*Lumnitzera litoria*	1	1.51	0.38	0.48	2.38	0.02	4.17
	Lumnitzera racemosa	14	9.09	11.02	0.48	20.59	0.17	58.33
Meliaceae	*Xylocarpus moluccensis*	1	1.51	1.25	0.48	3.24	0.02	4.17
	Xylocarpus rumphii	1	1.51	0.21	0.48	2.21	0.02	4.17
	Total	236	100	100	100	300	1.51	983.33

3.3. Biomass and C-stocks of Mangrove.

The mean of DBH, AGB, AGC, BGB, and BGC of mangrove stands in Labuhan Village is presented in Table 3. The average biomass of mangrove stands in Labuhan Village was 168.05 ± 35.34 Mg B ha^{-}

[1] which was slightly higher than that of in the previous study in the same area (163.96 ± 35.34 Mg B ha^{-1}) [9]. However, the former study encompassed only three plots in one research station while our study covered broader areas, 24 plots in six stations (Figure 1). The mangrove stands biomass was equivalent to 74.70 ± 15.58 Mg C ha^{-1} which can be assumed that the mangrove stands of Labuhan Village sequestered 274.15 ± 57.18 Mg CO_2 ha^{-1}.
The average of above-ground carbon (AGC) of mangrove stands in this study was 53.89 ± 10.65 Mg C ha^{-1} (Table 3), which was higher than that of North Sulawesi, Indonesia and Okinawa, Japan (30.8 and 40.25 Mg C ha^{-1} respectively) [25],[5]. Meanwhile, even though R. mucronata had highest number of mangrove (83 trees), in total R. apiculata had significantly higher biomass and carbon (63.69 ± 11.32 Mg B ha^{-1} and 28.67 ± 3.84 Mg C ha^{-1} respectively) (Tukey's multiple comparisons test $p < 0.05$). It was due to the facts that R. apiculata had bigger DBH than R. mucronata, and the wood density of R. apiculata was also higher than that of R. mucronata [15]. The mangrove forest of Labuhan Village not only harbour some mangrove species, but also most importantly provide for carbon sequestration. Therefore, local communities and authorities should preserve this mangrove habitat.

Table 3. The mean of DBH, AGB, AGC, BGB, and BGC of mangrove stands in Labuhan Village. The standard errors (±) represent six stations (n=6). Two-way ANOVA showed statistically significant differences ($p < 0.05$) in DBH, AGB, AGC, BGB, and BGC of mangrove stands among different species.

Species	DBH	AGB (Mg B ha^{-1})	AGC (Mg C ha^{-1})	BGB (Mg B ha^{-1})	BGC (Mg C ha^{-1})
R. apiculata	15.70 ± 4.93	45.53 ± 5.66	21.39 ± 2.66	18.16 ± 5.66	7.08 ± 1.18
R. mucronata	12.41 ± 4.75	29.97 ± 5.24	14.09 ± 2.46	17.74 ± 1.95	6.92 ± 0.76
B. gymnorrhiza	9.80 ± 2.51	15.15 ± 3.27	5.90 ± 1.27	7.06 ± 1.48	2.75 ± 0.57
B. parviflora	11.05 ± 5.20	5.33 ± 1.35	2.50 ± 0.63	2.25 ± 0.56	0.87 ± 0.22
C. Decandra	15.28 ± 5.52	1.87 ± 0.51	0.88 ± 0.24	1.02 ± 0.26	0.39 ± 0.10
L. litoria	12.74 ± 0.00	0.40 ± 0.16	0.18 ± 0.07	0.18 ± 0.07	0.07 ± 0.03
L. racemosa	16.58 ± 7.75	14.91 ± 5.76	7.01 ± 2.71	5.76 ± 2.20	2.24 ± 0.86
X. moluccensis	22.93 ± 0.00	1.33 ± 0.54	0.62 ± 0.25	1.11 ± 0.45	0.43 ± 1.17
X. rumphii	9.55 ± 0.00	0.19 ± 0.08	0.09 ± 0.03	0.09 ± 0.03	0.03 ± 0.01
Total	-	114.68 ± 22.68	53.89 ± 10.65	53.37 ± 12.66	20.81 ± 4.93

4. Conclusion

This study was the first comprehensive mangrove forest inventory in Labuhan Village. We employed random sampling design across mangrove forest of Labuhan Village as the basic of mangrove identification and DBH measurement. Mangrove allometry was used to solve the biomass and C stocks of mangrove stands. The mangrove forest of Labuhan Village harboured nine species of true mangroves in which *R. mucronata* and *R. apiculata* dominated the forest communities with an IVI of 100.40% and 80.46% respectively. *R. apiculata* also had the highest contribution to the mangrove trees' biomass and C stocks (63.69 ± 11.32 Mg B ha^{-1} and 28.67 ± 3.84 Mg C ha^{-1} respectively). Meanwhile, the average biomass and C of mangrove stands in Labuhan Village were 168.05 ± 35.34 Mg B ha^{-1} and 74.70 ± 15.58 Mg C ha^{-1} respectively, which was equivalent to 274.15 ± 57.18 Mg CO_2 ha^{-1} sequestered and stored in the biomass. A follow-up study on the soil C stocks of the mangrove ecosystem is recommended to quantity the total ecosystem C stocks of the mangrove forest.

5. **References**

[1] Mitra A, Banerjee K and Sett S 2012 Spatial Variation in Organic Carbon Density of Mangrove Soil in Indian Sundarbans *Natl. Acad. Sci. Lett.* 35 147–154
[2] Donato D, Kauffman JB, Murdiyarso D, Kurnianto S, Stidham M and Kanninen M 2011 Mangroves among the most carbon-rich forests in the tropics *Nat. Geosci.* 4 293–297

[3] Giri C, Ochieng E, Tieszen LL, Zhu Z, Singh A, Loveland T, Masek J and Duke N 2011 Status and distribution of mangrove forests of the world using earth observation satellite data *Glob. Ecol. Biogeogr.* 20 154–159

[4] Polidoro BA *et al.* 2010 The Loss of Species: Mangrove Extinction Risk and Geographic Areas of Global Concern *PLOS ONE* 5 e10095

[5] Murdiyarso D, Donato D, Kauffman JB, Kurnianto S, Stidham M and Kanninen M 2010 *Carbon storage in mangrove and peatland ecosystems: a preliminary account from plots in Indonesia* (Bogor: CIFOR)

[6] Wylie L, Sutton-Grier AE and Moore A 2016 Keys to successful blue carbon projects: Lessons learned from global case studies *Mar. Policy* 65 76–84

[7] Aslan A, Rahman AF, Warren MW and Robeson SM 2016 Mapping spatial distribution and biomass of coastal wetland vegetation in Indonesian Papua by combining active and passive remotely sensed data *Remote Sens. Environ.* 183 65–81

[8] Suhardjono S 2013 Hutan Mangrove Cagar Alam Pulau Sempu, Jawa Timur *J. Biol. Indones.* 9 121–130 [in Bahasa Indonesia]

[9] Asadi MA, Guntur G, Ricky AB, Novianti P and Andik I 2017 Mangrove ecosystem C-stocks of Lamongan, Indonesia and its correlation with forest age *Res. J. Chem. Environ.* 21(8) 1–9

[10] Hoekstra P 1989 *Hydrodynamics and depositional processes of the Solo and Porong Deltas, East Java, Indonesia*, in: van der Linden WJM *et al* (Eds.), *Coastal Lowlands: Geology and Geotechnology* (Dordrecht: Springer Netherlands) pp 161–173

[11] Giesen W, Wulffraat S, Zieren M, Scholten L 2006 *Mangrove Guidebook For Southeast Asia* (Bangkok: FAO and Wetlands International)

[12] Abino AC, Castillo JAA and Lee YJ 2014 Species diversity, biomass and carbon stock assessments of a natural mangrove forest in Palawan, Philippines *Pak. J. Bot.* 46 1955–1962

[13] Nguyen H, Lamb D, Herbohn J and Firn J 2014 Designing Mixed Species Tree Plantations for the Tropics: Balancing Ecological Attributes of Species with Landholder Preferences in the Philippines *PLOS ONE* 9 e95267

[14] Comley BWT and McGuinness KA 2005 Above- and below-ground biomass, and allometry, of four common northern Australian mangroves *Aust. J. Bot.* 53 431–436

[15] Komiyama A, Ong JE and Poungparn S 2008 Allometry, biomass, and productivity of mangrove forests: A review *Mangrove Ecol. – Appl. For. Costal Zone Manag.* 89 128–137

[16] Kauffman JB and Donato D 2012 *Protocols for the measurement, monitoring and reporting of structure, biomass and carbon stocks in mangrove forests* (Bogor: CIFOR)

[17] Xue B, Yan C, Lu H and Bai Y 2009 Mangrove-Derived Organic Carbon in Sediment from Zhangjiang Estuary (China) Mangrove Wetland *J. Coast. Res.* 949–956

[18] Hossain MD and Nuruddin A 2016 Soil and Mangrove: A Review *J. Environ. Sci. Technol.* 9 198–207

[19] Kathiresan K 2003 How do mangrove forests induce sedimentation? *Rev. Biol. Trop.* 51 355–360

[20] Mishra R 1968 *Ecology Workbook* (New Delhi: Oxford and IBH Co.)

[21] Cerón-Souza I, Rivera-Ocasio E, Medina E, Jiménez JA, McMillan WO and Bermingham E 2010 Hybridization and introgression in New World red mangroves, Rhizophora (Rhizophoraceae) *Am. J. Bot.* 97 945–957

[22] Naskar S and Palit PK 2015 Anatomical and physiological adaptations of mangroves *Wetl. Ecol. Manag.* 23 357–370

[23] Reef R and Lovelock CE 2015 Regulation of water balance in mangroves *Ann. Bot.* 115 385–395

[24] Spalding MD 2001 M*angroves* A2 - Steele, John H., in: Encyclopedia of Ocean Sciences (Second Edition)* (Oxford: Academic Press) pp 496–504

[25] Khan MNI, Suwa R and Hagihara A 2009 Biomass and aboveground net primary production in a subtropical mangrove stand of *Kandelia obovata* (S., L.) Yong at Manko Wetland, Okinawa, Japan *Wetl. Ecol. Manag.* 17 585–599

Acknowledgments
The authors would like to thank the rector of Brawijaya University and the dean of Faculty of Fisheries and Marine Science of UB for the publication support. The authors acknowledge the field work support from Ade Riizky Pratama and Muhammad Zuhal Fikri.

Measuring from Space the Efficiency of Local Forest Management: The Successful Case of the Indigenous Community of Cherán, Mexico

FJ Osorno-Covarrubias[1,2] S Couturier[1,2,*] and M Piceno Hernández[1]

1 Laboratory of Geospatial Analysis (LAGE), Institute of Geography, UNAM, Mexico City, Mexico
2 Research Center for Complexity Science (Centro de Ciencias de la Complejidad, C3), UNAM, Mexico City, Mexico
* Corresponding Author
E-mail: stephcamelo@mailcity.com

Abstract. Remote sensing-based forest cover change maps are usually elaborated for national level baseline information towards environmental conservation schemes. Community forest management is acknowledged as a key component of sustainable use of natural resources. In this research, we provide a means of technological appropriation for the community of Cherán, state of Michoacán, Mexico, to complement their local forest management initiative since 2012, after severe illegal logging occurred all along the previous decade. Participatory mapping, collaborative online cartography and global forest cover products were shared with the Community Council of the Commons. The environmental effect of the governance change occurred in 2012 was assessed using the Global Forest Cover Change (GFCC) product from 2000 to 2016. We find a significant reduction of deforestation in Cherán, due to local empowerment, which contrasts with overall trend in the region. We discuss the potential benefits of an alliance between communities and the academy through the construction of decentralized spatial data infrastructures in Mexico.

1. Introduction

In the context of global environmental change, deforestation critically contributes to greenhouse gas emissions and biodiversity loss. Since the early 2000s, the parties to the United Nations Framework Convention on Climate Change (UNFCCC) have encouraged mechanisms towards the reduction of deforestation worldwide[1]. Forest cover observation from space has become an effective means of measuring deforestation at national and global levels. The derived forest cover change maps have been widely employed by remote sensing specialists and consulting groups in order to document carbon stock baselines in the frame of UNFCCC proposed incentives in tropical countries [2]. However, an important component for the sustainability of forest conservation in the long run is local governance upon natural resources [3]. Specifically, for an effective reversal of ecosystem degradation it is argued that local communities should be considered an active part in the management of forests [4]. In Mexico, most of the managed forest cover is under communal or semi-communal property and much of the management is implemented according to traditional and indigenous knowledge systems.
In this research, an academic effort is intended to strengthen the local forest management system of the indigenous community of Cherán, state of Michoacán (Mexico) (figure 1).

2. Characteristics of the Study Area

Forestry is a prominent economical sector in the in state of Michoacán (Fig. 1b); 1st national position for resin production, 3rd for wood production. Yet, land use conversions to grazing land and avocado plantation are a direct deforestation threat in this region [5], and frequently occur with the intervention of criminal groups [6]. To this respect, the evolution of forest management in Cherán (Fig. 1b) is a peculiar, emblematic case of community-based management in Mexico: their communal land was allegedly affected by heavy illegal logging from 2001 to 2012, at which date the organized majority of Cherán inhabitants, after repelling the armed illegal loggers, took control over the territory [7-9]. Since then, community forest management has been implemented by the Community Council of the Commons (CCC), with the aim of restauration and reforestation [10].

Under these circumstances, the purpose of this research was twofold: 1) Identify the academic knowledge and technology which might be helpful to the indigenous people of Cherán in their local process of forest recovery; and 2) Bring evidence of any forest cover tendency over the alleged illegal logging process from 2000 to 2012, and then over the alleged forest recovery process from 2012 to 2016, using available global forest cover products.

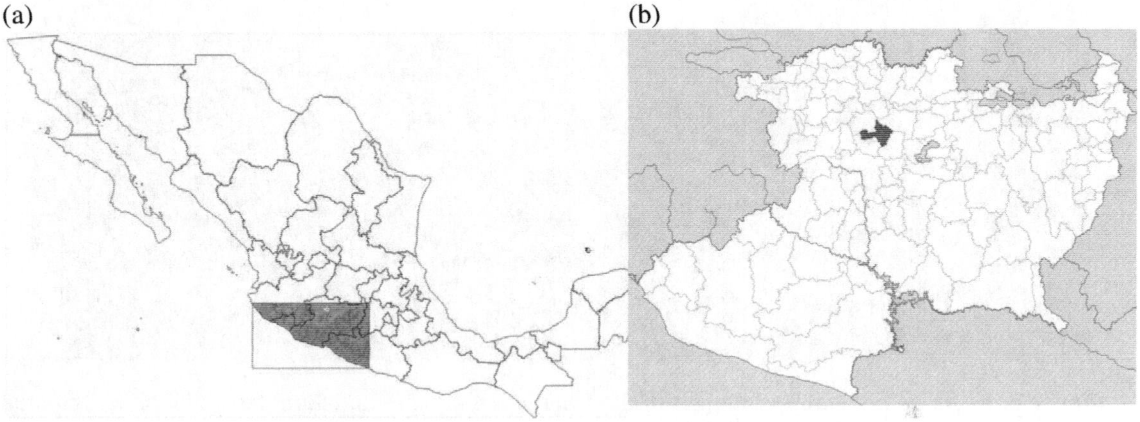

Figure 1. Location of the study area: (a) Purépecha Plateau (small rectangle) within the 10-20N lat., 100-110W long. quadrant. (b) Cherán community (red color) within the Michoacán state.

3. Methods

In order to efficiently accompany the organizational process of local forest management, a strategy of collaborative cartography (within the concepts of Citizen Science and NeoGeography [11]) was implemented, with three major elements: (1) Participatory mapping, (2) Construction of an online cartographic platform, and (3) spatial analysis of satellite – derived global forest cover products.

Following initial visits to the Community Council of the Commons (CCC) in the Cherán community, workshops of participatory mapping were planned, submitted to the CCC and authorized; they were implemented in Cherán in 2015 and 2016 [12]. Then, an online cartographic platform was designed with the open source software GeoNode, intended as a potential Collaborative Spatial Data Infrastructure (CSDI) for the Cherán community. Spatial materials derived from the participatory workshops, as well as freely available satellite imagery were selected and incorporated into the platform.

Finally, the Global Forest Cover Change (GFCC) [13], based on archive Landsat satellite imagery, was downloaded and analyzed. GFCC provides an estimate of the forest cover in year 2000 ("Tree canopy cover for year 2000" GIS raster layer, at 30m resolution, which states for each pixel the percentage canopy closure), as well as a yearly estimate of deforestation rate from 2000 to 2016 ("Year of gross forest cover loss event", or "lossyear" GIS layer). The latter layer is encoded as 0 (no loss) or a value ranging from 1 to 16, corresponding to deforestation occurrence in year 2001 to 2016. Deforestation trends were assessed according to GFCC for the Cherán community and for a surrounding portion of the Purépecha Plateau (Fig. 1a) before and after the governance change occurred; that is, from 2000 to 2011 and from 2012 to 2016.

4. Results

The participatory mapping workshops revealed the previous establishment of a local Geographic Information System (GIS) for forest management and the cartographic experience of specific sectors of the population. Forest degradation as well as restauration efforts had been precisely delineated in this GIS, under the supervision of the CCC. On the other hand, many Cherán participants expressed interest in integrating the spatial data collected in the workshop onto online cartography, to complete a collective memory on their territorial defense and establish educational materials (figure 2)

Figure 2. Interactive 3D web visualizer of the Cherán territory, included in the Collaborative Spatial Data Infrastructure (CSDI). Vector data shown were obtained through participatory mapping [12].

Additionally, options of payments for environmental services with international agents was perceived as an interest by CCC, while specific national incentives were perceived with distrust. Based on agreements derived from these workshop, a Collaborative Spatial Data Infrastructure (CSDI) was built, including local GIS information and color composites of satellite imagery (figure 3).

Finally, an analysis was conducted using the GFCC data. Total deforestation in the Purépecha Plateau was 34,143 ha from 2000 to 2016, setting an overall deforestation rate of 2,134 ha/year (see figure 4a). The accumulated forest loss within the traditional territorial boundaries of Cherán was 798 ha over the same period (see figure 4b) and deforestation rate had a peak in year 2011; that year the community confronted illegal loggers and gradually took territorial control within the traditional boundaries of Cherán. After this 2011 peak however, deforestation diminished to low levels in 2012-14, and then was almost null in 2015-16 (figure 4b). This pattern of deforestation reduction contrasts sharply with the increase observed in the rest of the Purépecha Plateau in the same period, reaching a new peak in 2016 for the 16 years observed period (figure 4a).

The spatio-temporal pattern of this process can be seen in figure 5. As of 2012, deforestation intensifies and occurs, almost entirely, outside the perimeter controlled by the Cherán community. This can be appreciated in the location and extension of the red pixels compared to that of the black pixels (deforestation before 2012).

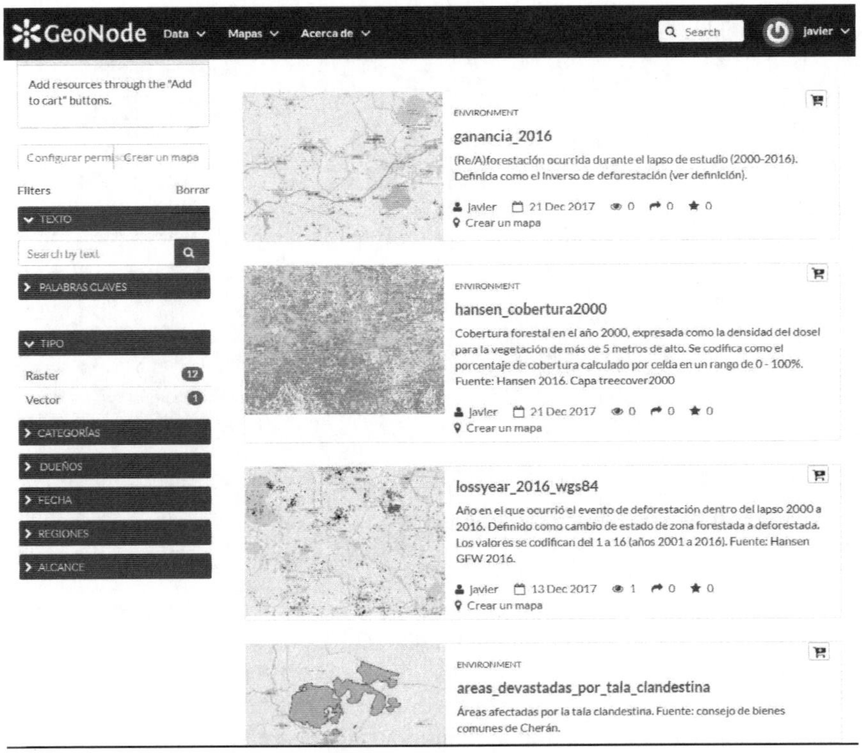

Figure 3. Collaborative Spatial Data Infrastructure (CSDI) proposed to support forest monitoring by Cherán K'eri authorities.

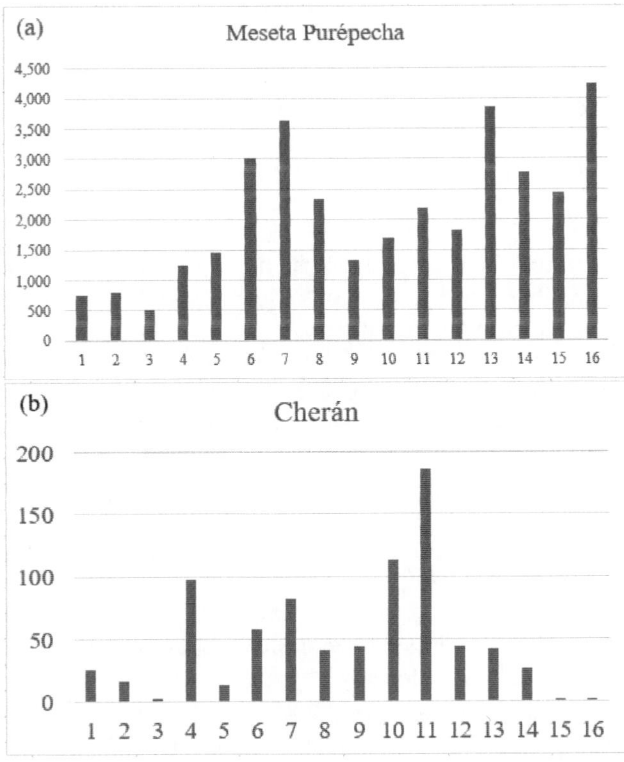

Figure 4. Deforestation rates (ha/year) from 2000 to 2016, derived from the GFCC global forest cover product: (a) Purépecha Plateau (b) Cherán.

Figure 5. Deforestation pattern in communities of the Purépecha Plateau in the vicinity of Cherán (white polygon), according to the GFCC global forest cover product [13]. Deforestation areas are depicted in black (between 2000 and 2011), yellow (year 2012), and red (2013-2016). After 2011, note the reduced deforestation in Cherán, which contrasts with extensive deforestation in neighbouring communities.

5. Discussion

Previous research from the remote sensing perspective has documented the severe environmental degradation in the Meseta Purépecha; detailed deforestation hotspots in Michoacán [14], and in Cherán before the political shift [7]. The corresponding situation of illegal logging in the Meseta Purépecha and the particular case of Cherán before the political shift have also been studied from a sociopolitical and territorial perspective [6-9, 15]. Within this context, this research builds upon results of a socio-territorial analysis of the emblematic political shift in Cherán [10, 12]. A Collaborative Spatial Data Infrastructure (CSDI) is proposed, which we argue may be in accordance with preparing grounds for *Community Mapping, Measuring, Monitoring* [4] in Cherán, in a framework of Citizen Science.

To support this view, the results of our participatory workshops pointed to a high level of local expertise towards a digital forest monitoring system, which is an asset for the feasibility of a Citizen Science framework [16]. Then, the identification of a few needs from the Community (e.g. the construction of a georeferenced database for a collective memory on territorial defense, the analysis of technical options in view of payments for the environmental service of forest conservation) makes possible the relevance and mutual use of information shared in a CSDI. A special feature of this CSDI was the preliminary analysis of forest cover cartography based on the global forest cover change product GFCC. This analysis proved to be in accordance with the Cherán K'eri version of the situation and provides up to date information on the regional situation around Cherán, potentially strategical to the CCC.

The participatory work with the Cherán Community is still incipient (limited to a diagnosis) and would need more efforts and incentives from the academy in order to offer capacitation in the future, perhaps with the perspective of a community forest management scheme such as the one of nearby Nuevo San Juan Parangaricutiro [17], if ever such perspective is an option in Cherán. Also, from the remote sensing perspective, our study should aim at greater data quality, using available higher resolution remote sensing data [14], because it offers further independent validation on the reliability of citizen science practice [16]. As a first step, we should aim at a rigorous accuracy assessment of the GFCC satellite estimate.

6. Conclusions

In the global context of uncontrolled, increasing human consumption rates, many peripheral territories are affected by severe social-environmental damage. Views from Neo-geography or Citizen Science are that the academy should make an alliance with local actors that suffer these damages and accompany their quest for more social and environmental justice, with the perspective that uncontrolled merchandise production be slowed down to a constrained, more sustainable pace. This is the case of illegal logging for unsustainable wood production in Michoacán, Mexico, and specifically in Cherán, where local defenders of the environment have lost their lives facing criminal organizations. Schemes such as Community Measuring, Mapping, and Monitoring (CMMM) have been proposed to strengthen local forest management systems, where the academy should complement local knowledge with technical / communication tools.

This research seeks to accompany the notorious governance shift undergone by the Cherán K'eri traditional authorities in 2011 towards an autonomous, sustainable resource use in the Cherán territory. For the purpose of helping in the implementation of CMMM in Cherán, we propose a set of methodological steps. First, potentially useful academic tools and knowledge were identified through participatory mapping. Second, a set of these tools was integrated into an online cartographic platform (a Collaborative Spatial Data Infrastructure - CSDI) and visualized in the perspective of *Community Mapping and Measuring*. Finally, we found that the freely available GFCC time series product could be a useful tool to set an external, preliminary reference to *Community Monitoring*. Our analysis of the GFCC product between 2001 and 2016 confirmed the environmental tendencies claimed by the Cherán K'eri authorities along this time period: the community was indeed affected by high deforestation rates between 2004 and 2011 (allegedly caused by the illegal logging criminal organization), and, from 2012 up to our days, deforestation has drastically reduced (because of the expulsion of the illegal logging organization from Cherán, the removal of the former formal government, involved in the illegal activity, and the reforestation policy of the Cherán K'eri autonomous authorities). Our findings also point to a contrasting situation, however, in the Purépecha Plateau, and specifically in communities next to the Cherán territory, where unsustainable logging could be increasing since 2012, suggesting the persistence of illegal logging organizations, warning for a potentially conflictive situation in the vicinity of the Cherán community.

Based on this experience, and under favourable alliance conditions, we suggest similar Collaborative Spatial Data Infrastructures be built to technically support *Community Mapping, Measuring and Monitoring* aimed at local forest sustainable management elsewhere in Mexico and in rural settings of the sub-tropics. Future works in Cherán may include the accuracy assessment and quantitative adjustments of the GFCC-based forest cover figures using the more accurate information of the local GIS developed by the Community Council of the Commons in Cherán, as well as the elaboration of more cartographic products (printed and online) which might help Cherán K'eri authorities document and appropriately inform about the forest cover tendencies shown in this research.

7. References

[1] Mollicone D, Freibauer A, Schulze E, Braatz S, Grassi G, Federici S. 2007 Elements for the expected mechanisms on 'reduced emissions from deforestation and degradation, REDD'under UNFCCC. *Environmental Research Letters*. 2(4):045024.

[2] Mitchell AL, Rosenqvist A, Mora B. 2017 Current remote sensing approaches to monitoring forest degradation in support of countries measurement, reporting and verification (MRV) systems for REDD+. *Carbon balance and management*. 12(1):9.

[3] Pelletier J, Gélinas N, Skutsch M. 2016 The place of community forest management in the REDD+ landscape. *Forests*. 7(8):170.

[4] McCall MK, Chutz N, Skutsch M. 2016 Moving from measuring, reporting, verification (MRV) of forest carbon to community mapping, measuring, monitoring (MMM): perspectives from Mexico. *PloS one*. 11(6):e0146038.

[5] Barsimantov J, Antezana JN. 2012 Forest cover change and land tenure change in Mexico's avocado region: Is community forestry related to reduced deforestation for high value crops? *Applied Geography*. 32(2):844-53.

[6] Del Conde A. 2015 Illegal Logging and its Dynamics of Violence Within the P'urhépecha Plateau. *Landscapes of Violence*. 3(1):6.

[7] España-Boquera ML, Champo-Jiménez O. 2016 Proceso de deforestación en el municipio de Cherán, Michoacán, México (2006-2012). *Madera y bosques*. 22(1):141-53.
[8] Fuentes Díaz A, Paleta Pérez G. 2105 Violencia y autodefensas comunitarias en Michoacán, México. *Íconos: Revista de ciencias sociales*. (53):171-86.
[9] Fuentes-Díaz A. 2014 Autodefensa y justicia en los márgenes del Estado. *Clivajes. Revista de Ciencias Sociales*. (2).
[10] Lariagon R, Piceno M. 2016 La lutte pour l'autonomie de la communauté de Cherán: reconfiguration territoriale et subjective. *Cahiers des Amériques latines*. (81):133-51.
[11] Osorno-Covarrubias J, Couturier S, Ricárdez M. 2015 El rol de la Geografía y sus hibridaciones recientes frente a la crisis de sustentabilidad global. *Boletín de la Asociación de Geógrafos Españoles*. (69):93-112.
[12] Piceno Hernandez M. La cartografía colaborativa como herramienta para la defensa del territorio y los bienes comunes en Cherán, Michoacán [Tesis de maestria]. México: Universidad Nacional Autónoma de México; 2017.
[13] Hansen MC, Potapov PV, Moore R, Hancher M, Turubanova S, Tyukavina A, et al. 2013 High-resolution global maps of 21st-century forest cover change. *science*. 342(6160):850-53.
[14] Mas J-F, Lemoine-Rodríguez R, González R, López-Sánchez J, Piña-Garduño A, Herrera-Flores E. 2017 Evaluación de las tasas de deforestación en Michoacán a escala detallada mediante un método híbrido de clasificación de imágenes SPOT. *Madera y bosques*. 23(2):119-31.
[15] Fuentes-Díaz A. 2015 Narcotráfico y autodefensa comunitaria en" Tierra Caliente", Michoacán, México. *CienciaUAT*. 10(1):68-82.
[16] Elbroch M, Mwampamba TH, Santos MJ, Zylberberg M, Liebenberg L, Minye J, et al. 2011 The value, limitations, and challenges of employing local experts in conservation research. *Conservation Biology*. 25(6):1195-202.
[17] Velázquez A. Las enseñanzas de San Juan: investigación participativa para el manejo integral de recursos naturales/comp. por Alejandro Velázquez, Alejandro Torres, Gerardo Bocco. México: Secretaría del Medio Ambiente y Recursos Naturales, Instituto Nacional de Ecología,; 203. 595 p.

Acknowledgments
The authors duly acknowledge the participation of numerous inhabitants of Cherán and their authorities (Cherán K'eri), through their Community Council of the Commons (CCC). Field work in Cherán and the elaboration of cartographic materials were supported by the DGAPA UNAM sponsored projects PAPIIT IA300515 and PAPIIT IN302417; The online collaborative platform was developed under the ASTIG (Social Empowerment on Geographic Information Technology) project, in the Research Center for Complexity Science (Centro de Ciencias de la Complejidad, C3) at UNAM; Thanks are also due to the CONACYT national public research funds for the Master's thesis scholarship of the third author.

A Study on the Proper Selection of Ecological Habitat for the Wild Radiation of Crested ibis(*Nipponia Nippon)*

Yoon HeeJung[1] and JiYoung Choi[2]

1 Ph D Student, in Department of Ecocreative Natural Sciences, Ewha Womans University, Seoul, South Korea
2 Ph D Student, in Department of Environmental Science and Engineering, Ewha Womans University, Seoul, South Korea
E-mail: velvetpaw19@gmail.com

Abstract. This study aims to select the radiated area for the wild grazing of Crested ibis(Nipponia Nippon) in 2017 and develop a best-fit models for the wild grazing candidate area and dispersed area based on the existing restoration center in Gyeongnam area. According to the restoration policy of the endangered species Crested ibis, the regions that have a high suitability index are selected. In order to increase genetic diversity and prevent massive bird flu outbreaks, designating protection area for Crested ibis is necessary and it is recommended to release them in a geographically dispersed manner. As a result, the data obtained from the Crested ibis releasing could be shared with other restoration centers and also used as a basic data for the conservation of other endangered species in the entire Korea Peninsula.

1. Introduction

Crested ibis is a specially conserved species registered as a Vulnerable (VU) in the Red List of Endangered Species in 1988 by the International Union for Conservation of Nature (IUCN). In Korea, Crested ibis is extinct and it has been on artificial breeding since 2008. According to the Ministry of Environment's propagation in July, 2008, the number of individuals of Crested ibis has been successfully increased from 2 to 313. Currently, Upo Restoration Center of Crested ibis cares for them. The geographical location of Korea allows to observe various species of birds for each season because of being in the middle of the East Asia-Australia route, that is one of the main routes of migratory birds. Especially, the Upo Swamp provides a resting place and a rich food resource for the birds and works as a representative wetland in Gyeongnam area.[1]

In celebration of 10th Meeting of the Conference of the Contracting Parties to the Convention on Wetlands, the re-introduction target for the wild grazing of Crested ibis was selected as Upo Swamp. It aims to preserve biodiversity through the restoration of Crested ibis, since the Upo Swamp and its surrounding paddies are properly managed under Ramsar Wetlands criteria. The preceded Crested ibis habitat suitability evaluation focusing on rivers and hills, that was conducted in China and Japan, excludes rice fields and their management. Hence, this study has a significance of providing a complimentary evaluation method to apply it in Korea.

Since Crested ibis is one of the most vulnerable species to avian influenza, its genetic diversity is difficult to be obtained. In this study, the habitat fit factors for Crested ibis were selected through expert questionnaire and the habitat evaluation index was selected considering the habitat environment of Crested ibis. This study aims to select the radiated area for the wild grazing of Crested ibis(*Nipponia Nippon*) in 2017, and develop a best-fit model for the wild grazing candidate area and dispersed area based on the existing restoration center in Gyeongnam area. If the population of Crested ibis(*Nipponia Nippon*) is increased in the future, the risk of collective accidents should be reduced and genetic diversity should be ensured by dispersing wild grazing in suitable habitats.

2. Materials and Methods

2.1. Subject of Study

Crested ibis was a common species prevalent in Northeast Asia until the end of the 19th century. It was distributed in eastern Russia, Japan, and China, and it was bred throughout the whole of Korea. It was wintering throughout the Korean Peninsula and Taiwan, but it started to decrease rapidly from late 19th century to early 20th century.[2]

One of the most recently endangered bird species in the world is the crested ibis (Nipponia Nippon; IUCN Red Data Book, BirdLife International 2001). It was once widespread across Northeast Asia, with a range encompassing China, Russia, Korea, and Japan until the 1860s. Suffering from over-hunting and habitat loss, the crested ibis populations had finally collapsed in the late 19th and early 20th centuries, to the extent that it was thought to be completely extinct from the wild.[3]

Taxonomically, Crested ibis belongs to the Animal, Vertebrate, Avis, Ciconiiformes and Threskiornithidae, Twenty-eight species have been recorded all over the world, but the Crested ibis (Reunion Flightless Ibis, Threskiornis solitarius) is estimated to have been extinct since 1705. There are 27 species of crested ibis .[4]

It was distributed all over the Northeast Asia region. In Russia, it was inhabited in the Amur River basin and Khabarovsk. As the amphibians, insects, etc. are exploding in summer during the summer, various birds have been used as breeding grounds as well as Crested ibis.

Crested ibis seemed to have been found mainly for breeding in summer, but the winter temperatures in Russia are so low that it is hard to find food, so when winter arrives, it travels to the Korean peninsula or southeastern part of China or Japan.

Except for a few species, Crested ibis has a very limited habitat, of which 7 species are in a state of extinction due to their reduced population. In addition to being designated as an endangered species, the loss of wetland behind and the lack of food resources in tofu are causing considerable threats to survival.

Although it is designated as National Natural Monument No. 198 in Korea, it is regarded as extinct because there is no record of observation since winter 1978.[5]

The International Union for Conservation of Nature (IUCN) lists Vulnerable (VU) species on the Red List of Endangered Species.

In China, the Crested ibis restoration project was initiated at the national level. As the artificial breeding project succeeded, reintroduction of Crested ibis from China was promoted in Korea and Japan where wild Crested ibis was extinct.[6]

The reintroduction of Crested ibis in China is known to return to the nature of artificially propagated Crested ibis, and the number of wild populations is now more than that of artificial rearing.

In Japan and Korea, it is necessary to introduce species that have already been exterminated in the wild from China and to expand artificial rearing in Korea, or in the case of Japan as an initial stage of wild spinning.

Currently, Crested ibis is artificially multiplying for restoration in Upo Swamp in 2008 and plans to naturally emit Crested ibis in 2017. Therefore, we will select the joint area considering the species for Crested ibis radiation.

When Crested ibis is inhabited in one area, the habitat evaluation index is set according to the high possibility of bird flu and the like.

2.2. Study Methods

The research progresses in the order of the following figure (Figure 1). The purpose of this study was to investigate the current status of the Crested ibis 's habitat and select the factors through expert opinion for more objective and realistic emission.

A study on the domestic preliminary research on ibis has been carried out in the selection of Crested ibis habitability factors in the DMZ as a military area.[7]

The restoration of Crested ibis was first performed in Japan and China, which are neighboring countries of Crested ibis distribution.

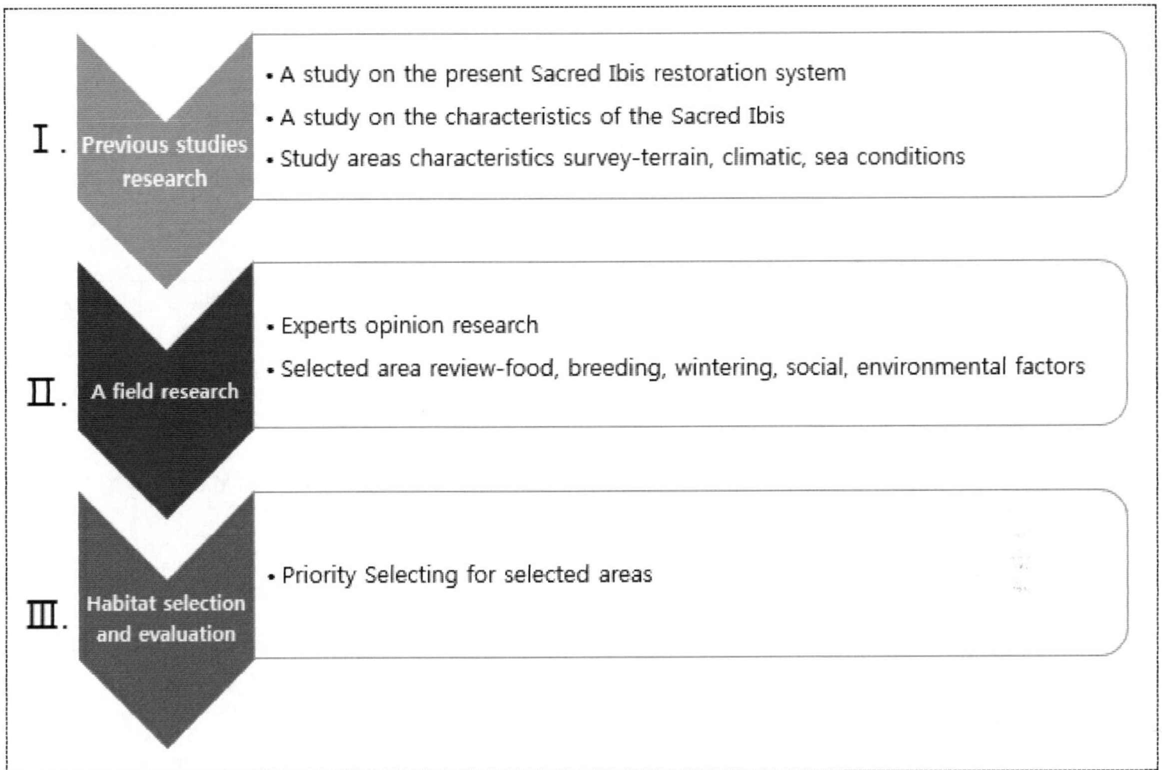

Figure 1. Study flow

3. Results and Discussion

3.1. Study Methods

For the selection of the suitability factors for the habitat, the environmental and geographical characteristics of the Crested ibis were selected from the preliminary study and the opinions of six domestic experts of Crested ibis.

The fit fitness factors of the selected Crested ibis were as follows (Table 1).

Crested ibis was classified into five major factors to consider when selecting habitat.

First, since the prey and habitat of Crested ibis are important factors, the feeding factors of Crested ibis and its approach should be easy.

The main food source of Crested ibis is loach, frog, insect, etc. Most of them live in rice fields.[8]

Second, the factors of breeding were selected as the evaluation items such as the distance of feeding grounds, breeding grounds, number of enemies, biodiversity index, and the labor and density of rice fields.

Table 1. Factors for habitat evaluation for Crested Ibies.

Categories	Evaluation factors	Value	Categories	Evaluation factors	Value
Food Factor	1. A rice paddy	Y/N	**Wintering Factor**	1. Depth of freshwater	20cm
	2. Small irrigation ponds	Y/N		2. Partial freshwater in paddies	Y/N
	3. Fallow ground biotope	Y/N		3. Safe feeding place	Y/N
	4. Eco-corridor	Y/N		4. Safe sleeping place	Y/N

IOP Conf. Series: Earth and Environmental Science **151** (2018) 012011 doi:10.1088/1755-1315/151/1/012011

	5. Two-crop faming	Y/N		5. Artificial feeding site	Y/N
Reproductive Factor	1. Distance from feeding and breeding place	100m	**Social Factor**	1. The public's approval	pros and cons
	2. A enemy species /population size	a lot/single		2. Local residents's willingness to participate	High/Low
	3. Diversity index	High/Low	**Environmental Factor**	1. Environment-friendly agriculture	Y/N
	4. Tree height	15m		2. Pollution source	Y/N
	5. Density of forest	High/Low		3. Urban development degree	High/Low

Based on the evaluation items of bird habitat suitability in the previous study, the priority item for the selection of the crested ibis wildfire area was selected and it can be used as an evaluation criterion for selecting the crested ibis area .[9]

In particular, the results of the existing research mainly focus on the forests, and the focus is on the distance from the surrounding habitat types and the area of the connected space.[10]

In this study, the breeding and wintering seasons, social factors, and surrounding environmental condition such as eco-friendly farming and pollution source. Biotope composition and double cropping are suggested as additional evaluation items.

Considering the food resources of Crested ibis, the establishment of a habitat for wild radishes is closely related to the behavior of local residents. Particularly, discussion management is more important because it plays a role of food habitat of Crested ibis. In this respect, consultation between local residents and the management is essential. Habitat management of the wildfire sites should actively encourage the participation of the residents adjacent to the target area and ensure that the habitat is sustainably managed through them and linked to various projects, including monitoring in the future.

3.2. Candidate Area According to Habitat Evaluation

According to the evaluation area, the top 5 sites in the area suitable for the Crested ibis habitat were selected from the 18 cities in Gyeongnam area where the Crested ibis restoration center is located (Table 2)

The site of Crested ibis radiation area in Gyeongsangnam - do was selected as the surrounding environment, food resources for wild spinning, and habitat conditions.

The 1st place is the Bugok eco industrial park in Changnyeong-gun. The region was selected as the most suitable region by subtracting 92.6 points from the total of 5 evaluation factors.

Bongha Village in Gimhae City was the second place with a total score of 89.4, and the third place was selected as Natural Ecological Farming Complex of Upo Swamp in Changnyeong County.

Table 2. Candidate area for wild Radiation

No	Site	Score	Fitness possibility
1	**Bugok eco industrial park**	**92.6**	**suitability**
2	**Bongha Village, Hallim-myeon**	**89.4**	**suitability**
3	**The IPO Wetland**	**87**	**suitability**
4	**Bu-ri, Chahwang-myeon, Sanhcheong-gun**	**85.4**	**suitability**
5	**Gyeongsangnam-do Forest Enviroment Research Institute**	**83.4**	**suitability**

Crested ibis wild grazing was firstly applied in the area where sufficient feeder and habitat were established. Among the 19 areas of Gyeongsangnam-do, there are eco-friendly agricultural complex area and Gyeongsangnam-do forest environment research area.
However, this area is a place where residents and local governments need to consult. Crested ibis has been ranked as the No. 1 place in the Bonggok Ecological Agricultural Zone in Changnyeong-gun, which is rich in food prefectures. However, this area needs to be discussed by residents and local governments. Respectively. In particular, Upo Wetland is the second most important wetland in the world and registered as a Ramsar Convention in 1998, and is regarded as an important habitat for waterbirds. Upo Wetland is a region with a size of about 700,000 pyong. Rice fields and fields formed around Upo Wetland can be suitable places for wild animals that rely on farming culture and aquatic ecosystem like Crested ibis. The reason why the restoration center is located in Upo Wetland is the reason for this evaluation.

4. Conclusion

This study evaluates the five habitat suitable elements to choose five regions in Gyung-nam area for releasing Crested ibis. The five factors for evaluation are 'Food Factor', 'Reproductive Factor', 'Wintering Factor', 'Social Factor', and 'Environment'.
The regions that have enough feeding grounds and breeding habitats are prioritized.
As a result, Buggok Ecological Farming Complex in Changnyeong-gun is selected as the first place with abundant feeding grounds and Bongha Village in Gimhae and Upo Swamp ecological agriculture complex are following. However, since Buggok Ecological Farming Complex requires resident agreement with local government, Bongha Village and Upo Swamp ecological agriculture complex are considered more suitable.
According to the restoration policy of the endangered species Crested ibis, the regions that have a high suitability index are selected. In order to increase genetic diversity and prevent massive bird flu outbreaks, designating protection area for Crested ibis is necessary, and it is recommended to release them in a geographically dispersed manner.
In condition of a successful Crested ibis releasing, it could be associated with ecotourism. For example, Japan runs the Crested ibis Forest Park and the Crested ibis Village around the protection area and China is currently developing the restoration program in collaboration between the government and local residents.
Crested ibis restoration and wild spinning is not a simple species restoration project. It must be accompanied by a national effort for biodiversity, awareness of environmental conservation in the local community, and education of community members who are stakeholders, as a place branding business.
Additionally, the data obtained from the Crested ibis releasing could be shared with other restoration centers and also used as a basic data for the conservation of other endangered species in the entire Korea peninsula.

5. References

[1] UNDP / GEF National Wetland Conservation Project Management Team. 2008. A Study on the Beneficial Value of Major Wetlands in Korea. 3: 76-86
[2] Changnyeong County Office. (2018). Distribution status .[online]Available at: http://www.cng.go.kr/tour/ibis/00001184.web [Accessed 17 Nov. 2017].
[3] Collar NJAV, Chan SAndreev. (2001). "Threatened Birds of Asia: The BirdLife International Red Data Book 5th edition." Cambridge
[4] Changnyeong County Office. (2018). Classification of Crested ibis. [online]Available at: http://www.cng.go.kr/tour/ibis/00001183.web [Accessed 17 Nov. 2017]
[5] Kang J HI K, Yoo S H, Kang T H, Baek W KKim. (2008). Current Status and Distribution of the Natural Monument Bird Species in Korea. "The Korean Society of Phycology", 5: 73-84
[6] Li X. H and D Li 1998. Current status and the future of the crested ibis (*Nipponia nippon*): A case study by population viability analysis. Ecological Research 13: 323-333.
[7] Rho P H and Han S H 2010 Habitat Evaluation for Selecting Candidate Restoration Site for Crested Ibis (*Nipponia nippon*) in Cheolwon, Korea. Kor. J. Orni. 17(3): 275-283

[8] Ma Z. J., C. Q. Ding, X. H. Li, B. Z. Lu, T. Q. Zhai and G. M. Zheng. 2001. Feeding site selection of crested ibis in winter. Zoological Research 22: 46-50.
[9] McGarigal K and B J Marks 1995. Fragstats, spatial pattern analysis program for quantifying landscape structure. Oregon State University, Corvallis, Oregon, USA.
[10] Li X. H., D. M. Li, Y. M. Li, Z. J. Ma and T. Q. Zhai. 2002. Habitat evaluation for crested ibis: A GIS-based approach. Ecological Research 17: 565-573.

Acknowledgements

This work was supported by a grant from "Crested ibis Survey of habitable areas based on Upo Crested ibis Restoration Center", funded by "GREF(Gyeongsangnamdo Ramsar Environmental Foundation)" of the Republic of Korea.

IOP Conf. Series: Earth and Environmental Science **151** (2018) 012012 doi:10.1088/1755-1315/151/1/012012

Evaluation of the Effects of Surface Slope in Discretization of Groundwater Models

D K S Y Klaas[1,2,3], M A Imteaz[1], A Arulrajah[1], I Sudiayem[3], E M E Klaas[3] and E C M Klaas[3]

1 Department of Civil & Construction Engineering, Swinburne University of Technology, Melbourne, VIC, Australia. dklaas@swin.edu.au
2 Department of Civil Engineering, Politeknik Negeri Kupang, Indonesia.
3 Department of Health, Science and Technology, Klinik Garuda, Malang, Indonesia

Abstract. A robust representation of hydrogeological properties of the examined study area in discretization step is essential in groundwater modelling to improve the accuracy and efficiency of the model. A number of studies have investigated the effect of grid cell size in model discretization using comparison of performances of different models representing different cell sizes. However, although grid size refinement effect on model performances was calculated, the impact of mean slope was not conclusively discussed. In this study, five models distinguished by five spatial discretization schemes; 10 x 10 m, 20 x 20 m, 30 x 30 m, 40 x 40 m and 50 x 50 m were constructed. Using PEST, hydraulic conductivity and specific yield values over a selection of pilot points were estimated. The effect of surface slope is discussed in order to recommend the most appropriate location for observation well placement in terms of topographical characteristic. It is confirmed that the deterioration of model performance is controlled by mean slope of the surface. Results reveal that model performance increases substantially for areas of low slope (< 3 %) and medium slope (3 ~ 10 %) for smaller a grid cell size. Therefore, to improve model performance, it is recommended that the observations wells are placed in areas of low and medium slopes.

1. Introduction

Discretization is one way to spatially and temporally assign physical properties of the real world to the model. Optimal choice of size and number of cells in discretization step would principally enhance the efficiency and accuracy of the model as well as reducing the complexity of model [1]. Spatial discretization comprises of spatially dividing model area into a finite number of cells to which model properties are characterised. In classic numerical methods, such as finite difference method, the nodes which are located in the centre of the cell represent the input components, such as precipitation and hydraulic conductivity. The node is also the place to which the numerical equation is solved to yield the output of the model, such as hydraulic head and discharge.

Figure 1 shows that the grid refinement progressively from 1-cell grid towards 16-cell grid resulting in smaller distance between node and observation well (d_n). This obviously reduces the head disparity (Δh_n) between simulated head (h_s) and observed head (h_o), which statistically improves model performance. However, on the other hand due to the additional computational tasks as a result of increased grid numbers due to grid refinement, time consumption would exponentially increase. In some cases, very finer grid may only provide an insignificant model improvement at the huge expense of computational time. Grid refinement is important in enhancing model performances and to some extent poor representation of the catchment surface attributed by coarser grid results in deterioration of model performance [2]. The aim of the study is to confirm that by assessing the effect of grid cell properties on model calibration we are able to efficiently select most appropriate cell size during the discretization step of model development; and identify the most suitable location for observation well placement in terms of topographical characteristic. Hence, a guideline to select observation well based on the effect of grid cell properties (size and distance between centre of grid cell and observation well) and the representation of topography is provided in this study.

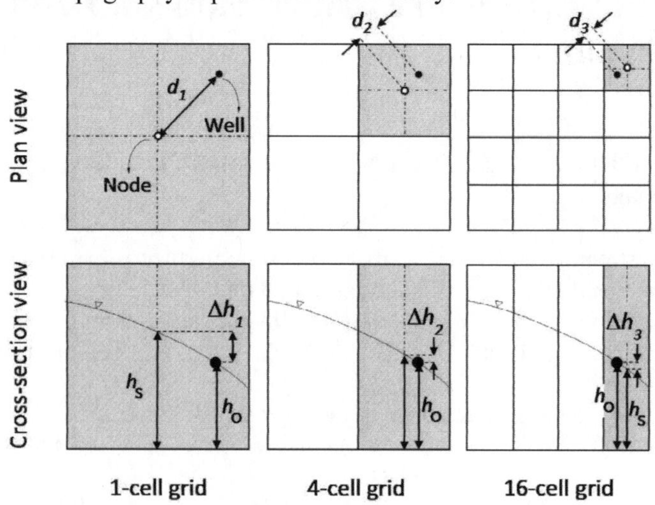

Figure 1. Horizontal conceptualization of three examples of model grid representing observed head (h_o), simulated head (h_s), head disparity (Δh_n) and distance between node and observation well (d_n). The shadowed cells are the observed grid cells. Here, the denser grid represented by bigger cell number yields smaller distance ($d_3 < d_2 < d_1$) and smaller head disparity ($\Delta h_3 < \Delta h_2 < \Delta h_1$). In modeling Δh_n is assigned as statistics criterion

2. Materials and Methodology

2.1. Study Area

This experimental area is located in Rote island, Indonesia, geographically located between latitudes 10°46'42.17"S ~ 10 ° 43'36.91"S and longitudes 123 ° 3'14.84"E ~ 123 ° 9'17.64"E. Having a topographically bounded surface-drainage basin area of 20.11 km^2, the Oemau spring is located around 3 km from Ba'a, the capital of the Rote Island [3].

In the rainy season, monthly rainfall amount reaches around 400 mm in February, while its intensity then usually decreases in subsequent months and may reach to only 4 mm/month in the dry season (August) [4]. The humidity increases during wet months (December – February) to around 92 %; and subsequently drops in the dry season in November to as low as 75 %. Geomorphologicaly characterised by low karstification degree and typified by the absence of preferential flow paths and conduits and, the area is classified as eogenetic karst [5]-[10].

2.2. Groundwater Model

This study utilised MODFLOW [11] developed by USGS operated in groundwater modeling system (GMS) user interface [12] to simulate the hydro-geological processes in the catchment. Under the steady-state and transient simulations, a no-flow boundary is used along the catchment boundary of the spring outlet assuming no flux flowing through the boundary. Specified head boundary was set for the small stream downstream the spring using Time-Variant Specified-Head (CHD) package provided in GMS MODFLOW.

The conceptual model consists of one horizontal layer which assumes the continuum representation of unconfined carbonate aquifer in the study area. The porosity was set as a constant value of 0.3 to represent the typical nature of the recharge area [13]. The drains were simulated using Drain (DRN) package [14] to model the surface drainage network of the recharge area.

The water balance method [15] was used to quantify the recharge. Using SCS-CN method [16], each land use was assigned the run-off value as the input for the recharge. The calibration and validation processes used the daily observed groundwater heads for a 16-month period collected from 7 dug wells scattered in the recharge area. The model was then spatially discretised by a grid system comprising of a finite difference mesh of cells. The grid cell of the model domain was homogenously discretised to 5 different sizes, i.e. 10 x 10 m, 20 x 20 m, 30 x 30 m, 40 x 40 m and 50 x 50 m, to assess the impact of spatial discretization properties (cell size and distance between observation well and centre of grid cell).

3. Results and Discussion

3.1. Grouping of Slope Cluster

The recharge area is dominated with a highly undulated topography with elevation ranging between 97.55 m and 339.55 m above sea level. Areas with mild slope are generally found in the middle of the catchment where rice farms are situated; and steeper slopes dominate the upper and lower areas of the catchment. Using ArcGIS the surface of the area studied was clustered into three slope groups (Figure 2). The clustering is intended to categorise the area into distinct slope distribution and to explain the influence of each mean area slope to the performance of the model, in relation to different grid sizes.

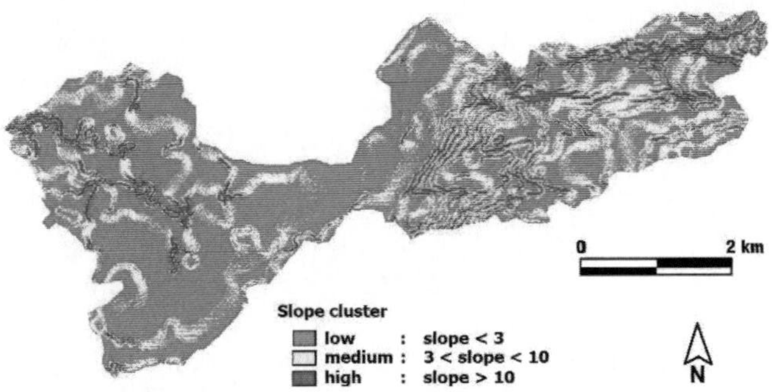

Figure 2. Distribution of areas according to assigned slope cluster

The clustering shows that the catchment is dominated by areas with low and medium clusters; 47.51 % and 41.44 % respectively, while high slope areas share only 11.04 % of the total area. Four wells (OW5, OW6, OW13 and OW15), which are mainly located near the rice farm areas, fall in low slope cluster; while other two wells (OW2 and OW9) fall into medium slope cluster. Well OW17, situated in the hilly area upstream of the catchment, is categorised to rest in high slope cluster.

3.2. Effects of Surface Slope of the Grid

Figure 3 presents the model performance at each observation well with respect to mean slope of the surface area. In general, the increase of slope in each well as a result of modification of grid cell size results in deterioration of model performance represented by increased $RMSE_h$ values. The performance decrease due to increased slope differs variably, with the most notable decrease appears at wells, OW6, OW9, OW13 and OW15 (between around 1250 and 2040 %) which are located in low and medium slope areas. In contrast, OW17 representing high slope area responds moderately by around 250% decrease to the increased slope. Meanwhile, the performance change is barely seen in OW2 and OW5, with only around 25 % and 40 % respectively, conceivably owing to their position which are adjacent to almost steady surface water elevation of a pool downstream the spring.

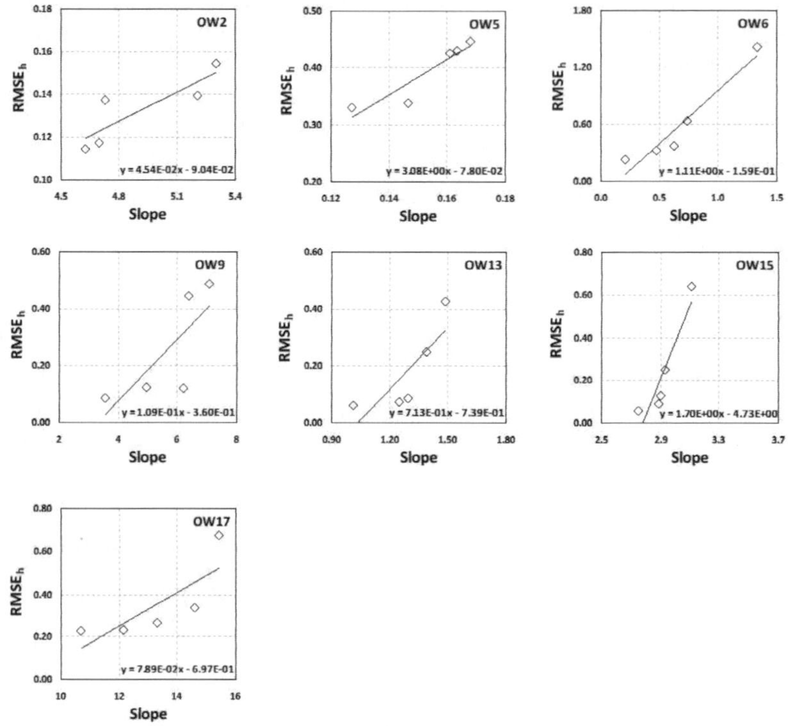

Figure 3. $RMSE_h$ results vs. mean slopes of the area in cell containing observation wells

Therefore, it can be concluded that generally in area of smooth topography the possibility of significantly enhancing model performance using grid refinement method is greater than that in high slope areas. The rationale behind this is that grid size coarsening resulting in higher mean slope gradients consequently generates smaller modelled recharge volumes. This at the end would decrease simulated head. The relationship between slope augmentation and simulated head confirms that the higher slopes contribute to lowered simulated head (h_s) and increased of ME_h values. The effect of grid cell slope shows that the selection of observation wells locations, with regard to their topographical gradient, to be included into a model is critical because it significantly contributes to the effectiveness of the model to achieve a better statistical performance while keeping a less computation time to do so.

4. Conclusions

In this study the effect of mean slope in discretization of groundwater models is obvious. It is found that generally model performance increases most significantly for areas of low slope ($\leq$ 3 %) and medium slope (3 ~ 10%). In these two areas, $RMSE_h$ greatly increases from approximately 7 to 19 times for smaller grid sizes. However, in high slope areas (>10%), model performance increases only a maximum of four times. Another important finding is that the model performance is controlled by the change in grid cell surface geometry represented by the mean slope of the respective cell. Steeper slope would generate smaller flux into the cell thus it decreases simulated head, which results in decreased model performance represented by higher $RMSE_h$.

The results obtained from the simulation of the five models assist us to improve our understanding of how they impact the model performance. This henceforth provides an improved insight and rationale of not only efficient grid cell selection but also the preference for selecting or placing observation well(s). The most optimal model performance was obtained in the low and medium slope areas given the cell size and slope are reduced. Hence, it is recommended to place observation wells within low and medium slope areas to help improving model performance.

5. References

[1] Klaas D K S Y and Imteaz M A 2017 Investigating the impact of the properties of pilot points on calibration of groundwater models: case study of a karst catchment in Rote Island, Indonesia *Hydrogeol J* **25**(6) 1703–1719 doi: 10.1007/s10040-017-1590-4

[2] Klaas D K S Y, Imteaz M A and Arulrajah A 2016 Evaluating the impact of grid cell properties in spatial discretization of groundwater model for a tropical karst catchment in Rote Island, Indonesia *Hydrol Res* doi: 10.2166/nh.2016.250

[3] Klaas D K S Y 2008. *Indigenous Water Management: Sustainable Water Conservation Strategies in a Karstic Dominated Area in Rote Island, NTT Province, Indonesia* (MEngSc Thesis). Monash University, Melbourne, Australia.

[4] Klaas D K S Y and Imteaz M A 2017 Development of a groundwater model for a tropical eogenetic karst aquifer in Rote Island, Indonesia *Proc. 37th Hydrol and Water Res Symp Queenstown, New Zealand* 228-235

[5] Klaas D K S Y, Imteaz M A and Arulrajah A 2017 Development of groundwater vulnerability zones in a data-scarce eogenetic karst area using Head-Guided Zonation and particle-tracking simulation methods *Water Res* **122**, 17-26 doi: 10.1016/j.watres.2017.05.056

[6] Klaas D K S Y, Imteaz M A, Sudiayem I, Klaas E M E and Klaas E C M 2017 Novel approaches in sub-surface parameterisation to calibrate groundwater models *Proc. 3rd Int Conf on Water Res and Environ (WRE 2017). IOP Conf. Series: Earth and Environmental Science* **82** 1-8 doi: 10.1088/1755-1315/82/1/012014

[7] Klaas D K S Y, Imteaz M A and Arulrajah A 2017 Application of head-guided zonation method to delineate groundwater vulnerability zones in a data-scarce catchment. *Europ Water* **57**:395-398

[8] Klaas D K S Y, Imteaz M A, Sudiayem I, Klaas E M E and Klaas E C M 2017 On the development of a new methodology in subsurface parameterisation on the calibration of groundwater models. *Proc. 1st Int Conf on Adv Environ Eng (AEE 2017). IOP Conf. Series: Earth and Environmental Science* **92** 012026 doi: 10.1088/1755-1315/92/1/012026

[9] Klaas D K S Y, Imteaz M A, Sudiayem I, Klaas E M E and Klaas E C M 2017 Assessing the impact of grid cell properties in spatial discretization of groundwater model. *Proc. 1st Int Conf Adv on Eco Eng Dev (ICEED 2017). IOP Conf. Series: Earth and Environmental Science* **109** 012037 doi: 10.1088/1755-1315/109/1/012037

[10] Klaas D K S Y, Imteaz M A, Sudiayem I, Klaas E M E and Klaas E C M 2017 On the development of a new methodology in subsurface parameterisation on the calibration of groundwater models. *Proc. 1st Int Conf on Adv Environ Eng (AEE 2017). IOP Conf. Series: Earth and Environmental Science* **92** 012026 doi: 10.1088/1755-1315/92/1/012026

[11] McDonald M G and Harbaugh A W 1988 *A Modular three-dimensional finite-difference ground-water flow model*, U.S.G.S., Techniques of Water-Resources Investigations 06-A1.

[12] Aquaveo 2011 *Groundwater Modeling System 8.0*. Aquaveo, University of Texas, Provo, Utah, USA.

[13] Johnson A I 1967. *Specific Yield – Compilation of Specific Yields for Various Materials*, US Department of the Interior Geological Survey

[14] Harbaugh A W, Hill M C and McDonald M G 2000 *MODFLOW-2000, the U.S. Geological Survey Modular Ground-Water Model: User Guide to Modularization Concepts and the Ground-Water Flow Process* (USGS Open-File Rep 00-92, USGS)

[15] Bras R L 1990 *Hydrology: An introduction to hydrologic science*. Addison-Wesley, Reading, Massachusetts.

[16] Viessman W and Lewis G L 2003 Introduction to Hydrology 5th ed (Upper Saddle River: Pearson Education)

Acknowledgments
The authors wish to acknowledge the Australian government for supporting the research through the Australia Awards Scholarship (AAS) and Allison Sudrajat Prize Award in the Swinburne University of Technology, Australia.

IOP Conf. Series: Earth and Environmental Science **151** (2018) 012013 doi:10.1088/1755-1315/151/1/012013

Numerical study of local scour around bridge piers

H Omara[1] and A Tawfik[1,2]

1 Egypt - Japan University of Science and Technology (E-Just), Environmental Engineering Department, P.O. Box 179 – New Borg El Arab City – Postal Code 21934 - Alexandria – Egypt Hewida.omara@ejust.edu.eg

2 National Research Centre, Water Pollution Research Department, 12622, Dokki, Giza, Egypt

E-mail: hewida.omara@ejust.edu.eg; ahmed.tawfik@ejust.edu.eg

Abstract. Numerical models have more flexibility avoiding the critical limitation in physical ones. The predictive capability of the hydro-morphological model around bridge piers in clear-water conditions have been investigated using mathematical modelling flow 3D 11.2 as there is a lack in using that model. The objective of the study to investigate the effect of cross section of the pier on the predictive capability of the model. Computations has been performed and the results are compared with several sets of experimental data available in the literature. The Van Rijn sediment transport model and (RNG) k-ε model are used. However, this study strongly demonstrates that a 3D hydro-morphological model can effectively predict the scour around piers with different cross section, there is significantly under predicts for the scour depth at the nose of the pier with different pier shapes. It is found that for the circular pier the capability of the model depending on the flow properties not only the geometry of the pier.

1. Introduction

Three-dimensional flows around vertical obstructions in natural rivers or streams are inherently complex because of generation of flow separation and turbulence structures in a wide range of scales [1]. Various analytical, numerical, experimental and data processing methods have been used to estimate the local pier scour. several three-dimensional numerical studies around a single vertical circular cylinder have been conducted to study the scour process and estimate the scour depth.

A fully 3D hydrodynamic model flow occurring at the base of a cylindrical bridge pier within a scour hole was used by [2,3] investigated mean flow and turbulence kinetic energy distribution inside the equilibrium scour hole around cylindrical piers. Computational studies by focusing on flow patterns around piers on a flat rigid bed have been reported by [4–6]. [5] used FLUENT to predict the 3-d flow field around a circular cylinder for rigid and erodible beds which gives satisfactory agreement in the bed shear stresses with those calculated from the experimental velocities near the bed. Also, [7] used different turbulence models and compared with several sets of experimental data of [8], Dargahi (1987) and [9]to investigate the flow around cylindrical pier. they declared that the standard and the RNG k- ε models can be used to calculate the flow field around the pier although they slightly overestimate the velocity near the bed. [10] applied the same DES approach to study the dynamics of the flow past the cylindrical pier mounted on a flat rigid bed studied experimentally by Dargahi (1987) and obtained results in good agreement with the experimental observations. [11] investigate the effect of pier bluntness on the dynamics of scour and the predictive capabilities of the model for three bridge piers (cylindrical, square, and diamond cross-sectional shape) with movable-bed channels developed hydro-morpho dynamic CURVIB method. They observed discrepancies between experiments and

simulations. Modelling of the free-surface flow around a pier in a fixed scour hole was assessed by [12] to determine shear stress distributions at the channel-bed using mesh-based numerical models. The bed shear stress from FEM agrees well with the measurements. Some discrepancies exist for locations just before water enters the scour hole. Prediction of local scour around a single square pier using numerical FSUM and physical models for suspended sediment concentration and bed morphology was investigated by [13]. They found that the model predicts a bigger scour depth in comparison to the measured scour depth in the physical model. Both numerical model and experiment results show that the maximum scour depth occurs at two front edges of the pier. [14]carried out the 3D numerical analysis of pier scour using the level set method. they simulated the scour evolution and free surface under the currents and wave. Results showed good agreement with experimental observation made by [15]. A numerical scour model was used by [16] to get approximation approach of the equilibrium scour depth around a mono-pile by FLUENT.
A few CFD models have been developed for computing flow patterns and bed-profile changes around hydraulic structures, e.g., SSIIM, Fluent and Flow-3D models. There is a lack for using flow 3 d to simulate scour around bridges piers. The objective of the present study is to investigate the capabilities in simulating scour around circular piers for piers with different cross section using 3-D computational fluid dynamics (CFD) solver Flow 3D 11.2.

2. Numerical Model

FLOW 3D can simulate bed and suspended load rate for steady or unsteady flow conditions. It is developed by Flow Science Inc., uses a non-hydrostatic finite difference model to solve the 3D Navier-Stokes equations, and has a powerful capability to deal with free surface flow and sediment transport issues [17]. The model can simulate the deposition and entrainment of sand, silt and other non-cohesive sediment, furthermore, it enables specifying multiple sediment species, and includes a bed load transport model, a nonlinear drift-flux model, and empirical equations to predict the entrainment and erosion of sediment. The continuity and momentum Navier-Stokes equations are applied:

$$\frac{\partial u_i}{\partial x_i} = 0 \qquad (1)$$

$$u_j \frac{\partial u_i}{\partial x_i} = \frac{-1}{P} \frac{\partial}{\partial x_i} (p z_{ij} + p u_i \, u_j) \qquad (2)$$

Where, P is the dynamic pressure, ρ the flow density, ui the i component of the local time-averaged flow velocity, δ the Kronecker Delta function and $pu_i\, u_j$ is the turbulence stresses. The numerical solution to complement, a turbulence model and renormalization-group (RNG) k-ε model has been used. Sediment scour requires accurate estimates of the near-wall shear stresses, so a good turbulence model should be selected for turbulent flows. The RNG model is known to describe more accurately low intensity turbulence flows and flows having strong shear regions. For computational domain, it was found that the upstream inlet should place at sufficient distance from the pier to ensure that the flow becomes fully developed 40D. Sufficient distance is provided between the pier and the downstream outflow to ensure that the flow returned to the undisturbed pattern approximately 12 times the pier diameter downstream of the pier [18]. So, the domain length used is, l=52D, with fixed bed at the inlet to prevent scour to prevent scour at the inlet, Figure 1. The pile is located at (x=0, y=0 and z= - 0.2) of the domain. The bed surface is at z=0.

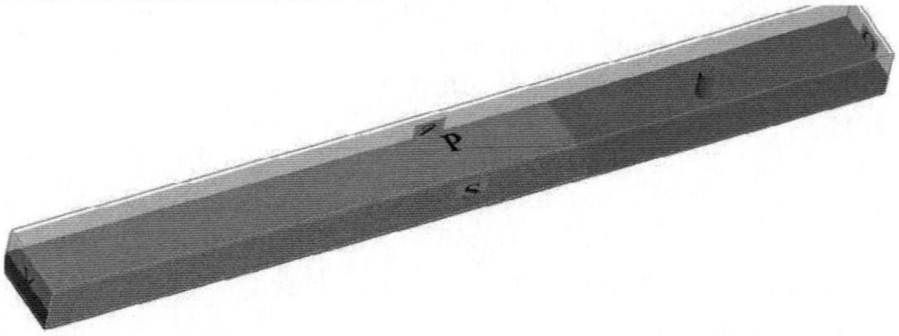

Figure 1. boundary condition of present study

Finer mesh is required near solid boundaries including the pier and packed sediment bed to resolve the flow details near the solid boundaries. A considerable effort has been put into optimizing the computational mesh for the best grid and time convergence. The size of the cells adjacent to solid boundaries is chosen to satisfy the limits of the wall unit distance 11.225<Z+<30 based on law of wall to avoid calculation in the laminar sub-layer as Viscosity damps turbulence near a wall and the value of Z+ should be smaller than 30 to ensure that the law of the wall is applicable. The wall unit distance Z+ is defined by the following equation:

$$Z+ = \frac{\rho u z_p}{\mu} \tag{3}$$

where ρfluid density; u is shear velocity; $z_p$distance from Point p to the wall; and μ dynamic viscosity of the fluid. Finer grid is also provided near the air-water interface to capture the small variation of the free surface [7]. It was used a mesh measuring 0.003 m near the pier and sediment bed surface and gradually increase to 0.025 m in the rest of the domain. The law-of-the-wall for the mean velocity modified for roughness is expressed as follows:

$$\frac{u_p}{u} = \frac{1}{k} ln \left[E \frac{\rho c_\mu^{\frac{1}{4}} k_p^{\frac{1}{2}} z_p}{\mu} \right] - \Delta B \tag{4}$$

where up mean flow velocity at point p; k von Karman's constant (0.418); u shear velocity related to the bottom shear stress τ ($\tau = \rho u^2$); ρ fluid density; $c_\mu$constant; kp turbulent kinetic energy at point p; and ΔB roughness function. Free surface can be included in the one-fluid incompressible mode. In FLOW-3D, A proper definition of the boundary conditions is important for an accurate capture of the free-surface dynamics. The volume of fluid (VOF) method is employed. The top surface is defined by stagnation pressure with the atmospheric pressure value. At the bottom, there is packed sediment layer with thickness 0.2 m. At the exit of the channel, the outflow condition is used so the water level at the outlet is calculated from inside the domain, Figure 2. At a stagnation pressure source in the inlet, fluid is assumed to enter the domain at constant velocity. The roughness height of the bottom is (2.5 D50).

3. Sediment Scour Model

This model assumes non-cohesive sediment species and estimates the motion of sediment by predicting the erosion, advection and deposition process. Soulsby-Whitehouse equation [Sou97] used to predict the critical shields parameter $\theta_{cr,i}$:

$$\theta_{cr,i} = \frac{0.3}{1+1.2d_{*,i}} + 0.055\left[1 - exp\left(-0.02d_{*,i}\right)\right] \quad \text{where} \quad d_{*,i} = d_i \left[\frac{\rho f(\rho i - \rho f)\|g\|}{\mu_f^2}\right]^{\frac{1}{3}} \tag{5}$$

ρi is the density of the sediment species i, ρf is the fluid density, di is the diameter, μf is the dynamic viscosity of fluid and ||g||is the magnitude of the acceleration of gravity. The local shields parameter is:

$$\theta_i = \frac{\tau}{\|g\| d_i(\rho_i - \rho_f)} \tag{6}$$

where τ is calculated using the law of the wall and the quadratic law of bottom shear stress for 3D turbulent flow and shallow water turbulent flow, respectively, with consideration of bed surface roughness. Nikuradse roughness of the bed surface ks is:

$$ks = \text{Crough } D_{50}, \text{ packed} \tag{7}$$

where Crough is coefficient and ks is the physical roughness height taken here and equal to the 2.5. The entrainment coefficient can be used to scale scour rates or fit experimental data where it found that 0.005 give good result. The volumetric sediment transport rate per width of bed Van Rijn [vanRijn84] which has been described as follows is used,

$$\Phi_i = \beta_{VR,i} d_{*,i}{}^{-0.3} \left(\frac{\theta_i}{\theta_{cr,i}} - 1\right)^{2.1} c_{b,i} \tag{8}$$

Where $\beta_{VR,i}$ is coefficients typically equal to 0.053. $c_{b,i}$ is the volume fraction of species i in the bed material, Φ_i is the dimensionless bed-load transport rate which is taken 0.05, and is related to the volumetric bed-load transport rate, $q_{b,i}$, by:

$$q_{b,i} = \Phi_i \left[\|g\| \left(\frac{\rho_i - \rho_f}{\rho_f}\right) d_i^3\right]^{1/2} \tag{9}$$

4. Case Study

The laboratory experiments reported by [9] which also used by [7,19] to validate their numerical model are simulated in the present study (flatbed to calibrate the model and equilibrium scour bed) for cylindrical pier, [20,21] for rectangular cross section pier and [22] for square cross section pier. All required information for physical experiment are presented in Table 1.

Table 1. Relevant Information on Various Experimental Conditions Modeled in the Present Study

Experiment data	Ahmed & Rajaratnam, (1998)	*Vaghefi & Ghodsian, (2015)* [23]	*Abdul-Hassan & Hadi, (2016)*[20]	*Fael et al., (2016)* [21]	*Ghiassi& Abbasnia, (2013)*[22]	Khosronejad et al. (2012)[11]
Pier cross section	cylindrical		Rectangular, square with around nose	Rectangular square nose	square	
Flume dimension	20/1.22 m		12/ 0.5m	14/2m	13/ 0.6m	10/0.45 m
Y (m)	0.182	0.18	0.12	0.2	0.088	0.139
U(m/sec)	0.2927	0.32	0.18	0.3	0.15	0.22
D(m)/pier length	0.089	0.05	0.045/0.2	0.05/0.2	0.05	0.165
D_{50}(mm)	1.8	0.6	0.71	0.86	0.53	0.85
Bed slope %So	0.02	0.002	0	0	0	0
ds Scour depth (m)	0.040	0.08	0.043	0.038	0.014	0.06
U* /U*c	0.62	0.86	0.56	0.96	0.6	0.73
Fr=u/√gh	0.21	0.24	0.17	0.22	0.16	0.13
R=uY/v (×103)	53.2	57.6	14.7	15.5	13	30

Note: Fr *is Froud number; and* R *is Reynolds number.*

5. Results and Discussion

Numerical simulations for experiment data were carried out in order to validate the applicability of the flow 3D model for evolutions of flow and bed elevation around a cylindrical bridge pier as shown in Figure 2 Moreover, the effect of different pier shapes on the predictive capabilities of the model was studied. The erosion process can be impeded by particles that are large relative to the size of the groove that forms at the base of the scour hole in front of the pier. The groove has been identified as the principle erosion zone for bridge pier scour [8,24,25] and forms under the action of the downflow that occurs in front of the pier.

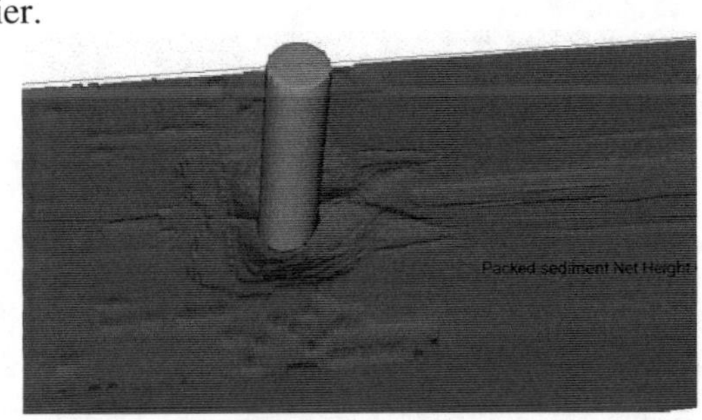

Figure 2. the scour hole shape using flow 3d

Figure 3 shows scour contours at (t=30 min) observed by Ahmed and Rajaratnam 1998 and simulated using the computational fluid dynamics (CFD) versus the model results. The computed scour contours by *Fael et al.* [26] used coupling of a stochastic model for sediment pickup and deposition using a momentum equation of sediment particles in order to account the temporal change in bed topography for the effect of nonequilibrium sediment transport. It is notice that the maximum scour depth is 0.040 ,

0.041, 0.0397 m for the experiment, [26]) and present model. From that, flow 3d gives the most agreement result with the experiment (0.040m) however, there is some disagreement observed for the scour at the area close to the pier nose; while the maximum scour depth locates at pier sides. [26] mention that this was because a double-sided sharp slope immediately close to pier nose in numerical results, which may collapse in local avalanches of sediment into the erosion zone under the action of turbulent flow in the test. Hence, the presented maximum scour depth agrees well with the test. However, the scour hole shape and location between the numerical simulation and the test are quite different the model gives scour depth with sufficient accuracy. Comparisons between the bed elevation simulated using flow 3D and the experiment of [27] at the plane of symmetry are shown in figure 4. It gives better result for the scour shape. It confirms that for cylindrical pier the model gives under estimation to the scour at the nose however it gives the maximum scour at the pier sides with good agreement with the paper. Also, [11] with hydrodynamic model declared that the numerical model they used gives overpredicted the growth rate of scour and failed to capture the region of deep scour at the pier nose For the cylindrical and square shapes.

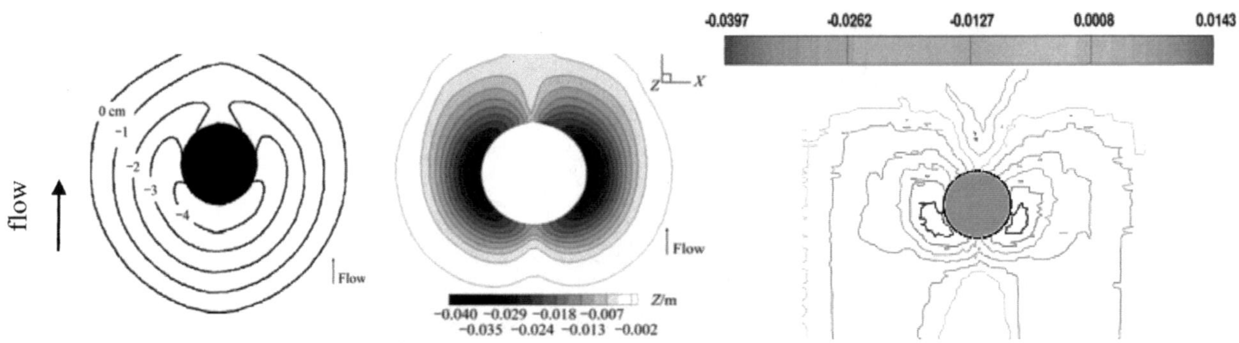

(a) (b) (c)

Figure 3. Scour hole contours at (t=30 min) (cm): (a) observed by [9]; (b) simulated by [26] (m); (c) model results(m).

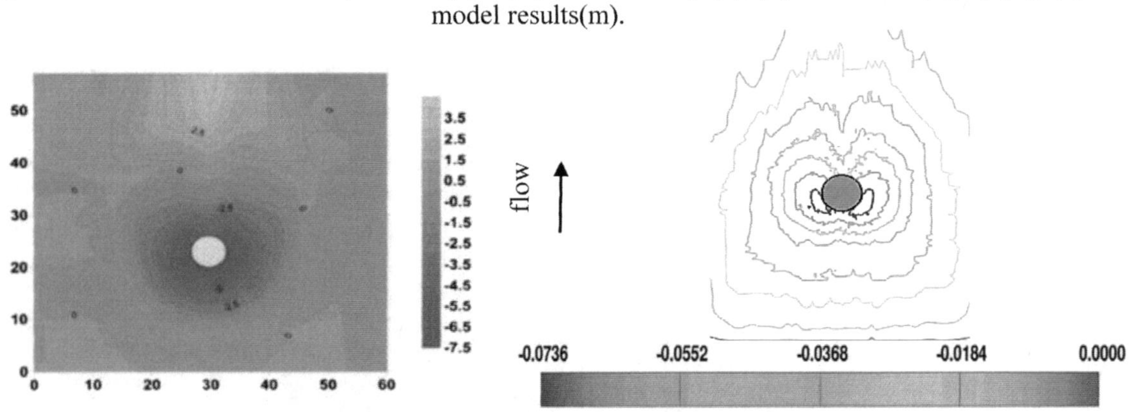

Figure 4. Bed contours observed by Vaghefi et al. [27] experimental (cm) and numerical (m)

For square shapes, Comparisons between the bed elevation observed and simulated by [22] and present results at the plane of symmetry are shown in Figure 5. From the figure, it is declared that the observed maximum scour depth locates at the front corners of the pier which is agree well with the results of the model. However, there is still disturbance at the nose but it can be neglected.

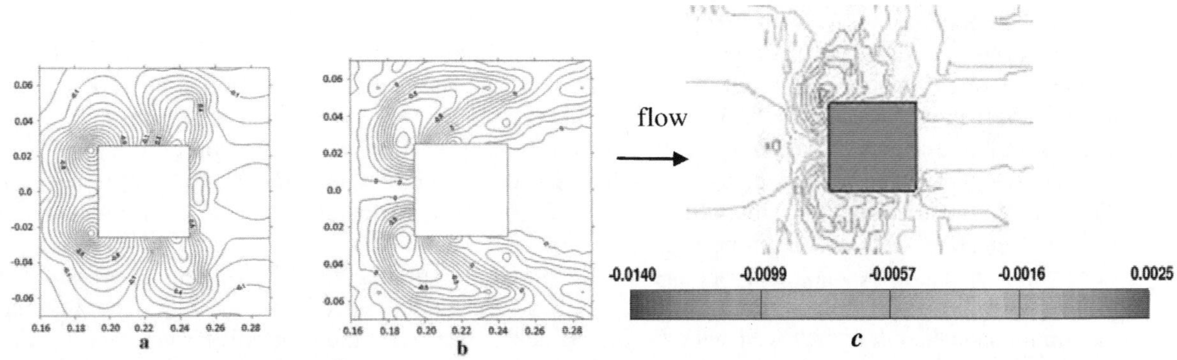

Figure 5. Scour hole shape (a) observed and (b) simulated [22] (cm); (c) model results(m) of the experiment of Ghiassi & Abbasnia [22].

For rectangular shape with both square or round, Figures 6 and 7 show comparison between the observed scour contours of [20,21] and simulated one. Figure 6 shows that the maximum scour depth observed is 0.038 for rectangular with circular nose and 0,042 m for square nose versus the maximum scour depth simulated is 0.0375 for circular nose and 0.043 for square one. Figure 6 also present the scour hole shape observed by [21] and present results, it give the same agreement with the maximum scour depth observed 0,06 and simulated one 0.0609 m. The model gave good agreement with the observed for the maximum scour depth and outer perimeter of the scour hole.

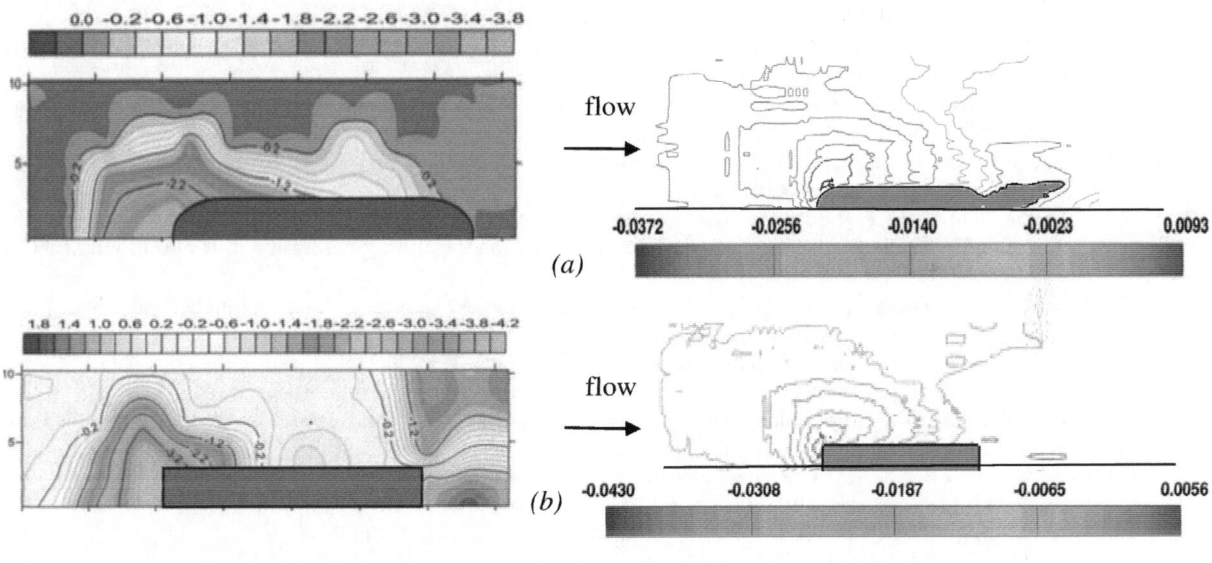

Figure 6. Scour hole shape for experiment of [20,21] (cm) and model results (m) (a) Rectangular circular nose (a) rounded nose (Oblong)

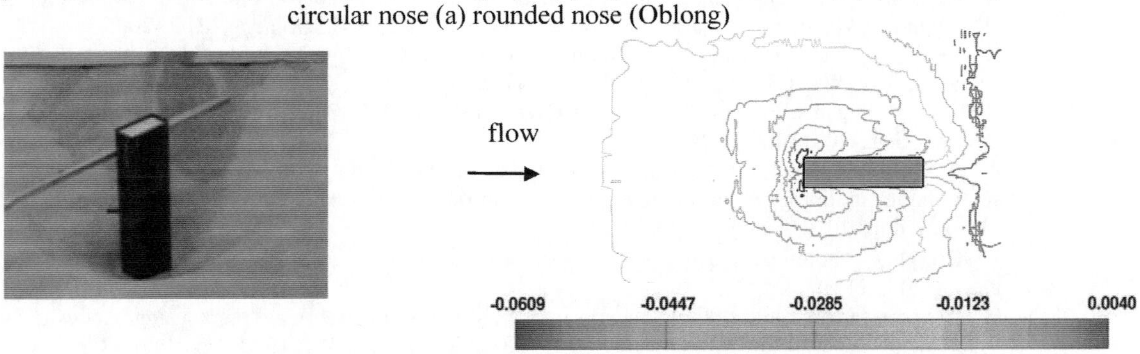

Figure 7. Scour hole shape of [21] experiment and present result

6. Conclusions

It should also be expected that 'perfect' results are difficult to obtain in view of the many factors involved, which cannot be modelled directly using a numerical simulation,[16]. The present investigation shows that the Flow 3D can be effectively used to estimate the maximum scour depth. The model simulates the maximum scour hole depth with sufficient accuracy on contrary with the prediction of the scour hole shape as it is failed to simulate the observed scour depth at the nose and the capability of the model to simulate the scour shape well depends on the cross sections shape and for the same cross section flow properties.

For cylindrical case, the maximum scour depth observed in experimental data located at the area close to the pier nose; while the maximum scour depth from the model situated at pier sides. The presented maximum scour depth agrees well with the experiments. However, the scour hole shape and location between the numerical simulation and the test are quite different. [11] also failed to capture the region of deep scour at the pier nose and declared that the predictive capabilities of the URANS hydro-morpho dynamic depend on the details of the geometry of the studied pier cross-section only. Although, it is found that for the circular pier the capability of the model depending on the flow properties not only the geometry of the pier.

7. References

[1] I.K. and Y.S. Hyung Suk Kim1, Mohamed Nabi2, Computational Modeling of Scours, Hydraul. Eng. (2014) 79–84.
[2] S. Dey, I.S.K. Bose, and G.L.N. Sastry, CLEAR WATER SCOUR AT CIRCULAR PIERS : A MODEL, J. Hydraul. Eng. (1995) 869–876.
[3] J.F.A.J.M. ANDREAS ROULUND1, B. MUTLU SUMER1, Numerical and experimental investigation of flow and scour around a circular pile, J. Fluid Mech. 534 (2005) 351–401. doi:10.1017/S0022112005004507.
[4] S.. Richardson, E.V. and Davis, Evaluating Scour at Bridges, HEC 18, Third Ed. Novemb. 1995 Metr. Report No. (1995).
[5] K.H.M. Ali, O. Karim, Simulation of flow around piers, J. Hydraul. Res. 40 (2002) 161–174. doi:10.1080/00221680209499859.
[6] L. Ge, S.O. Lee, F. Sotiropoulos, M. Asce, T. Sturm, M. Asce, 3D Unsteady RANS Modeling of Complex Hydraulic Engineering Flows . II : Model Validation and Flow Physics, J. Hydraul. Eng. (2005) 809–820.
[7] T.M. Salaheldin, J. Imran, M.H. Chaudhry, Numerical Modeling of Three-Dimensional Flow Field Around Circular Piers, J. Hydraul. Eng. Vol. 130, (2004) 91–100.
[8] B.W. Melville, LOCAL SCOUR AT BRIDGE SITES, phD. 1994 (1975).
[9] F. Ahmed, N. Rajaratnam, Flow around Bridge Piers_Ahmed.pdf, J. Hydraul. Eng. 124 (1998) 288–300.
[10] C. Escauriaza, F. Sotiropoulos, Initial stages of erosion and bed form development in a turbulent flow around a cylindrical pier, J. Geophys. Res. Earth Surf. 116 (2011) 1–24. doi:10.1029/2010JF001749.
[11] A. Khosronejad, S. Kang, F. Sotiropoulos, Experimental and computational investigation of local scour around bridge piers, Adv. Water Resour. 37 (2012) 73–85. doi:10.1016/j.advwatres.2011.09.013.
[12] M.N. Sakib, phD-CFD Techniques for Simulation of Flow in a Scour Hole around a bridge pier, Ph.D. (2013).
[13] N.V. Thanh, D.H. Chung, T.D. Nghien, Prediction of the Local Scour at the Bridge Square Pier Using a 3D Numerical Model, Open J. Appl. Sci. 4 (2014) 34–42. doi:10.4236/ojapps.2014.42005.
[14] H.S. Kim, M. Nabi, I. Kimura, Y. Shimizu, Numerical investigation of local scour at two adjacent cylinders, Adv. Water Resour. 70 (2014) 131–147. doi:10.1016/j.advwatres.2014.04.018.
[15] B.M. Sumer1, and J. Fredsøe, SCOUR AROUND PILE IN COMBINED WAVES AND CURRENT, J. Hydraul. Eng. (2001) 403–411.
[16] A.L.J. Pang, M. Skote, S.Y. Lim, A. Gullman-strand, J, N. Morgan, A numerical approach for determining equilibrium scour depth around a mono-pile due to steady currents, Phys. Procedia. 57 (2016) 114–124. doi:10.1016/j.apor.2016.02.010.
[17] Z. Xie, phD Theoretical and numerical research on sediment transport in pressurised flow conditions, Ph.D. (2011) 217.
[18] M. Akhtaruzzaman Sarker, Flow measurement around scoured bridge piers using Acoustic-Doppler Velocimeter (ADV), Flow Meas. Instrum. 9 (1998) 217–227. doi:10.1016/S0955-5986(98)00028-4.

[19] D.A. Lyn, Turbulence Models for Sediment Transport Engineering, Sediment. Eng. (2013) 763–825. doi:doi:10.1061/9780784408148.ch16.
[20] Abdul-Hassan, Zaid Hadi, EXPERIMENTAL STUDY OF BRIDGE PIER, Int. J. Civ. Eng. Technol. 7 (2016) 162–171.
[21] C. Fael, R. Lança, A. Cardoso, Effect of pier shape and pier alignment on the equilibrium scour depth at single piers, Int. J. Sediment Res. (2016) 1–7. doi:10.1016/j.ijsrc.2016.04.001.
[22] R. Ghiassi, A.H. Abbasnia, Investigation of Vorticity Effects on Local Scouring, Arab. J. Sci. Eng. 38 (2013) 537–548. doi:10.1007/s13369-012-0337-8.
[23] M. Vaghefi, G. and Masoud, The effect of circular bridge piers with different inclination angles toward, Sadhana. (2015). doi:10.1007/s12046-015-0443-x.
[24] R. Ettema, Scour at bridge piers, Rep. No. 216, Sch. Eng. Univ. Auckland, Auckland, New Zealand, 527pp, 1980. (1980).
[25] S.S. Shen, H. W., Schneider, V. R., and Karaki, Mechanics of local scour., U.S. Dep. Commer. Natl. Bur. Stan- Dards, Inst. Appl. Technol. Washington, D.C. (1966).
[26] Z.H.U. Zhi-wen, L.I.U. Zhen-qing, CFD prediction of local scour hole around bridge piers, J. Cent. South Univ. (2012) 273–281. doi:10.1007/s11771.
[27] M. Vaghefi, M. Ghodsian, S. Salimi, Scour Formation Due to Laterally Inclined Circular Pier, Arab. J. Sci. Eng. 41 (2016) 1311–1318. doi:10.1007/s13369-015-1920-6.

8. Acknowledgments

The first researcher is supported by a scholarship from the Mission Department for the Government of Egypt which is gratefully acknowledged. Second I am thankful to Egypt-Japan University of Science and Technology (E-JUST) and Japan International Cooperation Agency (JICA) for offering the tools and equipment needed for the research work.

9. Notation

D	*pier diameter/width;*
d_s	*equilibrium scour depth;*
D_{50}	*mean sediment size;*
gi	*gravitational acceleration in i th direction*
k	*von Karman constant;*
So	*longitudinal slope of flume;*
t	*time;*
U	*average flow velocity;*
u*	*shear velocity;*
Z+	*wall unit distance;*
zp	*distance from point p to wall;*
q	*volume fraction of qth phase in control volume;*
ΔB	*roughness function;*

IOP Conf. Series: Earth and Environmental Science **151** (2018) 012014 doi:10.1088/1755-1315/151/1/012014

The Xique-Xique Network towards sustainability and solidarity

M Dantas[1,3,] and G Silva[2,]

1 Concordia University, Montreal, QC, Canada
3 Research sponsored by Science without Borders scholarship by CAPES - Brazil
E-mail: mo_arauj@live.concordia.ca
2 IFRN, Natal, RN, Brazil
E-mail: gerlane_rn@bol.com.br

Abstract. The industrialized food system holds considerable power over some of the most important aspects of our societies, such as health, environment, economy, culture and politics, though their main focus is private profit. In response to issues related to the production and distribution of goods alternative initiatives are emerging all over the world related to the creation of a food systems that promotes solidarity, diversity, democracy. This paper looked at an organization emerged from a group of women members of the Landless Movement called Rede Xique-Xique. The study presents data from a recent interview conducted with a community organizer involved with the Network, also includes records from previous unpublished research, consisting of a participant observation conducted in 2011. The study found that the network organization brought improvements to the living conditions of farmers and voice to a rather invisible portion of the population.

1. Introduction

The Northeast is a region of Brazil that has been marked by a longstanding history of exclusion history., The last Census from the Instituto Brasileiro de Geografia e Estatística-IBGE [1], revealed once more that the lowest wages in Brazil comes from this region. The reason for these regional disparities will not be discussed in this paper as they are complex and are rooted in the Brazilian colonization and industrialization processes, the last of which was took place when with national efforts to buy a ready to be implemented industry was concentred, and resulted in the industrialization of only a few regions in the southeast supporting inequality. Some areas in the countryside of the northeast have suffered from severe water shortage for decades. According to Almeida, Tavares and De Jesus et al [2], this is due to a natural desertification process of the area, accompanied by excessive activity of exploitation and depletion of natural resources. The lack of infrastructure and public policies to support better living conditions and to deal with these long periods of droughts in the semi-arid regions have contributed to the continuous impoverishment of this area.

In spite of the odds, something is booming in the region, the mapping of solidarity economy enterprises published by the IBGE [3] reveals that the Northeast accounts for 40% of all the solidarity economy organizations in the country. In addition, the study shows that the majority of these organizations are located in rural areas, suggesting solidarity economy as an alternative answer to struggling communities.

This case study aims to look at the aspects of how community organization happened in response to the effects of exclusion and capital accumulation in the industrialized food system, focusing in the grassroots organization that emerged from a group of women member of the Landless Movement called *Rede Xique-Xique de Economia Solidaria* Xique-Xique Network of Solidarity Economy - XXN located in the semi-arid region of Brazil in the state of Rio Grande do Norte – RN.

2. Methodology

This paper is composed of a literature review to introduce the the effects of the globalized food industry particularly in Brazil and the solidarity economy alternative models that emerged and developed in response to crises. The study presents data from an interview conducted via Google hangout with a community organizer and popular educator currently involved with the XXN through a rural development agency called Association for Assistance of Peasants Communities - AACC. In addition, also includes records from previous research, which consisted of a participant observation and a survey conducted in 2011 [4], with family farmers members of the Xique Xique network, from the settlement of São Miguel do Gostoso also located in the state of RN.
This paper is presented in five parts: The introduction, the methodology, the Thesis Statement, the Literature review, The Data and Fieldwork and Final Considerations.

2.1. Thesis Statement

The 'Food' topic has become very popular in the media, market adds, academia, environmentalist discourses and lately even the first lady Michelle Obama's agenda, making its significance obvious. In fact, if I could name one system that holds considerable power over some of the most important aspects of our societies, such as health, environment, economy, culture and politics, it would be our complex and highly industrialized and globalized food system.
In this sense this article observed the effects of the development of the agroindustry sector in the world and more specially in Brazil and how the solidarity economy alternative models have been responding to the issues in the northeast of the country, particularly regarding sustainability in the family farming practices of a group of farm workers from the Landless Movement - MST called Rede Xique-xique.

3. Literature Review

A need for a globalized food system is usually sold as the only way to feed a rapidly growing population. Based on monoculture, by high technology and large scale production they can achieve high profitability and sell products at lower prices. However, at what social, moral, economic and environmental costs?

3.1. Effects of a Globalized Food System in the Economy, Health, Environment, Culture and Politics.

"The agro-industrial farming system has been spectacularly successful at encouraging uniformity, destroying diversity, polluting soil and water, corroding human health and impoverishing farm labour" [5].
Sandy Brown and Christy Getz [6] described the cruel contradiction within California agribusiness in which hunger affects the very people who work in the production of food of the most productive state in the United States. Inequality and scarcity have nothing to do with the amount of production, it is the distribution that causes hunger to even the food-producing workers and other low income population. Brown and Getz [6] concluded that these conditions are the result of the agrarian capital accumulation that has been constructed within our economic and political system though neoliberal trade policies such as free trade agreements and immigration politics.
Food Alienation is also a characteristic of the society in the globalized food era. Where does food come from? The supermarkets! Human disconnection with the food cycle was promoted by the market with the help of government authorities. Alworth [7] discourses about the surge of the 'supermarket centers' as part of the 'Marshall Plan' after World War II, the period of "the industrialization, fragmentation, and anomie of modernity" [7], The supermarket was designed and continually evolved to became a place that promoted entertainment and emotionally satisfying activities in order to trigger more consumption and time expenditure, reorganizing the social interactions of production, distribution and sales of goods.

Water shortages, soil exhaustion, social assistance dependence caused by under-employment, inadequate access to food triggering both undernourishment and overweight health issues. These are just some of the issues affecting even the richest of developed countries.

When it comes to developing countries the situation worsens. For instance, the production of food as fuel can be considerate morally wrong when roughly 70 per cent of the 1.4 billion extremely poor people in the world live in rural areas [8]. The commercialization of food as fuel rises food prices in places where a slightest variation could greatly aggravate the poverty problems. Besides the biofuel may not be the large scale solution to our energy issues in the first place, the FAO suggested that even if we use all the world's crops to produce biofuels, it would represent at most 13% of the global primary energy needed, it also would require the use of 85 per cent of the world's freshwater resources [9].

Brazil is one of the largest food producing countries of the world, in contradiction according the country remains between the most unequal nations in the world. Inequality is one of the biggest problems Brazil currently faces. Land distribution is no different as it has been historically concentrated in the hand of a few. Although many governments have admitted the social economic importance of an agrarian reform and even created policies to tackle the problem, none really complied with their promises, thus an expansive agrarian reform never took place in Brazil. In order to partially fulfill the constitutional and legal provisions on Agrarian Reform, the Brazilian government gradually acquires land or expropriates unproductive country estates for land redistribution, in an inefficient manner and usually only in response to popular manifestation.

The country suffers from infrastructure problems especially outside the major cities and that gives place to precarious agricultural working conditions that contributes to the rural exodus and increases social inequality in the cities. The agribusiness sector does not contribute to poverty reduction or to improve food and nutrition security in the country, since agribusiness in Brazil adopts large scale approach and export orientation [10]. Family farming in the other hand, employs almost 75% of the rural labor and is responsible for Brazilians' food safety, as it provides the basic food products consumed by most [1]. According to the 2006 Census the agribusiness held 15.6% of the total number of farm establishments that occupied 75.7% of the cultivable area in the country. Family farmers presented similar inverted numbers, they accounted for 84.4% of the total number of establishments and occupied 24.3% area of arable area of the country. The average size of family farms was 18.37 ha, and the traditional, of 309.18 ha. The results of Census showed the concentrated agrarian structure.

3.2. Alternatives Approaches to the Industrial Food System

In response to crises and in an attempt to to regain social connections, related to the production and distribution of goods, small farmworkers, indigenous people and other members of civil society all over the world have been building alternative food systems that promotes solidarity, diversity, democracy and food justice, although they are still largely ignored by policymakers and mainstream media and other agencies who deal with agricultural and environmental issues they continue to grow and gain attention as grassroots movements.

Singer [11] talks about popular movements for agrarian reform that had been on going for a long time in Brazil, it was heavily repressed and lost power during the dictatorial coup that took place between 1964 and 1985 and regain strength between the latest 1970s and the beginning of the 1980s as the military regime weakened, as a result in 1984 the Landless Workers Movement - MST was created initially with the main focus of resolving the land distribution issues on individual bases, through land settlement, however they soon realized that turning landless workers into small holders of land in a isolated family farmer unit, made production, sales and even subsistence difficult, thus not enough to improve the lives of the rural workers, and some of the settled had to give up their land. Thereafter, in 1989 started promoting organization and market of settlements thru formation of associations and cooperatives, subsequently the MST deemed settlement organization as important as land distribution. Presently the rural settlements in Brazil, especially the ones organized by the MST, represents spaces for the adoption of agro ecological principles and practices as as they have an agrarian social structure based on the family unit, associative and cooperative practices and ecological-environmental concerns.

4. Case Study Presentation and Discussion

The *XXN* is an solidarity economy association, founded on the semi-arid region of Northeast Brazil. The network started in 1999, with a group of women from a land settlement called Mulungunzinho in Mossoro, Rio Grande do Norte - RN, it began as many solidarity initiatives, brainstorming ideas on how to generate income through collective production to overcome hardship. Agriculture was part of their daily lives and traditions, thus they opted for the production and market of organic goods. According to the association online portal, the network today has urban and rural workers from settlements across the RN state, organized in more than fifty production groups articulated in 19 municipal market centers (Org. Portal: Rede Xiquexique).

4.1. Xique-xique Network in the Farmer'S Voices

In 2011, I and a fellow popular educator Gerlane Silva [4] conducted a research aimed at identifying the social impacts of the participation of farmers in the XXN from the farmers' point of view, at one of the network's center, the São Miguel do Gostoso Center, in a small rural municipality of RN which in 2008 the census assessed of having a population of 9,093 inhabitants [1].

We talked to members of small production groups from the Xique-Xique Network, one of the groups we approached worked with catering services and told us that the political support of the network ensured the quota of 30% of the municipal public school meal program be provided by members of the of network. By law at least 30% the municipal public school meals, when available, should be provided by local producers, however often times the local producers are not aware of the law or how to contract with the public authorities and the program does not fulfill its purpose.

Another member shared the experience of continuous learning through the meetings between members of the network and attributed her daughter's admission in the State University of Rio Grande do Norte - UERN to her participation in these meetings, she added that her daughter was the first in her family to ever attend an University. A group formed by family farmers stated that been part of the network had enabled them to sell their produce directly to the customers in the Xique-Xique farmers market. Separately their production would be to small and they had been accustomed to sell the surplus of production to middlemen.

In one of the conversations, a farmer talked about feminism and how the participation in the network changed her life as she could now actively participate in the community, in her own words, "We are happy because now we know ourselves… we are real women who did not know what we were before". One of the Xique-xique's network principles is feminism, and it reaches remote areas where women still ask the permission of their husband to talk in public. Apparently the discovery of respect and of women's rights brought a search for independence and recognition to some feminine members of this community.

Overall the participants declared that the network brought improvements to their living conditions, however, they still facing many issues related to basic infra-structure and access to health and education they also complained about the lack of technical support in some areas, for instance in the process of producing and saving organic seeds.

4.2. Xique-xique Network from a Development Agent'S View

The second part of this case study is based on an interview conducted on April 12th 2015, with Lidiane Freire a popular educator graduated in social economy that gives support to the XXN as a community development agent, she is commissioned by the AACC – Association for Assistance of Peasants Communities, a non profit organization that has partnership with the federal government, they obtain funds from public and private sources. The interview was conducted in Portuguese, then translated and edited to allow for better understanding.

She was asked about her experience as an agent of community development and solidarity economy within the XXN. Lidiane Freire started by recalling how the terms popular education and social economy seemed to be strange for the farmers. "They do not use these terms, we have begun to discuss them in workshops just recently". However, in the agent's view their construction of knowledge comes from sharing experiences and collective work. The management process is horizontal, not because they know the concepts of popular education and social economy, but because of their constitution as working owners. "It is just their cultural attitude" (my own translation of quote from the interview with Lidiane Freire).

Lately Freire have been working with the National Register for the Development of Solidarity Economy, as the agency AACC have recognized the need to give an identity to small groups, both formal and informal, so they can be recognized by the government and have access to new social policies focused at the solidarity economy sector. She also facilitates workshops with the communities involved in the XXN groups. These workshops involve reflections about their quotidian, the demands to be solved throughout collective development of strategies to overcame the issues. "They invite me to facilitate workshops about solidarity economy, based on work I have done with other production groups, sometimes, I feel like they expect I will bring salvation." (my own translation of quote from the interview with Lidiane Freire)

When Lidiane started working with the network some members said they had never heard about the term solidarity economy. However, as they talked about their experience and the ways they relate to each other, She showed them that they already knew, because they rely on each other, they unify to overcome community demands, and that is the basis of solidarity economy. For Lidiane "Solidarity economy does not come from rich people, it comes from people who do not feel included in the capitalist system, although they do not make this analysis, and do not appoint capitalism itself as a problem, they seek collectively to overcome the difficulties that resulted from their exclusion of this economic organization." (my own translation of quote from the interview with Lidiane Freire). In addition, in relation to capitalism there is a difference between the groups located in cities and the ones in rural areas. In the cities there is an oscillation between employment rates and the solidarity economy when there is higher unemployment there is a boom in social economy. In the urban areas, the groups are usually composed of young participants who are unemployed for a period, they seek solidarity economy as an immediate remedy for the lack of resources. However, as soon as opportunity arises for a job, even if it is not a good one, they usually go for it. She reported: "We lost almost all groups in the capital Natal, because the cost to do business is too high and people get scared." Solidarity economy is seen as a tool for generating immediate income. In Freire's opinion a down point is there is still no political understanding that Solidarity is simply a better way to organize society, a way to establish a kind of relation with the economy, which rescues principles of justice that were denied and hidden by capitalism. However she points out that this issue occurs mostly in urban groups, as in rural communities the members are naturally organized in solidarity, "I think that is because the situation of exclusion of small farmers is historical" (my own translation of quote from the interview with Lidiane Freire).

The network started with a group of women from the Mulunguzinho settlement they decided to produce greens and the network was thought as a space to commercialize the production, they sought articulation with other production groups to gather resources to make it a possibility. Later on, they saw the need and the opportunity to have more variety on the same market place, therefore they invited groups from other MST settlements with different type products and became a network. The majority of the participants are women, especially in the small production groups.

Within the network cultural differences and knowledge are generally welcomed. However, when a participant's practices go against the network's principles, it generates conflict. Usually it happens when one family decides to produce in a conventional way to decrease loss by using pesticides, or when someone decides to lower their prices generating unfair competition between members. The problems are brought into the round table discussion at the local meetings at municipal center. After the discussion, the center deliberates time limits for the violations to be corrected, and perhaps attend to the needs and support the participants in solving the problems cause. Lidiane concludes that what happens in these cases is that the participant either leaves the network or corrects the infractions, thus decisions are always made at the round table.

This year the XXN completed ten years of operation and according to Lidiane Freire, the feelings of the participants at the anniversary forum were of optimism. They have grown stronger, and the network is now a reference of solidarity economy in the northeast because of its democratic form of organization

5. Final Considerations

The creation and development of the globalized industrial agricultural system in much of the last hundred years has made people disconnected to nature and the cycle associated with the production and distribution of food. A competitive market for profit and the accumulation of capital at any cost is no longer the means to an end but humanity socio-economic goal itself. In this process many groups have been excluded and some sought to create alternative to this form of organization.

Unlike many top down policies intended to support family farmers, but unable to reach the most remote communities, in a larger country such as Brazil, the Xique-xique Network's reach, appear to have given voice to people that otherwise would be invisible, specially women. However, this reach is limited. The settlement visited was still visibly marginalized and uses rudimentary tools and no machinery in their trade, some families still have limited access to electricity. They complained about the lack of technical support related to production, specially to produced organic seeds. Small scale family farming in organic agriculture needs as much access to technology and infrastructure as the industrial system does and it is available today but inaccessible to most. I agree with Halweil, when he states that "organic farming is a sophisticated combination of old wisdom and modern ecological innovations... It's heavily dependent on technology just not the technology that comes out of a chemical plant."[12].

The farm workers from Xique-xique network, reached to solidarity economy as a last resource to withstand hardship and exclusion, driven by a long period of oppression, prejudice and disinformation. However, it seems that the people observed in this study, see their practices as a way to overcome their marginalized condition, but not so much as a logical, even honorable and sustainable way to organize societies and natural resources. This may be due to the absence of entrepreneurship and the fear of co-responsibility. Self-management sometimes does not seem as easy task. Thus without understanding that solidarity is not only a emergency alternative, some prefer to go back to the dependence of a superior entity, even if they do not have their best interest in mind. Historically, most have been educated to have employers and to serve them as they need it. Thus, without the understanding that solidarity economy is not just a temporarily fix it up, is in fact a better way to participate in society, I am afraid the alternative will never become the main stream.

Although this case study does not represent an generalization of the profile of members of the MST, is worth mentioning the fact that the production groups from XXN came from land settlements of the MST, an characteristic itself that presents difficulties and possibilities. Difficulties because without means of production land acquisition itself is just the first step of a long journey into self-determination, therefore the government and civil society have to play a large role for its successful realization. The possibilities in other hand are related to the nature of the agricultural reform pursued by the the MST and observed in this case study. One of the conditions for the redistribution of the parcels of land to families is that the land must be used for agriculture purposes. Also, these parcels are usually part of a larger acreage that is then divided between families from the same social movement [13]. Thus by default, this process of agrarian reform is creating communities of small-scale producers that in some way share the MST values, which includes solidarity economy and agro-ecology, two very important features to create a sustainable food system. According to the MST online portal, so far, three hundred and fifty thousand families have been granted land. This number signifies the importance of the success of this land reorganization to a reconfiguration of the communities in the society.

Food systems are a social whole and are the forefront of our society organization, capable of bringing both the wellbeing and the demise of the human race and the environment. "Food is not simply the outcome of the agricultural system, is rarely just body fuel; rather, it is something which drastically separates us and fundamentally brings us together" [14].

6. References

[1] Brazil Gov. Instituto Brasileiro de Geografia e Estatística – [Brazilian Institute of Geography and Statistic] 2013. *2010 Censo [Census].*

[2] De Jesus E Almeida D Tavares D 2012. *Uma percepção do processo de desertificação no semiárido nordestino. I Seminário Nacional de Geologia e Planejamento Territorial.*

[3] Brazil Gov. Instituto Brasileiro de Geografia e Estatística – [Brazilian Institute of Geography and Statistic]. *Censo Agropecuário [Agricultural Census] 2006.* Brazil: 2009.
[4] Dantas M and Silva G 2011 Rede xique-xique de agricultura familiar - análise da dinâmica no núcleo de são miguel do gostoso, RN in *Conference Poster at the VI CONNEPI, IFRN,* Brazil.
[5] ETC Group 2013 Putting the cartel before the horse… and farm, seeds, soil, peasants, etc. *Who Will Control Agricultural Inputs.*
[6] Brown S and Getz C 2011 Farmworker food insecurity and the production of hunger in California. *Cultivating Food Justice: Race, Class, and Sustainability* pp. 121-146.
[7] Alworth D J 2010 Supermarket sociology. *New Literary History 41(2),* pp. 301-327.
[8] McKay B and Nehring R 2014 *Sustainable Agriculture: An Assessment of Brazil's Family Farm Programmes in Scaling Up Agroecological Food Production.*
[9] Afiff S *et al*. 2013 Biofuels and food security. a report by the high level panel of experts on food security and nutrition of the committee on world food security.
[10] Rocha C Burlandy L and Maluf R 2012 Small farms and sustainable rural development for food security: The brazilian experience. *Development Southern Africa 29(4),* pp. 519-529.
[11] P. Singer 2002 A recente ressurreição da economia solidária no brasil. *Produzir Para Viver: Os Caminhos Da Produção Não Capitalista.Rio De Janeiro: Civilização Brasileira 2*pp. 81-129.
[12] Halweil B 2006 Can organic farming feed us all? probably-but that may not be the right question. World Watch 19(3), pp. 18
[13] Movimento dos Trabalhadores Rurais Sem Terra [Landless Movement] – MST.
[14] Crewe L 2001 The besieged body: Geographies of retailing and consumption. *Prog. Hum. Geogr. 25(4),* pp. 629-640.

IOP Conf. Series: Earth and Environmental Science **151** (2018) 012015 doi:10.1088/1755-1315/151/1/012015

An Analysis of Tourist Participation Restoration-Ecotourism through Systems Thinking

S R Kim[1] and S D Lee[2]

1 Interdisciplinary Program of EcoCreative, The Graduate School, Ewha Womans University, Seoul, 03760, Republic of Korea
2 Department of Environmental Science and Engineering, College of Engineering, Ewha Womans University, Seoul, 03760, Republic of Korea
E-mail: lsd@ewha.ac.kr

Abstract. The purpose of this paper is to suggest a new form of Ecotourism, Tourist Participation Restoration-Ecotourism, based on the system thinking perspective using causal loop analysis. Analysis target area is Dadaepo which is not only ecologically important place but also famous tourist site. Three causal loop diagrams are developed, and a mutuality among variables are found. From these, variables that management harmful effects are elicited. The results are as follows. Release some marine seedlings with tourists and reduce the damaging of environment. The obstacle of ecological tours is the disruption of ecosystem. Give an education to the tourists about the environment in advance and limit the number of seedling that is released through the quiz, thereby preventing secondary damage to the ecosystem. Managing the decline of biodiversity is a establishment of standards. If the diversity falls below the standard, prevent the release of seedlings for maintain the biodiversity. This can be care for ecosystem of tourist site and satisfy the needs of tourists. The conclusion provides some research implications and future research direction.

1. Introduction

One of the benefits of the restoration of the mudflat ecosystem is the enhancement of cultural services. Ecotourism incorporates various types of trends such as bird watching, science research, photography, and tracking. It can be viewed as a subset of natural tourism activities, and can lead to increased incentive for natural environment experience and preservation. In ecotourism, it is important to balance the environment with maintaining the diversity of the environment and achieving economic benefits using the ecosystem. In particular, the most important means of sustaining sustainability is to limit the number of tourists to prevent damaging the ecosystem[1]. But controlling demand is a significant loss in terms of economic gain through the tourism industry. Thus, the pursuit of tourist control and the pursuit of economic profit is the biggest dilemma of ecological tourism.

Also, when selecting ecological tourist sights, places that are worth protecting, especially those with beautiful views, are main targets. However, the areas designated as protected zones are difficult to experience and do activity other than observing natural landscapes[2]. However, experience in tourist attractions plays an important role in the satisfaction of tourists and the visits to the revisit[3]. Also, eco-friendly and cultural experiences increase the attractiveness of ecotourism[4]. However, it is a great disadvantage to the existing ecological tourism sites to limit such activities. In addition, where the landscape is damaged it is difficult to develop as eco-tourism destination is also an unfortunate part of the existing eco-tourist destination selected points.

Therefore, for sustainable ecological tourism, tourists must have a good influence on the environment. It is also important to find ways to reduce control over the number of tourists. Also, tourists should provide satisfactory activities without having to compromise on the ecosystem. Lastly, it should be possible to develop eco - friendly tourist sites despite, the fact that ecological values have already decreased. This can extend the width of eco-tourism site and contribute to the popularization of ecological tourism. In this paper, to suggest a new form of Eco-tourism, Tourist Participation Restoration-Ecotourism. Analysis target area is Dadaepo which is not only ecologically important place but also famous tourist site. Through the problems and tourism issues of eco-tourism, we looked at how such problems could be offset.

2. Advanced Research

2.1. Eco-tourism

Since Korea hosts a tourism program centering on visitors, it has increased the risk of developing harmful environments to tourists, such as noise and pollution from visitors. However, the ecological tourism program minimizes the risk of ecological tourism by operating a strategic ecological tourism program that provides visitors with a chance of exploring nature and cultural resources, including wildlife and aquatic biology[5]

To induce people to be attracted to eco-tourists, it is not simply a sight seeing place, but the place that can be developed as experience variety of programs[6]. The more beneficial the experience is, the better the satisfaction of tourism, the better the organization of tourists, and the greater the change in attitude of tourists. As eco-friendly and cultural experiences increase, the attractiveness of eco-waving and tourism increases, and the quality of tourism is also increased through the protection of marine resources[4]. Plus, environmental responsibility of eco-tourist site has significant indirect effect on tourists' loyalty through cognitive and affective images[7].

Furthermore, it is important to deepen the understanding of environmental conservation and use it as a venue for environmental education, considering the importance of environmental conservation. Also, it is important to receive environmental education through expert advice or guidance from eco-focused tours. As for environmental preservation, it is essential to prevent environmental pollution, so it is regarded as one of the most economical measures for economic and social members, as it is one of the most economical measures for the economy[8].

In this study, we intend to contribute to developing eco - friendly programs by suggesting ecological rehabilitation activities that can reduce the ecological effects of tourists, as well as experience the ecological effects of tourists.

2.2. Fish Seeds Release

Restoration of marine ecosystems in Korea is largely divided into 'restoration of marine ecosystems' and 'restoration of the fishery ecosystem'. The fishery seed discharge project is a project to directly discharge the seeds of the fishery seed or shellfish, and is designed for restoration of fishery profits rather than restoring the ecosystem[9]. The fishery seed discharge project is also intended to increase fishing profits, and it is not intended to be a practical idea for reviving the ecosystem because it is intended to increase fishery profits rather than reviving the ecosystem. In this study, the general public wants to offer ecological restoration tours, which allow ordinary people to participate in the ecological system and enhance the quality of the ecosystem, not to mention other goals.

3. Analytical method

3.1. Research Area

Dadaepo Beach is located on the southwestern coastal area of the Nakdong-River, and it is the only beach that consist of a mud flat. The freshwater and salty water is mixed, and the sand mudflats are formed by deposits of sediment from the Nakdong-River. And there is a wide range of sandy beaches, 900 meters wide and 100 meters wide.

They rent tools for ecological activities like they can collect marine things and release them again, but they are not management well. Moreover, prior education is not doing well for tourists so they become confused. Figure 1. shows that the development of new tourism programs is needed here. In 'A study on the Users' Awareness of Waterfront in Busan's Main 7 tourist area', unlike other places, they replied with 'Unsatisfactory' or 'Very unsatisfactory'[10].

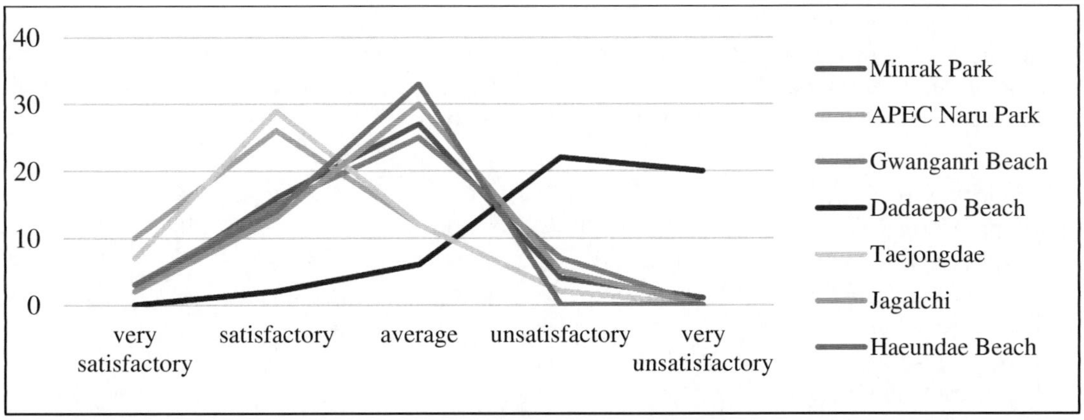

Figure 1. Overall recognition and satisfaction of the famous waterfront tourist area

3.2. Research Method

The system dynamics understands and explains the phenomenon in terms of dynamic and cyclical causation. Also, build a computer model based on this understanding. This is a method of experimenting on how a complex phenomenon of a complicated causal phenomenon varies dynamically into the computer. It is also a view of the phenomenon and the framework of accidents. Casual Loop Diagram can help determine the cyclic causation of variables that make up the system. If the relationship between variables is equal, they have 'positive causation'. In contrast, there is a 'negative causal relationship'. The feedback relationship is divided into 'reinforced loops' and 'balanced loops'. The reinforced loop is continuously progressed in a single direction unless any relationship has been severed. And the balanced loop indicates a steady state[11]. This method was a proper method due to the review of the overall changes over time and the causal relationship between variables.

First, a theoretical consideration was given to existing tourism patterns, damage, ecological tourism and ecological restoration, based on academic journals, related statutes, government agencies and newspaper articles. Second, individual causal loops were developed for the environment in which the existing environment and ecological restoration of the Dadaepo area were introduced. Finally, based on the individual causal loops, the integration causal loops are made. On that basis, proposed a sustainable ecological tourism plan at Dadaepo Beach. The study found that variables were not affected by an open circuit, but rather by interacting with causal relationships[12]. This will give you a systematic, integrated approach that will give an effective and practical approach.

4. Result

4.1. An analysis of Factors Affecting the Quality of Dadaepo Beach

Figure 2 shows the system casual loops diagram of the Dadaepo area. This is currently analysing the environment of Dadaepo Beach. 'Regulating service' and 'cultural services' represent ecosystem services. Regulating services refer to the value of the role in which the organisms remove and hold harmful pollutants such as phosphorus, industrial wastes, and heavy metals emitted by human activities. Cultural services are valued at leisure, tourism and education, and can be obtained without direct consumption or destruction of the ecosystem[13].

Since the Dadaepo beach is the place that meets the rivers and oceans, there is lots of nutrients. Therefore, many organisms inhabit here, and they provide organic matters to the next feeding class. When the creatures dig through the holes in the mudflats, the water penetrates through the holes and supplies oxygen to the sediment. It is showing positive feedback as shown in L1 and L2. Also, the purification of the water quality through the regulating services and the supply of organic matter and oxygen through Benthic organisms have 'delay effect'. There is an urgent need for preventive measures for long-term conservation, as the construction of Dadaepo Port has been planned and the pollution through the Nakdong-River is steadily draining. Accordingly, the positive reinforce loop on L1 is in danger of breaking.

In L3, as the quality of the landscape increases, the value of tourism rises and the number of tourists grows, therefore, it seems that there is only a positive effect. However, it is possible to see the decline in the ecosystem of the mud flats because of the inflow of tourists.

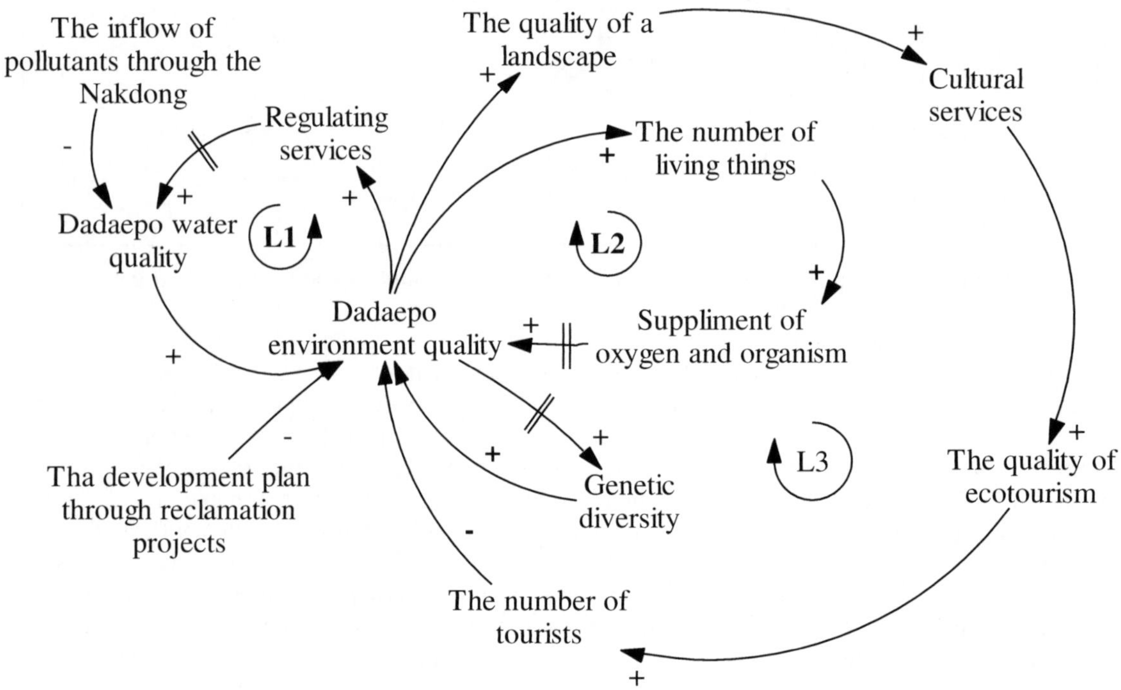

Figure 2. Casual loops diagram of Dadaepo Beach

4.2. Prediction Casual Loops Diagram during the Tourist Participation Restoration-Ecotourism

Figure 3 shows the prediction system casual loops diagram of the Tourist Participation Restoration-Ecotourism. The main activity here is the release of marine plants and seeds. To increase biomass of marine organism, it is a way to release the seeds of fish and shellfish. In the case of a practical project, there is a positive effect on the cultivation of fishing income, as well as increasing the quality of the resource management awareness and increasing the supply of high-quality fisheries products. Such activities with tourists can lead to a decrease in destruction of the environment by them. It is S1 that represents this part. However, there is a danger of ecological disturbance as it is an activity that allows organisms to be released directly into the ecosystem. S2 represents these problems. Provide prior education for the complexities of ecosystem and the importance of restoration. With this context, make a quiz and obtain a certain amount of marine seeds. This can prevent inadvertent discharge of releasing beyond environmental capacity. In addition, one can enjoy the educational effects of biodiversity, the importance of mudflow, and the animals observable at the Dadaepo beach. This curriculum is very important, because discharging the certain types of seed, not considering the environmental capacity of the oceans, is concerned about the disturbance of ecosystem and decrease in fishery resources[9].

Marine seedlings are cultured from the laboratory, and there is a risk of lowering the genetic diversity of organisms[9]. This part of the story is S3. Make a thorough preparation to prevent a reduction in

IOP Conf. Series: Earth and Environmental Science **151** (2018) 012015 doi:10.1088/1755-1315/151/1/012015

genetic diversity by establishing various criteria and avoiding inadvertent deviations from inadvertent standards.

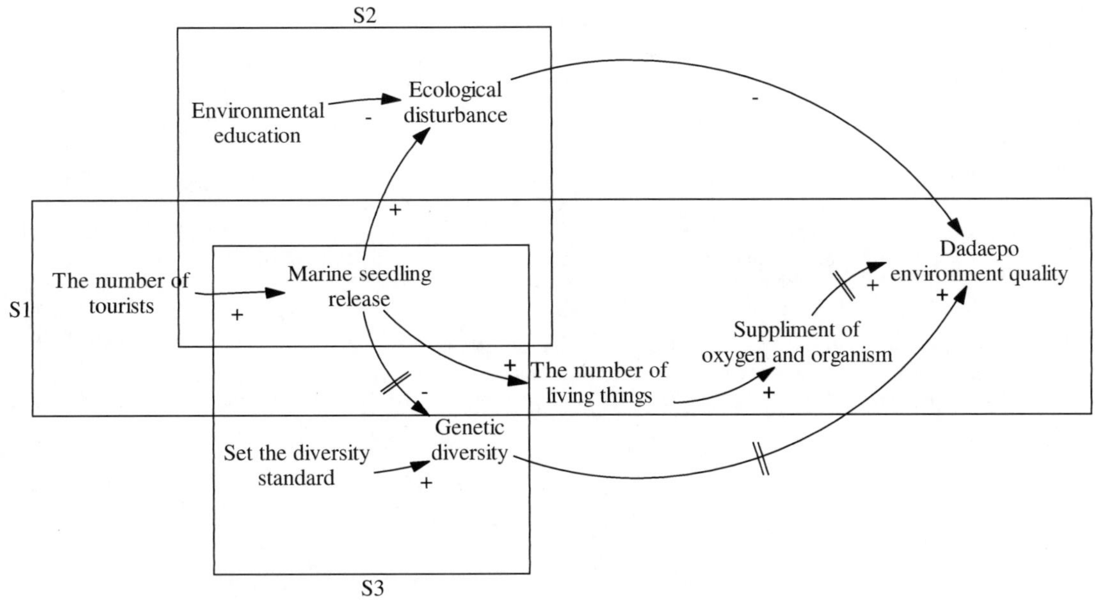

Figure 3. Casual loops diagram of Tourist Participation Restoration-Ecotourism

4.3. Integration Casual Loops Diagram

Figure 4 shows a integration diagram of individual casual loops that discussed earlier. Combine the Figure 2 that shows the existing environment and a casual loops of Figure 3 which can be kept sustainably managed, prediction casual loops diagram about the Tourist Participation Restoration-Ecotourism is made. The ecological damage problem caused by the increasing number of tourists is offset by the outflow of marine products. Also, the problems expected to occur due to restore activities will also be managed together in various ways, resulting in sustainable ecological tourism, even if there is no need to excessively limit the number of tourists.

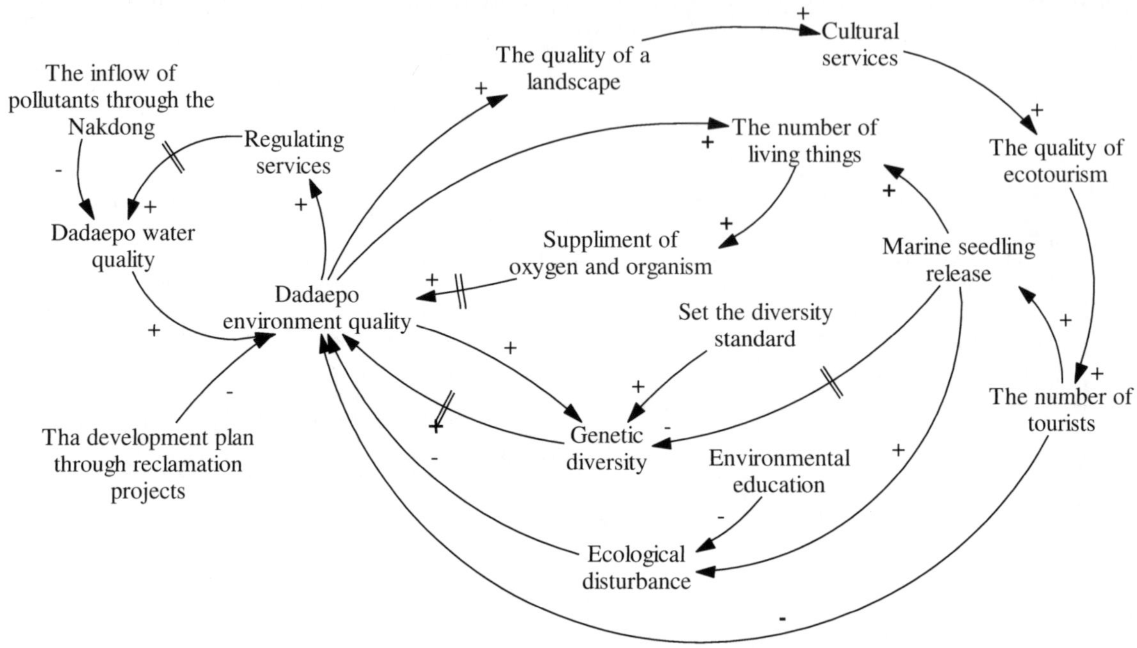

Figure 4. Integration casual loops diagram

However, the inflow of pollutants through the Nakdong-River or development through reclamation projects is a large-scale project that can not be prevented through the ecological tourism project. In the meantime, it was revealed that the restoration project of the government should be carried out at a certain level in order to develop an ecological rehabilitation tourism program. Further, continuous follow-up management should be continued to prevent secondary damages caused by the release of marine products, and the decrease in genetic diversity[14].

5. Conclusion and Suggestion

The study focused on the 'Tourist Participation Restoration-Ecotourism' at the Dadaepo beach, which is threatened by reckless exploitation in spites of the importance of mud flats, that can decrease the bad influences by tourists. Through this study, visitors will be able to provide experience that satisfy themselves, while reducing the risk of ecological disruption. Furthermore, Dadaepo Beach will turn over to the site for sustainable ecological tourism. However, since tourists are directly affected by the ecosystem, special follow-up management is required. Also, it will be more effective to work with the government unless it can be resolved with the tour program. It is a program that aims to solve the major dilemma of ecotourism, seeking a profit and protecting the ecosystem at once. This implies that both tourists and working-level employees have proposed a sustainable form of ecological tourism and confirmed their feasibility.

Dadaepo is geographically close to 'Eulsukdo Ecology Park', habitat of migratory birds, and 'Nakdong-River Estuary Eco Center', so it is easy to develop an ecological tour course. If it can be proposed by a concrete species that can be released through subsequent studies, it is expected to be reborn as an ecological tourist site representing Busan.

6. References

[1] Kim D K 2001 *Korea Tourism Research Institute* 1 1-146
[2] Stephen W, John N, Chae Y B and Kim H J 2012 *Ecotourism impacts, potentials and possibilities* HANOL (Korea)
[3] Jang H W 2012 *Korean Journal of Tourism Research* 27(1) 251-73.
[4] Gwak H G, Kim N O and Kim J 2008 *International Journal of Tourism and Hospitality Research* 22(2) 227-43.

[5] Choi Y E, You S J, Ham E K, Song K H and Chon J H 2014 *Journal of Tourism Sciences* 38(5) 59-78
[6] Cho J H and Kim S B 2007 *JOURNAL OF TOURISM&LEISURE RESEARCH* 19(1) 27-43.
[7] XU X, Lee M J and Hwang Y S 2016 *Tourism Study* 31(3) 1-17
[8] Park J W 2009 *Hanyang Law Review* 26(3) 179-200
[9] Fishery Resources Policy Division 2013 *Long-term Development Plan of Releasing Fish seedlings* Ministry of Oceans and Fisheries report pp1-27
[10] Yang D C, Lee M K, Kon M and Ahn C S 2012 *Proc. Int. Conf. on Journal of navigation and port* (Gobe, Japan)
[11] Jung J U and Kim H S 2007 *Korean System Dynamics Review* 8(1) 97-115
[12] Moon T H 2002 *Korean System Dynamics Review* 3(1) 61-77
[13] Richard B. Primack, Lee S D, KangH S, Kang H J and Kim J K 2014 *A Primer of Conservation Biology* WORLDSCIENCE (Korea)
[14] Jeong D S, Park C J and Jeon C Y 2008 *Korean Journal of Fisheries and Aquatic Sciences* 41 (6) 446-50.

Acknowledgments
This study was supported by the BK21 Plus Program (Creative Academy of Ecoscience, 31Z20130012990) funded by the Ministry of Education and National Research Foundation of Korea.

IOP Conf. Series: Earth and Environmental Science **151** (2018) 012016 doi:10.1088/1755-1315/151/1/012016

The Research on the LDC's Sustainable Development

Liu Bo[1], Jiang Xinpei[2*], Jin Ting[3], Li Bingyao[1], Ma Hui[1], Zheng Zheng[1],Li Junpeng[1]

1 School of water conservancy and hydroeletric power, Hebei University of Engineering, Handan, China
2 School of civil engineering, Hebei University of Engineering, Handan, China
3 School of economics and management, Zhejiang A&F University, Hangzhou, China
E-mail: jiangxinpei2007@163.com

Abstract. After the predatory primitive accumulation of development, the world development target is turn to the sustainable development. This paper aims to develop a sustainable development plan to create a sustainable world. A sustainable development model is constructed based on AHP, and the Index System of Sustainable development is designed. In terms of analysis about sustainability of policy, we chose Talent Policy combining with three Sustainable Indexes. When a country's comprehensive evaluation index is greater than 0.60, we define that country is sustainable. Therefore, the model in this question can state when and how a country is sustainable development. By using the model and study to determine the degree of Burmese sustainability development, we consider the Grey Forecast Model to analyse the changes population, natural resources(as water resources), economics, social and political environment. Therefore, the environment factors are negative to Myanmar's sustainable development.

1. Introduction

The rapid development of science and technology enhances the ability of human to transform nature and brings unprecedented prosperity of human society. And it also prepares the necessary material and technical conditions for further development in the future. However, from the mid-20th century, environmental issues frequently plagued mankind, the traditional development model was facing serious challenges. In this way, sustainable development strategy emerged. Sustainable development is defined by the 1987 Brundtland Report[1] as "development that meets the needs of the present without compromising the ability of future generations to meet their own needs." The core idea of sustainable development is that healthy economic development[2] should be based on the ecological sustainability, social justice and people's active participation in decision-making of their own development.

Establishing a model, this model should provide a measure method to distinguish more sustainable country and policy from less sustainable one. The sustainable development factors may include human health, food security, access to clean water, local environmental quality, energy access, livelihoods, community vulnerability and equitable sustainable development. According to the model, make a 20-yesars sustainable development plan for a LDC(Least Developed Country).

2. The Sustainable Development Model of the LDC

A sustainable development model is required to establish that can clearly define when and how a county is sustainable or unsustainable[3]. To solve this problem, we analyse the sustainability of a country by the use of sustainable development index system. The establishment of index system is equivalent to the complex objects hierarchize. So we choose Analytic Hierarchy Process as the country's sustainability development index system of comprehensive evaluation method. Combining the three sustainability development indexes, we can judge the sustainable development degree of a country. We use the same evaluation' model, just change the sustainability development indexes[4], that can explain the sustainability development strength of a policy.

2.1. Sustainable Development Model Based on AHP

2.1.1. An evaluation method for sustainable development. The sustainable development index is a complex system, which well composed of a series measure indicators. And it absolutely measures main level and the main aspects of the country's economic development, which dictates the use of multi-index comprehensive evaluation method[5] in sustainable development of the national economy. The comprehensive evaluation of sustainable development should solve the following three problems. First, construction of evaluation indexes system and selection of evaluation index; the second is the determination of index weight in evaluation indexes system; the third is method selection of comprehensive evaluation.

Sustainable development is a very complex research, describing the index system is equivalent to process hierarchically. After obtaining the index data of each player, the weights related to the goal of sustainable development will be determined[6]. Then, we can get a comprehensive evaluation result. Therefore, we choose AHP as our means to analyse the sustainable development system of the country.

2.1.2. The principle of index selection. Many factors[7] from three aspects influence the sustainable development. First aspect is some environment factors, like economy, resource, environment quality, human health, food security, livelihood, the vulnerability of the society, peace, and many other elements. The second aspect is the time factor involved contemporary and future generations. The last one is the space factor between different countries.

2.1.3. Model Establishment and Solution. Taken together, we selected the following indexes as the evaluation indexes of sustainability development in the Table 1.

Table 1. Evaluation Indexes

First-Level Index	Second-Level Index	Third-Level Index	Fourth-Level Index
Degree of sustainable development	Economy A	Economic aggregate A_1	Per capita GDP（dollar）
			GDP growth rate
		Economic benefit A_2	Funding grants
			External debt
	Society B	Population development B_1	Rate of population growth
			Related to combat deaths
		Living standard B_2	Tuberculosis treatment success
			Unemployment rate
	Environment C	Resource degradation C_1	Per capita ratio of water resources
			energy consumption
		Environmental pollution C_2	PM10
			Carbon dioxide emissions

2.1.4. The steps of AHP.

- A hierarchical hierarchical structure model is established[8].
- A comparative judgment matrix is constructed by 22 comparison methods.
- Determine the relative importance of the elements and test the consistency

2.2. Sustainable Development Index

2.2.1. Coordinated index of sustainable development. The basic characteristics of sustainable development are coordination of economic, social and environmental resources[9]. In order to study the sustainable development quantitatively, we introduced the coordination index CI to evaluate the coordinated development amount economy, society and environment. The mathematical meaning is that economic, social and environmental change rate balance each other. Meanwhile, any one of those will decrease the benefit. The formula is:

$$\mathrm{CI} = \frac{X + Y + Z}{\sqrt{X^2 + Y^2 + Z^2}} \qquad \mathrm{CI} = \frac{X + Y + Z}{\sqrt{X^2 + Y^2 + Z^2}} \tag{1}$$

In the formula, CI for coordination index (the minimum value of CI is -1.732, the maximum value of 1.732); A1, A2 for The sustainable development index of economic growth in different periods; B1, B2 for sustainable development index of social development in different periods; C1, C2 for sustainable development index of resource and environment in different periods.

When 1 <CI≤1.732, it indicates that economic growth, social development and environmental protection are moving in the direction of sustainable development and advancement in the development process, which we call strong sustainable development.

When 1 <CI≤1.732, it show that one of economic growth, social development and environmental protection deteriorates because of some reasons in the development process. But that could be offset by the development of others and also move toward the direction of sustainable development, which we call weak sustainable development.

When -1.732≤CI <0, show that one of economic growth, social development and environmental protection deteriorates because of some reasons in the development process. What's worse, that cannot be offset by the development of others and move toward the sustainable development direction, which we call it sustainable development. Among them, when -1.732≤CI≤-1, we call strong unsustainable development; when -1<CI<0, we call the weak unsustainable development. When CI = 0, we call sustainable development stagnation.

2.2.2. Sustainability index of sustainable development. One of the basic characteristics of sustainable development: the bigger economic change, social change and resource environment, if the rate of change is larger (ie, faster development rate), the sustainability of the stronger in the process of sustainable development. If development rate is slow, it is called weak sustainability of development; if the development rate is negative, it is called unsustainable. For this reason, we have introduced the sustainability indexes to measure speed of each factor on sustainable development.

According to the development characteristics, the three basic elements of development composed of economic factors, social factors and environmental factors. Let A sustainable economic index, B for social sustainability index, C for resources and environmental sustainability index. Through standardized processing[10], index numerical ranges between 0 and 1. In other words, each factor is always between in the worst state (value 0) and an optimum state change (value 1). The formula of sustainability index DI:

$$\mathrm{D} = \begin{cases} \frac{X_2 - X_1}{1.732 - X_1} & X_2 > X_1 \\ \frac{X_2 - X_1}{X_1} & X_2 < X_1 \\ 0 & X_2 = X_1 \end{cases} \qquad \mathrm{s.t.} \begin{cases} X_1 = \sqrt{A_1^2 + B_1^2 + C_1^2} \\ X_2 = \sqrt{A_2^2 + B_2^2 + C_2^2} \end{cases} \tag{2}$$

In the formula, DI for sustainability index; A1, A2 are different economic sustainable development index; B1, B2 for different social sustainable development index; C1, C2 for different environmental sustainable development index.

When 0<DI≤1, the development of the country moves towards sustainable direction, we called it the sustainable. If DI is growing, the sustainable development of the country will become stronger, which indicates that the country's development is scientific and rational. It gives policymakers revelation that move towards this direction and she state will sustainably develop. If DI gets smaller, the sustainable development of the country will become weaker, which indicates that the development of the country be unsustainable. It will remind policymakers of changing change the mode of development rapidly to enable the sustainable development of the country.

When -1≤DI<0, the development of the country have been marching to the sustainable direction, which we call it unsustainable. If DI get smaller, developing countries are increasingly unsustainable.

It will give policymakers a serious warning to reverse this adverse situation. Otherwise, the consequences could be disastrous.
When DI =0, the state development is stagnant, which we call sustainable development stagnation. It shows the state development is between sustainable and unsustainable. At this time, policymakers should analyse the situation and make sustainable development strategy to promote sustainable development.

2.2.3. The comprehensive evaluation index of sustainable development. Based on the determining of each single index' weight on the system, and dimensionless processing the index data, we can calculate the comprehensive evaluation index to evaluate the sustainable development, by using the linear weighted method. Comprehensive evaluation index Z specific calculation formula is as follows:

$$Z = \sum_{i=1}^{n} W_i \times U_i (i = 1,2,\cdots,n) \quad (3)$$

Z: comprehensive evaluation index; W_i: weight of evaluation index; U_i: the value of single index after the dimensionless processing, means the standard value of index; n: the number of evaluation index.
Using the domestic and foreign classification methods of comprehensive index as reference, we design a classification standard for the evaluation index and give qualitative description of the classification. It is the quality description for the comprehensive evaluation index[11]. Sustainable development is divided into five levels, the range of index of each levels are shown in Table 2.

Table 2. Sustainability Development Classification

Classification	Index Value	Evaluations
1	>0.9	Very strong sustainable development
2	0.80~90	Strong sustainable development
3	0.60~0.79	Basically sustainable development
4	0.30~0.59	Weak sustainable development
5	<0.30	Very weak sustainable development

So we can determine the strength of a country's capacity for sustainable development through comprehensive evaluation index Z and we also can know when and how a county is sustainable. If comprehensive evaluating index Z is greater than 0.6, we considered the state with the sustainable development, which can be invested.

3. Model Application of a Least Developed Country

For this problem, we need to make a country's 20-year sustainable development plan including projects, policies and aid[12]. First, according to the requirements of the problems, we chose a Least Developed Country (LDC) as our study object. Based on the model that we established in the Task One, we consider establishing Energy and Resource Sustainability Development Model and a Gray Forecast Model[13]. We analyse changes in economic conditions, political, social, etc. in the country. The analysis can provide guidance and evidence for the development plan[14].

3.1. The Framework Design of Index System

We chose Myanmar as research subjects. We designed a Myanmar's sustainable development index system based on general framework of sustainable development index system and its economic and social development in recent years. This design is shown in Table 3.

Table 3. Burma's Sustainable Development Index System Framework

target hierarchy A	Rule hierarchy B	Index hierarchy C	Variable hierarchy D
Degree of sustainable development	Economy B_1	Economic aggregate C_1	Per capita GDP（dollar）D_1
			GDP growth rate D_2
		Economic benefit C_2	Funding grants D_3
			External debt D_4
	Society B_2	Population development C_3	Rate of population growth D_5
			Related to combat deaths D_6

	Living standard C_4	Tuberculosis treatment successD_7
		Unemployment rateD_8
Environment B_3	Resource degradation C_5	Per capita ratio of water resources D_9
		energy consumption D_{10}
	Environmental pollution C_6	PM10 D_{11}
		Carbon dioxide emissions D_{12}

3.1.1. Evaluation of sustainable development about Burma. After the establishment of sustainable development system, we can evaluate Myanmar's economy by specific indexes. The step is shown Figure1.

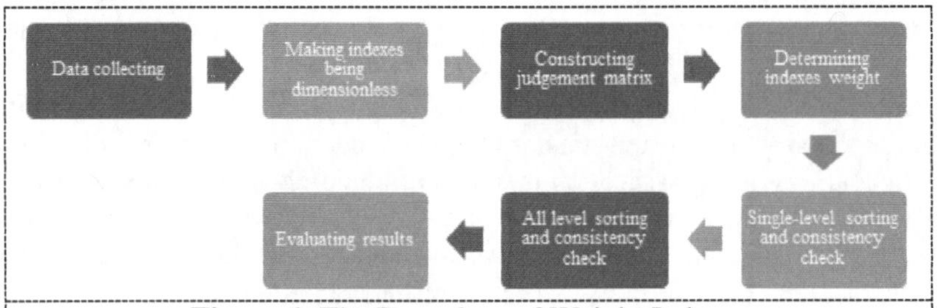

Figure 1. The Procedure of Weight Indexes

3.1.2. Data Gathering. To evaluate Myanmar's economic level of sustainable development and find the problems, we chose the raw data from 2000 to 2010 as our original data.

3.1.3. Single-level sorting and consistency check. According to scale method, construct judgment matrix at all hierarchy gradually. Then, calculate the weight of each index and check consistency. The single index weight of each hierarchy is show in Table 4.

Table 4. Weight of Each Index

Target H-ierarchy A	Rule Hiera-rchy B	Weight	Index Hierarchy C	Weight	Variable Hierarchy D	Weight
Degree of sustainable development	Economy B_1	0.540	Economic aggregate C_1	0.750	D_1	0.333
					D_2	0.667
			Economic benefit C_2	0.250	D_3	0.360
					D_4	0.640
	Society B_2	0.297	Population development C_3	0.800	D_5	0.667
					D_6	0.333
			Living standard C_4	0.200	D_7	0.800
					D_8	0.200
	Environment B_3	0.163	Resource degradation C_5	0.333	D_9	0.500
					D_{10}	0.500
			Environmental pollution C_6	0.667	D_{11}	0.167
					D_{12}	0.833

3.1.4. General level sorting and consistency check. To sort a certain level, we must use the previous level of hierarchy and single Ordering at this hierarchy[15]. The single ordering of second hierarchy is the general ranking of the second hierarchy. But in this way, general ranking must be sorted from the top to the lowest hierarchy by hierarchy. According to hierarchy general ranking, we can calculate the weight of general hierarchy ordering in sustainable development index system and check consistency.

3.2. Evaluation result for the sustainable development of Myanmar

3.2.1. Coordination Index. According to the formula 1 of Coordination Index, we can know the Coordination Index of Sustainable development about Myanmar from 2002 to 2010. The indexes are shown in Table 5.

Table 5. Coordination Index of Sustainable Development

Year	2002	2004	2006	2008	2010
Coordination Index of Sustainable Development CI	1.02	1.18	1.27	1.36	1.42

According to table 5, the coordination index of sustainable development is between 1.09 and 1.67 from 2002 to 2010, which is consistent with $1<CI\leq1.732$. It indicates that in the national development process, its economic growth, social development and environmental protection are coordinated and the structure of state is reasonable. That is coordinated development. Notably, there is a certain fluctuation on coordination level, which should be paid enough attention.

3.2.2. Sustainability Index. According sustainable index DI formulas 2 and index data each yea, we can obtain the Burmese sustainable index from 2002 to 2010, as shown in Table 6.

Table 6. Sustainability index of sustainable development

Years	2002	2004	2006	2008	2010
Sustainability index DI	0.19	0.17	0.26	0.34	0.49

From Table 6, 2002 to 2010, Myanmar sustainability index $0<DI\leq1$, the country's development is sustainable. And the greater the value of DI, the stronger sustainable development will be in Myanmar, which indicated that Myanmar's development is scientific and reasonable.

3.2.3. Comprehensive Index. According sustainable index DI formulas (2) and index data each year, we can obtain the Burmese comprehensive index of sustainable development from 2000 to 2010, as shown in Table 7.

Table 7. Burmese Comprehensive Index of Sustainable development

Index Type	2000	2002	2004	2006	2008	2010
Sustainability Index of Economic Growth A	0.38	0.41	0.53	0.67	0.73	0.81
Sustainability Index of social development B	0.43	0.46	0.62	0.73	0.74	0.83
Sustainability Index of Environment C	0.34	0.39	0.48	0.59	0.7	0.79
Comprehensive Index of Sustainable development Z	0.39	0.43	0.55	0.68	0.72	0.80

According to table 6, the Burmese comprehensive evaluation index of sustainable development is between 0.40 and 0.79 from 2000 to 2008. Table 7 indicate the level of sustainable development in Myanmar is changing over to basically sustainable development from weak sustainability development. Myanmar, in 2010, just reaches strong sustainability development country.

Referring to the analysis of the model we established in the sustainable development model of the LDC, Myanmar is a country with sustainable development and has a strong capacity for sustainable development in 2010.

3.3. Prediction of sustainable development index of Burma based on Grey Theory

Aiming to the development of Myanmar, we use GM[16] to predict the sustainability development index of Myanmar and made a future development plan.

- Step 1 Implementation

a. Column Generation

In order to weaken the randomness of the original time series, we need to accumulate the original time series. The time series accumulated is called Generated Column.

b. Grey Time Series Prediction

Use time series reflecting the characteristic of predicted object to construct Grey Prediction Model. In this way, we can predict the characteristic value at a certain time in the future.

- Step2 Calculating Steps

a. Generating the first order accumulated sequence according to the original data sequence in previous years.

b. Determining data matrix

Substitute sum into B and estimate parameters with Least Square Method.

- Step 3 Establishing the Forecasting Model of Per Capita Emissions In 2020 and solving the first-order linear differential equation.

Sequences have exponential growth law. However, the solution of first-order differential equations exactly is the solution of exponential growth in the form. Therefore, so we can assume that the sequence satisfies the following first-order linear differential equation model.

Obtain GM (1,1) time response function (k=1, 2, …)

- Step 4 Model Checking

Residual Check: Calculated by the time response function $x^{(1)}(k)$, According to the formula, Using the formula

$$x^{(0)}(k) = x^{(1)}(k) - x^{(1)}(k-1) \tag{4}$$

The calculation of the data reduction, and calculate the period of residual value $q(k)$ relative error value $e(k)$. The calculation formula of the relative residuals:

$$q(k) = \frac{x^{(0)}(k) - \hat{x}^{(0)}(k)}{x^{(0)}(k)}, k = 1,2,\cdots,n. \tag{5}$$

The relative error calculation formula:

$$\mathrm{e}(k) = \frac{q(k)}{x^{(0)}(k)} \tag{6}$$

Based on the Internet to find data, based on the grey prediction theory above, establishes a grey prediction model level, finally forecast the future trend of some indicators of sustainable development.

3.4. Prediction Results and Analysis

The prediction of foreign debt is shown in Figure 2 by using the grey prediction model. The prediction of population, prediction of Electric Power Prediction and prediction of Power Consumption are shown in Figure 3, 4, 5 respectively. And projections for other indicators is shown in Table 8.

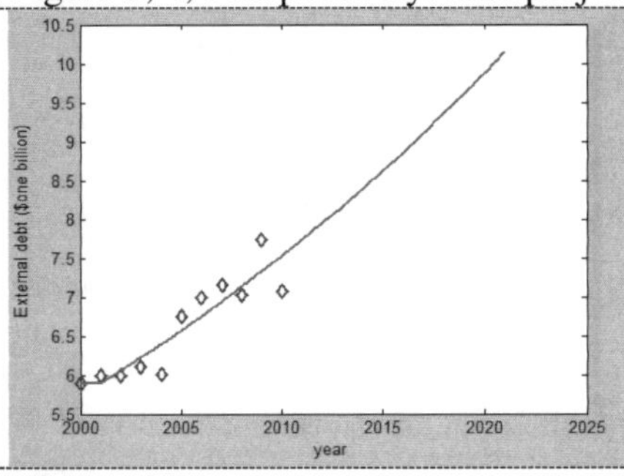

Figure 2. Prediction of External Debt

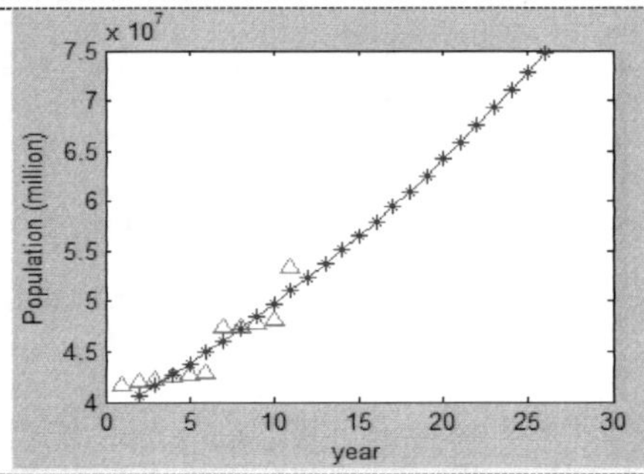

Figure 3. Population Prediction

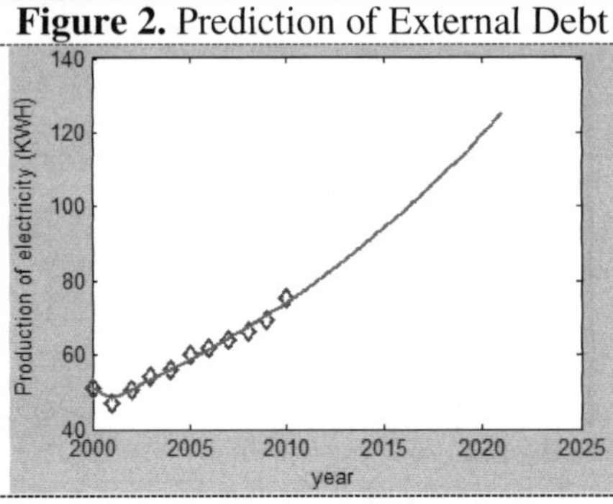

Figure 4. Prediction of Electric Power Production

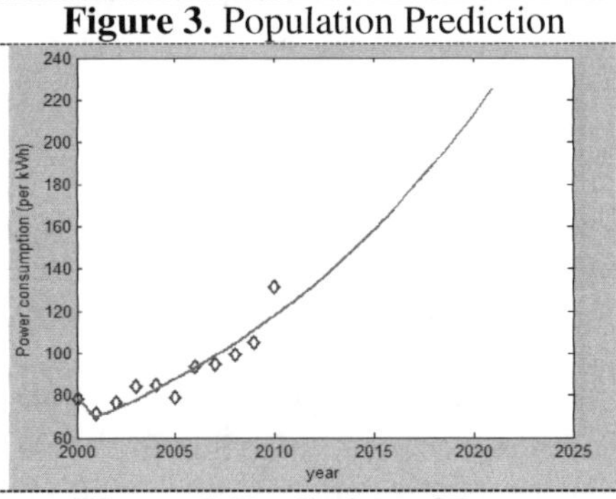

Figure 5. Prediction of Power Consumption

Table 8. Prediction of Each Index

	2011	2012	2013	2014	2015	2016	2017	2018	2019	2020
Population (Million Dollars)	52.4	53.7	55.1	56.5	58.0	59.5	61.0	62.6	64.2	65.8

Unemployment Rate (%)	7.2	7.5	7.8	8.1	8.4	8.7	9.0	9.3	9.7	10.0
Population Growth Rate	1.1	1.2	1.3	1.4	1.5	1.6	1.8	1.9	2.1	2.3
Combat Related deaths	380	463	565	688	839	1024	1248	1522	1855	2262
GDP(Billion Dollars)	60.5	59.1	57.7	56.1	5.2	53.9	52.7	51.5	50.3	49.2
Per Capita GDP(Dollars)	1367	1333	1300	1267	1235	1204	1174	1145	1116	1088
Financial Aid (Million Dollars)	546	679	845	1051	130	1626	2023	2516	3130	3893
Carbon Dioxide Emissions	13629	14125	14639	15172	15724	16297	16890	17505	18142	18803
Power consumption(kW/h)	125.1	132.7	140.8	149.4	158.4	168.1	178.3	189.1	200.6	212.8
Energy Depletion(Million Dollars)	2705	2976	3274	3602	3963	4360	4797	5277	5806	6387

According to the analysis of figure 2-5 and table 8, the following three points are obtained.

a. Population and population growth rate is rising year by year. Until 2020, the population will have reached to 650 million. Population growth may result in a reduction of per capita resources and employment difficulty. In other words, the unemployment rate increased year by year.

b. The increase of unemployment rate and reduction of per capita resources will result in declining of national economy. That indicates that the index per capita GDP be decreasing gradually.

c. The increase of population causes another increase of CO^2 emissions, which lead to the aggravation of environmental pollution. That means the resource consumption is too large and the index of energy depletion increases.

4. Conclusions and Suggestions

Based on the analytic hierarchy process, the sustainable Development Model Index System of the least developed countries is established, then the Sustainable Development Coordination Index CI and sustainable development Index DI are introduced. And the comprehensive Development index model of sustainable development is obtained. According to the value of the comprehensive evaluation Index Z, the sustainable development level of the developed countries can be judged, which is divided into 5 grades, they are Very strong sustainable development, Strong sustainable development, Basically sustainable development, Weak sustainable development and Very weak sustainable development.

Then Burma is selected from the LDC as the study object. According to the results of the model, Burma's sustainable development coordination index is between 1.02-1.42 from 2002 to 2010, that can be judged for sustainable and coordinated development country. the comprehensive index of sustainable development in Burma is between 0.39-0.79 from 2002 to 2010, which can be seen that Burma's sustainable development level is going from weak sustainable development to basically sustainable development.

According to above prediction, it can be learned that these indexes of population, population growth rate, energy depletion amount, CO^2 emissions, power consumption, war deaths and the unemployment rate will continue to increase in the coming years. Meanwhile, the indexes of financial assistance, GDP and per capita GDP are gradually decreasing.

It is supposed to advocate environmental protection, increase utilization of water resources, economize resources and establish the civilized ecological concept; the government is supposed to offer better social benefits, provide more job opportunities and improve the production and life condition. Only under this kind of condition, the sustainable development will progress better and better.

5. Reference

[1] Marková V, Lesníková P. Utilization of Corporate Sustainability Concept at Selected Enterprises in Slovakia [J]. Procedia Economics & Finance, 2015, 34:630-637.

[2] Mukhopadhyay M. 7. Situating Gender and Citizenship in Development Debates: Towards a Strategy[J]. Fortschritte der Neurologie-Psychiatrie, 2007, 63(8):297-302.
[3] Wu S. A Mathematical Sustainable Development Model[C]// International Conference on Education, Management and Computing Technology. 2015.
[4] Poortinga W, Darnton A. Segmenting for sustainability: The development of a sustainability segmentation model from a Welsh sample[J]. Journal of Environmental Psychology, 2016, 45:221-232.
[5] Deng C X, Xie B G, Yong-Xing W U, et al. The Quantitative & Comprehensive Appraisal of Urban Agricultural Sustainable Development in Shanghai[J]. Journal of Natural Resources, 2010.
[6] Leng J. Modeling the warning system of tour sustainable development with gravitational search algorithm support vector regression[J]. Artificial Intelligence Research, 2014, 3(2).
[7] Guo C Z. Empirical Analysis of the Ability of Urban Sustainable Development and Influencing Factors[J]. China Population Resources & Environment, 2010, 20(3):143-148.
[8] Peng Y. Application of AHP in regional sustainable development evaluation[J]. Journal of Anhui Agricultural Sciences, 2012.
[9] Ding L, Shao Z, Zhang H, et al. A Comprehensive Evaluation of Urban Sustainable Development in China Based on the TOPSIS-Entropy Method[J]. Sustainability, 2016, 8(8):746.
[10] Huang Lu. China's urban development and planning sustainability assessment: a approach based on sustainable science and index system [D]. Zhejiang University, 2015.
[11] Niu Xinguo and Li Yuebin. A preliminary study on the evaluation index system of urban sustainable development [C]// for the development of science and technology and social and economic development in twenty-first Century. 1999:21-23.
[12] He X, Wu Y, He X, et al. City sustainable development evaluation based on hesitant multiplicative fuzzy information. Mathematical Problems in Engineering, 2017(1) pp 1–9
[13] Conke L S, Ferreira T L. Urban metabolism: Measuring the city's contribution to sustainable development. Environmental Pollution, 2015, 202.146
[14] Wei Y, Huang C, Lam P T I, et al. Using urban-carrying capacity as a benchmark for sustainable urban development: An empirical study of Beijing. Sustainability, 2015, 7(3) pp 3244–3268
[15] Cafuta M R. Open space evaluation methodology and three dimensional evaluation model as a base for sustainable development tracking. Sustainability, 2015,7(10) pp 13690–13712
[16] Lu Yu, Wu Cao dissolved, Li Aimin. Analysis of grey prediction and the factor of border city economic development -- a case study of Chenglan railway 8 city as an example and based on [J]. technology industry, 2017, 17 (3): 10-13.

Assessment of Long-range Transport Contribution on Haze Episode in Northern Thailand, Laos and Myanmar

N Sirimongkonlertkun[1]

1 Civil Engineering Department, Rajamangala University of Technology Lanna, Chiang Rai, Thailand
E-mail: nionsiamgis@gmail.com

Abstract. The objective of this research is to study the hotspot occurrence pattern in January to April and Long-range Transport Contribution on Haze Episode in the areas sharing borders between northern Thailand, Laos and Myanmar. Hotspots or active fire detected by Moderate Resolution Imaging Spectroradiometer (MODIS) onboard of Terra and Aqua satellites, provided by NASA's Fire Information for Resource Management System (FIRMS) were used to represent open burning locations in the region. The daily backwardtrajectories in March 2014 to 2016, obtained from PM_{10} measurement station in Chiang Rai province was brought toanalysis using the Hybrid Single Particle Lagrangian Integrated Trajectory (HYSPLIT4) model. In the result, the hotspot occurrence has similar patterns. Highest number of hotspot was found in March, April, February and January, or 51, 31, 12 and 5 percent of the total number, respectively. The highest hotspot number was found in Myanmar, then Laos and Thailand, or 51, 28 and 21 percent, respectively. Backward trajectoriespatterns mainly moved across the south-western part of the study area, where hotspots were frequently occurred. Consequently, smoke haze problem in the province was influenced by burning at the upwind regions where PM_{10} was steadily accumulated until reaching the station. Obviously, the smoke problem in northern Thailand and its neighbours is boundless. Therefore, solving this regional problem by reduce burning activities in one country is insufficient. Collaboration between countries is necessary. Active sustainable land use, reduce burning activities after harvesting and maintain the sustainability of the natural resources and environment should be conductedIntroduction.

1. Introduction

Thailand recently experiences one of the worst air pollution episodes. For example, the elevated smoke haze plus particulate matter (PM_{10}) significantly blanketed Northern Thailand during March, 2007. In the meantime, Laos and Myanmar also faced the same situation. Smoke haze pollution regularly occurs in this region in March annually coinciding with the annual dry season when forest fires and agricultural burnings occurred intensively[1]. The unprecedented smoke haze that covers all areas in Northern Thailand was the problem that local people have to endure every year since 2007. The smoke haze situation directly contributes poor air quality in many areas causing adverse health impact to the residents. Usually, the main causes of this smoke haze are forest fires, open burnings and exhaust pollution from vehicles Kruasuwan*et al.*, 2014, [2]. According to Rayanakorn [2], it was found that 50 – 70 percent of small dust particles comes from open burning which included forest fires and agricultural burning or the burning off at farmlands while 10 percent comes from diesel engines, and the remainder is dust that blown over from another source. In addition, it is found that some areas have been affected by air pollution originated by forest fires occurred in neighbouring countries [3], [4] and [5]. This problem has direct impact to air quality that cause the value of particles with aerodynamic diameters less than 10 μm (PM_{10}) has value exceeds the standard value of the Pollution Control Department (PCD) which is 120 microgram per m^3. This high PM_{10} value has direct impact to the health of local people in particularly to the children younger than 5 year old and elders older than 60 years old. The value exceeds the standard has direct impact to respiratory system of the local people [5], [6]. However, the smoke-and haze problem that occurred in dry season (February - April), in particularly in March yearly has the source from the open biomass burning in high-elevated area [7] and[8] has conducted a research on the relationship of biomass burning with the elevation in Chiang Mai province using the hotspots that represents open fires, detected by MODIS as the open burning data. The result show that the area that high number of hotspots was found is the high-elevated area with the elevation of 401-600 m from the mean sea level. This result matched with the research conducted by [9] that high number of hotspots was found in Chiang Rai province is in the high-elevated area within logged forest area at the elevation of 401-600 m from the mean sea level. The logged forest area was used land preparation of the maize production. Burning in high elevated area may not impact to the urban area. However, long-range transport of air pollutants refers to the atmospheric transport of air pollutant within a moving air mass in a distance greater than 100 kilometres [10].

Among GMS countries, Thailand is the only country that monitors PM_{10} levels under the supervision of the Pollution Control Department (PCD), Ministry of Natural Resources and Environment. Monitoring stations were established at 8 Northern provinces, including Chiang Mai, Chiang Rai, Mae Hong Son, Lampang, Lamphun, Phrae, Nan and Phayao. Therefore, Thailand is seriously aware of the change of PM_{10} concentrations, especially during smoke haze episodes. Once it was found that, every PM_{10} monitoring station in the northern region of Thailand had 24-hour average PM_{10} levels that exceeded set standards for several days. As a result, Thailand's authority initiated to seek for mutual cooperation among 5 GMS countries to solve the problem of trans-boundary smoke haze. Then a working team on forestry and haze was also established for the Greater Mekong Sub-Region. The team was comprised of representatives from Cambodia, Laos, Myanmar, Vietnam, and Thailand. Importantly, these GMS countries agreed to work by aiming at the target of hotspot reduction. For instance, the goal to have cumulative hotspot counts not exceeded 50,000 (based on the data of 2006) was set to be achieved by 2015.

However, long-range transport of air pollutants refers to the atmospheric transport of air pollutant within a moving air mass in a distance greater than 100 kilometres [10]. The air pollutants emitted at a location could circulate the globe within a few days to weeks depending on meteorological conditions. In many cases, air pollutants were found thousands of kilometres away from their emission sources. While air pollutants were travelling around the globe, their impacts on human could be ranged from local to global. Therefore, through the trans-boundary transport phenomena, the air pollutants emitted within a state or country could evidently introduce adverse effects on other states or countries. Some research show that the relationship between hotspots and PM_{10} found that increase in hotspot increases PM_{10}, while long-range transport of pollutant air masses also impact local PM_{10} levels the relationship between hotspots and PM10 found that increase in hotspot increases PM_{10}, while long-range transport of pollutant air masses also impact local PM_{10} levels [1], [11] and [12]. Most of the researchers use Hybrid-Single Particle Lagrangian Integrated Trajectory (HYSPLIT) model from the Air Resources Laboratory (ARL) of the National Oceanic and Atmospheric Administration (NOAA) (the web-based version is available at http://www.arl.noaa.gov/ready/hysplit4.html [13]. HYSPLIT is one dispersion model type, which can compute the advection of air parcels, or its trajectories. When a source location emits pollutant parcel into the atmosphere, the trajectories of those can pinpoint how pollutants flow through the atmosphere. Most of the researchers use HYSPLIT as tool to study on the possibility of the movement of PM_{10}. This is also an important tool for studying the possibility of the movement of PM_{10} from distant source to the urban area as [14] used the HYSPLIT Model to determine the dispersion path way of smoke from biomass burning in Pathumthani. The resulted showed the pollutant from Pathumthani could transport to neighbouring area especially, the Bangkok Metropolitan Region. Sathavarodom [15] showed the possibility of Asian Dust reaching Thailand using ambient air monitoring data (PM_{10} and SO_2), the 10-day backward air mass trajectory from HYSPLIT4 model, 3 MODIS imagery during November 2001 and April 2002. The resulted show that the air mass from

China was not transported to Thailand or other Southeast Asia countries during those days. Manomaiphiboon*et al.*, [16] used Trajectory Modelling for examination of potential fire emissions for a March 2007 haze episode in UNR of Thailand. Pongkiatkul and Kim Oanh[17] attempt to use both trajectory analysis and monitoring data to assess the potential contribution of long-range transport to particulate air pollution (PM_{10}) in the Bangkok Metropolitan Region and [18] identify back trajectories of air masses arriving in Chiang Mai and Bangkok into distinct transport patterns by cluster analysis. Two-day backward trajectories at an altitude of 500 m in Chiang Mai and Bangkok were calculated between June 2008 to May 2009 using HYSPLIT Model developed by the Air Resources Laboratory of National Oceanic and Atmospheric Administration (NOAA).By the way, the air pollutants emitted at a location can normally circulate the globe within a few days to weeks depending on meteorological conditions. In many cases, air pollutants were found thousands of kilometres away from their emission sources. While air pollutants were travelling around the globe, their impacts on human could be ranged from local to global. Therefore, through the trans-boundary transport phenomena, the air pollutants emitted within a state or country can evidently introduce adverse effects in other states or countries.

Based on the review of the existing research listed above, the idea of this research is to study the pattern and trend of open burning conducted in Greater Mekong Sub-Region countries, particularly in the study period of 2014, 2015 and 2016, by using Thailand, Laos, and Myanmar as a case study. Hotspot data detected by MODIS sensors on board of Terra and Aqua satellites and provided by NASA's FIRMS website was applied in this research to represent burning locations in the region. The Spatial data analysis by GIS was used as the main tool for analysing the location of burning at study sites. And use the HYSPLIT model for study the possibility of PM_{10} movement to measurement station in Chiang Rai or the possible of the long-range transport contribution on haze episode in northern Thailand. The result from this study will be obvious information for planning and management of the smoke and haze problem due to open biomass burning in high-elevated area with concerned parties.

2. Methods

Hotspots or active fire data counted in this research were gathered from FIRMS website (https://earthdata.nasa.gov/earth-observation-data/near-real-time/firms). Each hotspot or detected active fire represented a 1-kilometer pixel, which might have contained one or more active burning fires in a certain period of time, either short or long-lived fires. As MODIS satellite images for Thailand are normally available only 1 or 2 times per satellite (either Terra or Aqua) and per day or night time, then the short-lived fires happened at other time during a day would not be detected. Moreover, the fire detection could be affected by cloud cover as well. Accordingly, the hotspot counting based on the available satellite images could be underestimated. Nevertheless, the hotspot counting was still considered as an effective way for monitoring the burning sites over the large study area.

Active fire or hotspot data obtained from the FIRMS website is in the format of spatial data which included several physical parameters designed by NASA, in particularly, latitude, longitude, detection date and time, brightness temperature, fire power and fire detection confidence which range from 0 to 100 percent. Thus, in order to apply the downloaded active fire data, it was necessary to create a spatial database for the data which are monthly and yearly grouped for each country. The process of the database development was conducted using Geographic Information System (GIS) tools. For analyzing long-range transport contribution on haze episode, the Hybrid Single-Particle Lagrangian Integrated Trajectory (HYSPLIT4) model, available at http://www.arl.noaa.gov/ready/hysplit4.html, was used. This online tool calculates the backward trajectories from the PM_{10} monitoring station. The output could be used to roughly represent the travel pathways of smoke from the origin of open burning. The model was run by using the "Final Run" meteorological data archives (FNL) of the Air Resource Laboratory, National Oceanic and Atmospheric Administration, USA. In principle, the HYSPLIT model is used for long-range transport study and the starting level should be in the free atmosphere. For this research, the starting time was selected at 12:00 am Thailand Local Standard Time, taking into account that the people normally burn during 10:00-14:00, and total run time was about 24 hours.

IOP Conf. Series: Earth and Environmental Science **151** (2018) 012017 doi:10.1088/1755-1315/151/1/012017

3. Results and Discussion

3.1.Pattern and Trend of Hotspot Occurrence in Three Countries

The general scenario of hotspot occurrence in three countries during the study period is described as following. Total of 747,477 hotspots were found. These countries has similar occurrence pattern in term of hotspot number, in which the highest number of hotspot was found in March, April, February and January, or 51, 31, 12 and 5 percent of total number of hotspot, respectively.

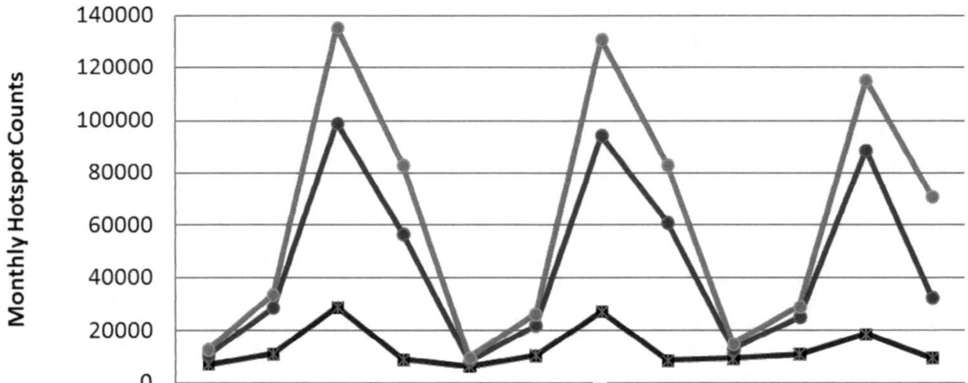

	Jan-14	Feb-14	Mar-14	Apr-14	Jan-15	Feb-15	Mar-15	Apr-15	Jan-16	Feb-16	Mar-16	Apr-16
Laos	2336	5132	36639	26412	1199	4318	36762	21866	2200	4154	26744	38386
Myanmar	3658	17462	70036	47695	2193	11465	66917	52517	3602	14200	70181	23221
Thailand	7146	11221	29093	9093	6314	10578	27346	8666	9425	11032	18631	9637

Figure 1. Hotspot occurrence number in Thailand, Laos and Myanmar in the period of January – April of 2014-2016

Fig. 1 shows the hotspot occurrence numbers in country level. The highest number of hotspot was found in Myanmar, following by Laos and Thailand, or 51, 28 and 21 percent, total number of hotspot, respectively as shown in Fig. 2.

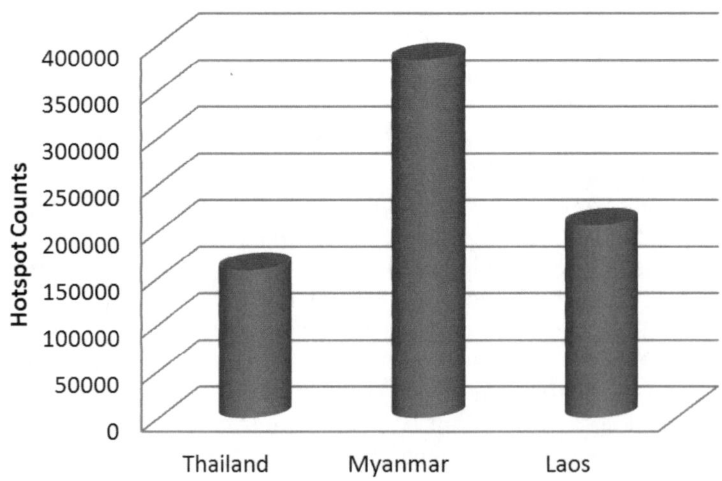

Figure 2. Hotspot occurrence pattern in Thailand, Laos and Myanmar in the period of January – April of 2014-2016

The important factor for the similar pattern of hotspot occurrence in this regional level is the agricultural activities of the local people in the Greater Mekong Sub-Region countries [18]. Therefore, it can be concluded that the obvious decreasing number of hotspot in three countries from January of each year, people usually begin to collect harvests. After the harvest season, most agriculturists will conduct burning to eliminate crop residues, in order to prepare lands for the next cultivation session. As a result, hotspots increase to highest number to the most during this period each year. In Northern Thailand, the majority of hotspot generally occurs at altitudes of 400 – 600 meters from mean sea level [18]. Those are mostly highlands and forest areas where local people often conduct agriculture, especially corn. Normally, agricultural burning is begun in February, so the increasing rate of hotspot from January to February will be higher comparing with other months during the burning season. Burning activities are conducted regularly in March, when the highest number of hotspots is found, with the lack of making fire breakers resulting in fires.

3.2.Long-rangetransport Pathway of Maize Straw Burning Smoke

The daily backward trajectories produced by HYSPLIT4 model for March of 2014-2016 are presented in Figs. 3 - 5. It can be seen that, most of backward trajectories patterns were moved in southwest direction across the southwestern part of Chiang Rai Province such as Mae Suay and Wiang Pa Pao districts before moving to PM_{10} measurement station located in urban area. The study also shows that most of the air mass movement moved across the area having intensive fire region. It is possible that PM_{10} that exceed standard is due to the movement of the air mass over the area and consequently moved to the urban area. This study is consistent with the study made by Yasanga*et al.*[18] that related to the study of the smoke and haze problem in Chiang Mai province, which show that PM_{10} exceed standard value was were influenced by open burning from the long-range upwind regions.

Another study by [19] also showsthat when PM_{10} peaked, the air mass arriving at the Songkhla and Phuket passed over the intensive fire region in Sumatra Island. Therefore, we can see that the smoke and haze problem or the problem of PM_{10} exceed standard value in Chiang Rai city has high possibility due to air mass movement rough PM_{10} fromburning area in elevated area to Chiang Rai city [20]. Therefore, the solution to the smoke-and-haze problem in Chiang Rai province need to be conducted together with the utilization of the land in high elevated area to reduce the problem sustainably.

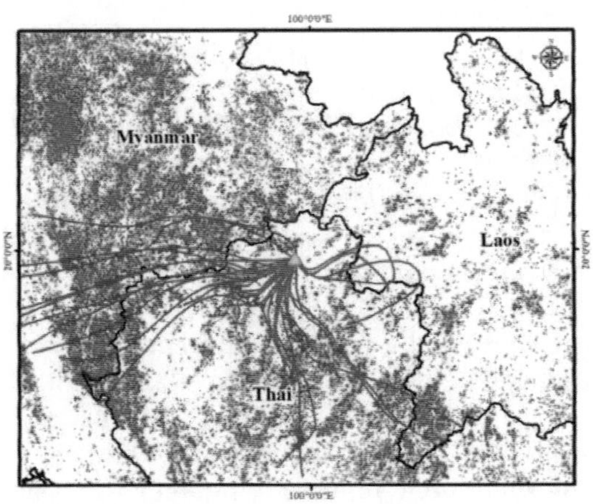

Figure 3. Backward trajectoriespatterns in March 2014

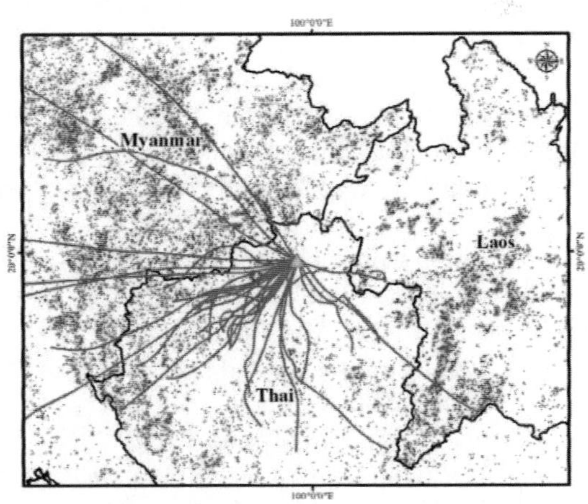

Figure 4. Backward trajectoriespatterns in March 2015

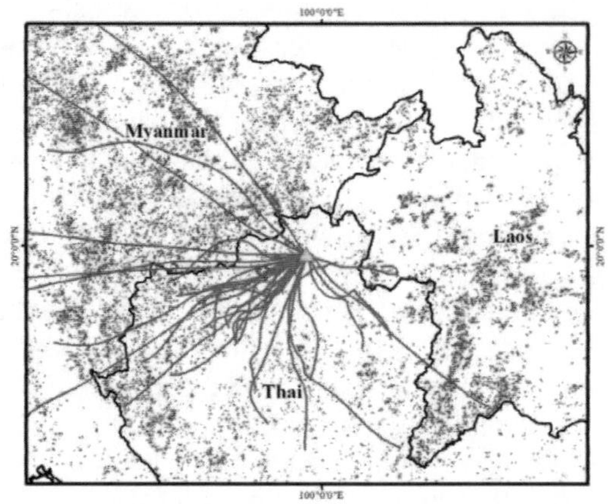

Figure 5. Backward trajectoriespatterns in March 2016

4. Conclusions

The pattern of open burning occurrence in Thailand, Laos and Myanmar is identical with the significant increasing rate of hotspot from January to February. Hotspot in these countries is also found peak in March annually. Therefore, the government should develop the policy aiming to lessen open burning, especially in February to March, which is considered as a main factor that causes smoke haze problem in the region.

5. References

[1] Bach N LandSirimongkonlerkunN2011 Satellite Data for Detecting Trans- Boundary Crop and Forest Fire Dynamics in Northern Thailand.*Int.J. Geoinformatics*, **7** 47-54

[2] Rayanakorn, M 2014 Hazeand Air Pollution in Chiang Mai http://www2.mahidol.ac.th/ks/vicharn/Haze-Airpollution.pdfLast Access: July, 2014

[3] Kim N TOanh and Leelasakultum K 2011 Analysis of meteorology and emission in haze episode prevalence over mountain-bounded region for early warning *Int.JThe Science of the environment***409** 2261-71

[4] Manomaiphiboon K 2007 Examination of Potential Fire Emissions for a March-2007 Haze Episodein Upper Northern Region of Thailand Using Trajectory Modelling, *Asian-Pacific RegConf. onPractical Environmental Technologies*KhonKaen Thailand

[5] KruasuwanJ, Chot-amornsak C,Ratjiranukoon P and Wiranvejayan O 2014 Climate Change Simulations for Thailand by the MM5-Regional Climate Model. http://elibrary.trf.or.th/project_content.asp?PJID=RDG5030045, Last Access: August, 2014

[6] Epidemiology& Intelligence Section2017 Report of Medical and public health http://odpc1.ddc.moph.go.th/epidpc10/list_report.php?item_id=6#**,** Last Access: January, 2017

[7] Pollution Control Department (PCD) 2005 *National master plan on open burning control*, Pollution Control Department, Bangkok

[8] Donthri S *et al* 2011 Prioritizing of Burned Areas Using Multi-source Spatial Data for Open Field Burning Surveillance and Prevention in Chiang Mai Province, Smoke-Haze Conf. Chiang Mai Thailand

[9] Sirimongkonlerkun N and Ponglirat S 2013 *Open space risk map from multiple sources in Chiang Rai province*, Research Report, Chiang Mai Public Policy Promotion Plan: Chiang Mai University.

[10] Glossary of Environment Statistics 2016 Studies in Methods, http://stats.oecd.org/glossary/detail.asp?ID=1558, Last Access: April 22, 2016.

[11] Sanchez-Ccoyllo O A, Silva DL, Andradse da Fatima MS and Freitas R 2006 Determination of O_3-CO- and PM_{10}-transport in the metropolitan area of Sao Paulo, Brazil through synoptic-scale analysis of back trajectories, *Meteoand Atm Physics***92**83-93.

[12] Riccio A, Giunta G and Chianese E 2007 Analysis of meteorology and emission in haze episode

prevalence over mountain-bounded region for early warning*Int.JThe Science of the environment***376** 198-214

[13] Hybrid Single Particle Lagrangian Integrated Trajectory (HYSPLIT) https://ready.arl.noaa.gov/HYSPLIT.phpLast Access September 13, 2016

[14] Tipayarom D2004 Source characterization for air pollution emission from open rice-straw burning in Thailand*Master research studyAsian Institute of Technology* Bangkok

[15] Sathavarodom T 2005 Study possibilityof long-range transport of ASIAN dust to Thailand usingmodelling and satellite data*Master research study Asian Institute of Technology* Bangkok

[16] Manomaiphiboon K, Pengchai P, Surapipith V, Hepanondh S, OnchangR, Wiwatwattana N, Tanpipat V and Garivait S 2009 Smoke-Haze Forecast Modeling for Upper North Thailand during March-April 2008, 3rd *Int. Conf. on Sustainable Energy and Environment* Bangkok Thailand 2009

[17] Pongkiatkul P and Kim N T Oanh2007 Assessment of potential long-range transport of particulate air pollution using trajectory modelling and monitoring data *AtmosphericResearch***85** 3-17

[18] Yasanga N, Traisathit P and Prasitwattanaseree S 2010 Identification of backward trajectory air transport patterns to Chiang Mai and Bangkok using cluster analysis, *11th Stat. and Applied Stat. Conf.* Chiang Mai Thailand

[19] Pentamwa P and Kim N T Oanh2008 Air quality in Southern Thailand during haze episode in relation to air mass trajectory *Songklanakarin J. Sci. Technol.* **30** (4) 539-546

[20] Health Information and Statistics, Haze situation in northern Thailand, http://www.hiso.or.th/hiso/health_event/ghealth_event5.php, Last Access: February 13, 2016

Acknowledgments
The authors would like to convey thanks to the National Aeronautics and Space Administration (NASA) for the readily available fire data from the Fire Information for Resources Management System (FIRMS). Also, the authors wish to thank the Air Resources Laboratory (ARL) of National Oceanic and Atmospheric Administration (NOAA) for the provision of the HYSPLIT transport, dispersion model, and READY website (http://www.arl.noaa.gor/ready.php) used in this publication.

Developing a New Activity Pollution Index for Emissions Quantitative Assessment in Projects Construction Phase: Case Study of an Administrative Building, Egypt

Islam Elmasoudi[1], Mona G. Ibrahim[2, 3], Wael Elham Mahmod[4, 5] and Emad Elbeltagi[6]

1 PhD student, Environmental Engineering Dept., School of Energy Resources, Environmental and Chemical & Petrochemical Engineering, Egypt-Japan University of Science and Technology, E-JUST, Alexandria,
2 Environmental Engineering Dept., School of Energy Resources, Environmental and Chemical & Petrochemical Engineering, Egypt-Japan University of Science and Technology, E-JUST, Alexandria
3 Environmental Health Department, High Institute of Public Health, Alexandria University
4 Environmental Engineering Department, School of Energy Resources, Environmental and Chemical & Petrochemical Engineering, Egypt-Japan University of Science and Technology, E-JUST, Alexandria, Egypt
5 Civil Engineering Department, Faculty of Engineering, Assiut University, 71515 Assiut, Egypt
6 Structural Engineering Dept., Mansoura University, Mansoura 35516, Egypt,
E-mail: islam.elmasoudi@ejust.edu.eg; mona.gamal@ejust.edu.eg; wael.elham@ejust.edu.eg; wdpp2006@aun.edu.eg;eelbelta@mans.edu.eg

Abstract. Quantitative assessment of emissions related to construction projects should be performed during the planning phase of the projects. This is significant to spot the high values of pollution during the construction phase. In this study, a model is developed to estimate pollution resulting from Buildings construction activities. The model calculates the generated pollution for each activity involved in the project as a result of dust, gases and noise emissions. A new index is developed namely Activity Pollution Index (API) which expresses the amount of total pollution for each activity during the project construction phase. Also, the developed model is able to display the resulted total pollution distribution throughout the project life that corresponding to the planned scheduling. An actual case study in an administrative building construction in Egypt is selected to demonstrate the practical use of the proposed model. The results show that the peak and minimum values of total pollution were occurred during the excavation activity and the formwork erections and steel fixing of the second segment of the building with values of API equal to 69 and 2, respectively.

1. Introduction

Air pollution has been and remains to be seen as a serious problem that affects human health and welfare. Construction industry carries a major threat to the environment as it has various impacts on the environment. Construction of buildings in Egypt is considered as one of the large sectors of the construction industry. Integrated environmental protection at the project planning phase should be performed to avoid or minimize pollution. To minimize the generated pollutants, construction

IOP Conf. Series: Earth and Environmental Science **151** (2018) 012018 doi:10.1088/1755-1315/151/1/012018

engineers and planners need to evaluate and optimize their resource utilization decisions and selections. For each activity in the project, these decisions include: a) selecting construction methods and raw materials, and b) identifying equipment's and labors crews' configurations and size. For example, using more productive equipment or using high engine power machines may save time, but it often increases air pollution. Construction equipment contributes a significant portion of air pollutions in urban areas during construction works such as site preparation, foundation works, road construction and maintenance [1].the main types of emissions in building construction are direct and indirect emissions. Direct emissions are defined as those emissions that produced from the major process considered while indirect emissions are emissions at upstream processes of the major process [2] .

A lot of research studies have performed that are concerned with environmental issues. For instance, Chen et al. [3] proposed an approach for quantitative assessment during the planning phase; this approach is based on the calculation of a "Pollution Index" for each of the construction activities. The index is simply a multiplication of the impact magnitude by the duration of the activity. In another study, Soderman [4] concluded that the quantitative risk analysis provides an assessment of environmental safety, the link between quantitative risk analysis and environmental management is fundamental for mitigating environmental impact. Chen et al. [5] used an analytic network process (ANP) to construct two decision models for environmental construction planning .Guggemos et al. [6] used the usage hours of the construction machines to estimate their emissions.

Hummer et al. [7] presented an optimization model for selecting construction equipment fleets that minimize total equipment cost and total emissions, or that minimize total equipment cost while limiting emissions. In another study, Arocho et al. [8] made a study on four street and utility construction projects and calculate emissions from the utilized non-road equipment. This study revealed that demolition and earthwork activities have a major effect on the total emissions during street construction. Baratia and Shen [9] developed a model to reduce the quantity of emissions produced per travelled distance for on-road construction equipment. Thy considered in their study three main operational factors that effect on emission rate. These factors are speed, road slope and payload.

This paper focuses on estimating emissions generated from buildings construction projects. The proposed environmental model aims at estimating the amount of pollution produced from construction activities. This model mitigates construction pollution by calculating the pollutants released from each project activity due to dust, harmful gases and noise, further it calculates the overall activity pollution index.

2. Methodology

The Activity Pollution Index model is developed to calculate the amount of air pollution generated from each activity in Building construction project. Further, it calculates the overall pollution of the project. Figure 1 illustrates the procedures followed to calculate the Activity Pollution Index (API). The proposed environmental model is divided into three sub-modules; dust, gases and noise.

Emission factors published by United States Environmental Protection Agency (US EPA) is used to estimate emissions from construction equipment usage. The major reason for applying US EPA emission factors for emissions estimating is that these factors are established according to the machine power and its deterioration and hence, give real emission pattern. This provides a unique emission factor for every machine which reflects the practical emission pattern better.

2.1. Estimating Dust Emissions

Dust arises from mechanical disturbance of granular material exposed to the air. It is termed fugitive because it isn't discharged to the atmosphere in confined flow stream (EPA). The main sources of fugitive dust in Buildings construction are; traffic on unpaved roads, materials handling, bulldozing and grading. Dust sub-module uses the empirical equations that presented by EPA ([10]–[13]). Figure 2 shows the flowchart for dust sources and factors affecting each one.

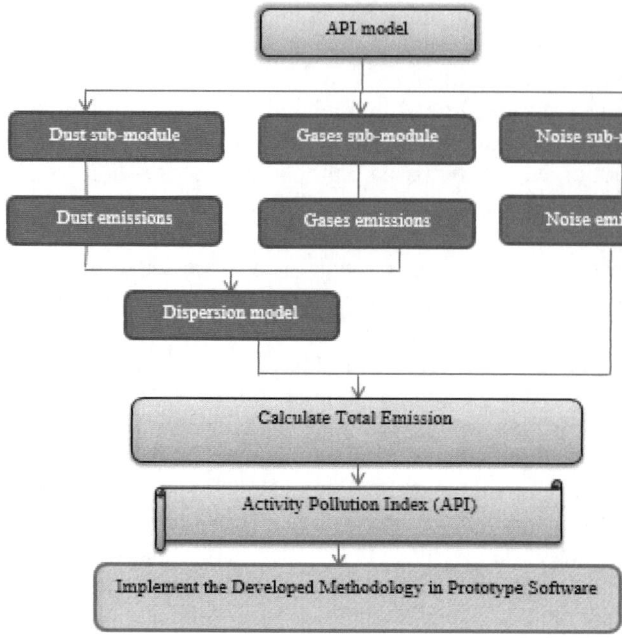

Figure 1. Flow chart of activity pollution index (API) model

2.2. *Estimating Gas Emissions*

Construction vehicles and plant machinery are the main sources of harmful gases emissions. This module covered most major pollutants released from combustion engines of equipment's during the buildings construction. These pollutants are Carbon monoxide (CO), Hydrocarbon (HC), nitrogen Oxides (NOX), sulfur oxides (SOX) and Particulate matter less than 10μm in aerodynamic diameter (PM10), Carbon dioxide (CO2), Methane (CH4) and nitrous oxide (N2O).. Methane (CH4) and nitrous oxide (N2O) are neglected in this study due to unavailability of data which is required to calculate them.

Gas emission from each equipment (Ei) can be evaluated using the general equation given below:

$$Ei = EFi * P * LF \quad (1)$$

Where: EFi is the emission factor for the emission i in g/(hp-hr); P is the power of the considered equipment in HP; LF is the operation load factor of the considered equipment.

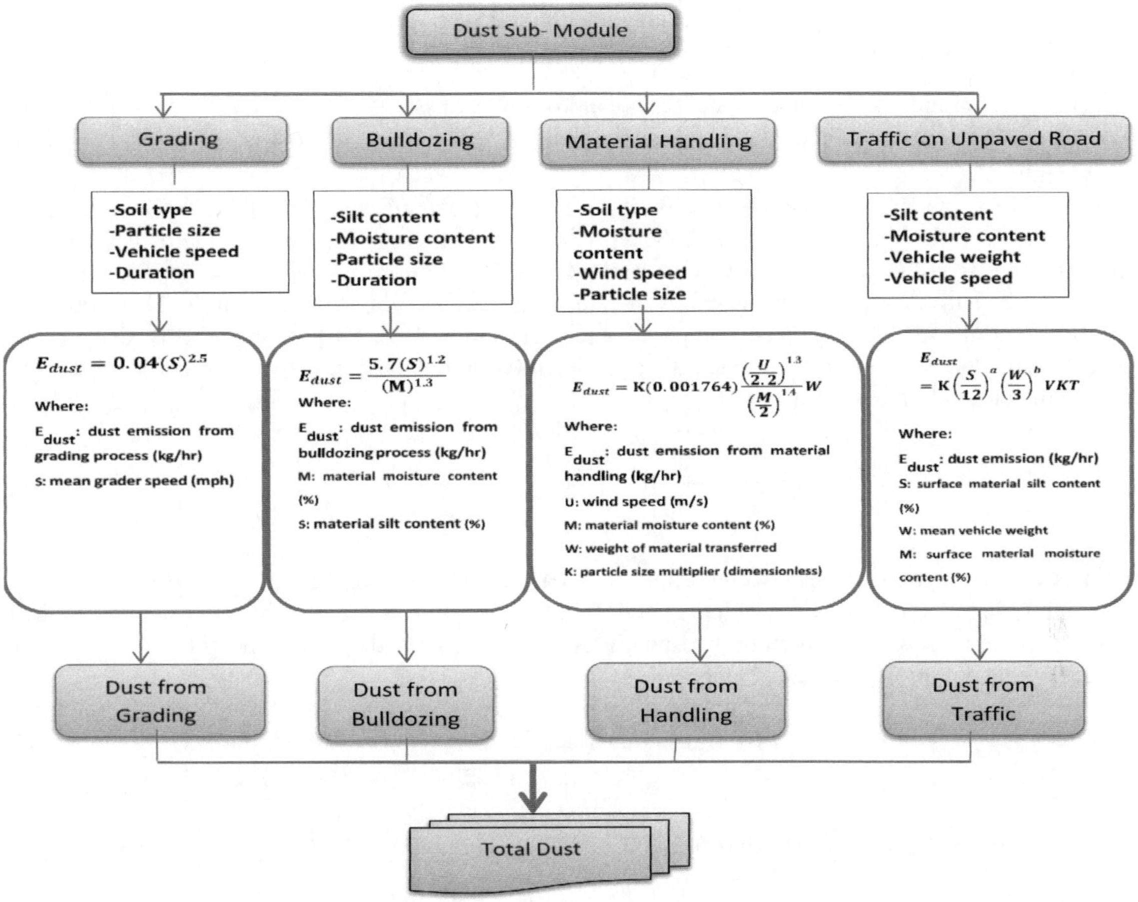

Figure 2. Flow chart of dust sub- module

2.3. *Estimating Noise Emissions*

The main source of noise in buildings construction projects is construction equipment. In addition, some activities cause noise during their execution. By determining the manufacturer, model type and power of equipment, noise component estimates the noise level. Noise is measured with a sound level meter and usually in terms of Decibels (dB). When two or more pieces of construction equipment are producing sound simultaneously, the total sound level isn't the summation of sound produced from each equipment. Because Decibels are geometric values, they can't be added or multiplied directly, but that can be done by calculating the equivalent sound level (Leq) from equation (2).

$$L_{eq} = 10Log\frac{\sum_{i=1}^{n} 10^{\frac{Li}{10}}}{N} \quad (2)$$

Where: Leq is the equivalent sound level; Li is the sound level for source i and N is the total number of sources.

2.4. *Calculating Activity Pollution Index (API)*

Total activity pollution index is calculated by normalizing the amount of dust, gases and noise emissions for each considering activity. The normalization process is done by determining the maximum threshold limit of pollutant. Such maximum threshold limits can be determined from environmental regulations and Acts in Egypt such as Law No.9 (Promulgating the Environmental Law 2009) [14]. Activity Pollution Index (API) can be estimated according to the following equation:

$$API_i = \frac{E_{i\,dust}}{E^m_{dust}} + \sum \frac{E_{i\,gases}}{E^m_{gases}} + \frac{E_{i\,noise}}{E^m_{noise}} \quad (3)$$

Where:
$E_{i\,dust}$, $E_{i\,gases}$ and $E_{i\,noise}$ are dust, gases and noise emissions from activity (i) respectively and E^m_{dust}, E^m_{gases} and E^m_{noise} are the maximum threshold limits for dust, gases and noise emissions respectively.
Table 1 shows the maximum threshold limit of pollutant according to environmental regulations applied in Egypt, especially Law No. 9 (2009). It is noticeable that the Egyptian standards describe the concentration of pollutants units in weight per volume. In addition, it is obvious that the emission values resulted from pollution module are in rate units (i.e. weight per time units). Therefore, the pollution module values cannot be compared with the standards. This problem is solved by using dispersion model which is based on Gaussian diffusion (equation 4). The dispersion model capable to convert the pollution rate to concentration.

$$C_X = \frac{Q}{\pi \times U \times \sigma_Y \times \sigma_Z}\left[e^{-\left(\frac{H^2}{2\sigma_Z^2}\right)}\right] \quad (4)$$

Where: C_X is the ground level concentration at distance x downwind (g/m3); Q is the quantity of pollutant (g/sec); U is the mean wind speed (m/sec); σ_Y $and \sigma_Z$ are the horizontal and vertical standard deviations of wind direction respectively (m) and H is the source height (m) (the value utilized by the pollution module is 1m as default value).

Table 1. The environmental law applied in Egypt law No.9 (2009)

Pollutant	Harmful Gases						Dust	Noise
	CO	NO2	PM10	SO2	HC	CO2		
Maximum Limit (μg /m3)	30000	400	400	350	600	35000	10000	90 (dB)

2.5. *Model Implementation*
The developed API model is coded and implemented using Visual Basic for application (VBA) on Microsoft Excel software. Friendly user interfaces are designed to facilitate entering data, sorting data, calculating the gases, dust and noise emissions for each activity and hence, calculating API and generating output reports. Therefore, the user enters each activity involved in the project.
The equipment required for each activity can be added. The gas, dust, and noise emissions are defined for each equipment which required for an activity. Finally, after entering the parameter required for each emission, the results will appear in output interface.

2.5.1. Gas sub *module.* All the machines utilized by the construction activities should be defined. Gas sub-module estimates the emissions of six main pollutants, carbon monoxide (CO) sulfur oxides (SOx), nitrogen oxides (NOx), particulate matter (PM), Hydrocarbons (HC) and carbon dioxide (CO2). The total gas emissions for all equipment utilized by each construction activity can be calculated.

2.5.2. *Dust sub- module.* Dust Sub-Module calculates the dust emissions generated from four sources, material handling, traffic on unpaved roads, bulldozing and grading.

2.5.3. *Noise sub-module.* The user selects the equipment and enter the manufacturer, type, power and the noise level of each different equipment utilized by each activity. Once the user enters these data, the noise sub-module calculates the total noise for all equipment utilized by the construction activity.

3. Case Study

3.1. *Case Description*
This case example considers the construction of the new administrative building of Egypt Japan University for Science and Technology (EJUST) in new Burg Alarab city-Alexandria Governorate in

Egypt.it is three stories building in four segments with total area of 2500 m2. Table 2 lists the main activities and the associated utilized equipment of the case study.

Table 2. The main activities and utilized equipment of the case study

Activity	Utilized Equipment	Number
Excavation	Loader (Caterpillar 966H (6m3))	2
	Truck (Mercides Actros 4040 (18 m3))	4
Backfilling	Loader (Caterpillar 966H (6m3))	1
	Truck (Mercides Actros 4040 (18 m3))	2
Soil Replacement- Compaction	Loader (Caterpillar 966H (6 m3))	1
	Truck (Mercides Actros 4040 (18 m3))	4
	Compactor (Caterpillar CS533E)	1
Formworks	Truck (Mercides Actros 4040)	1
Reinforcement	Truck (Mercides Actros 4040)	1
PC pouring for foundation	Truck mixer (Volvo FM400 (8 m3))w1	7
	Pump (Schwing S43SX)	2
RC pouring for foundation	Truck mixer (Volvo FM400 (8 m3)	7
	Pump (Schwing S43SX)	2
	Vibrator (Honda)	4
RC pouring for columns	Truck mixer (Volvo FM400 (8 m3)	3
	Pump (Schwing S43SX)	1
	Vibrator (Honda)	2
RC pouring for Beams & slabs	Truck mixer (Volvo FM400 (8 m3)	3
	Pump (Schwing S43SX)	1
	Vibrator (Honda)	2

4. Results and Discussion

The daily emissions generated from the main activities as a result of dust, gases noise and activity pollution index per each considered. It is worth to note that the factors that influence the activity pollution index (API) are: the construction method, the number of utilized equipment, the equipment percentage of usage and their power. Using the developed modules above, we can get the emissions generated from main activities as shown in Table 3.

Table3. Calculated emissions generated from main activities

Activity	Pollution Rate (gm/hr)						Dust	Noise	API
	HC	CO	NOX	PM	SO2	CO2	(gm)	(dB)	
Excavation	342.04	2442.28	5171.4	424.76	580.36	632743.64	41672.9	109	9.82
Backfilling	158.22	971.54	2044.9	170.78	290.18	316413.42	41672.9	109	5.23
Soil Replacement- Compaction	214.31	1391.96	3045.51	245.14	347.51	378873.22	93616.23	109	6.53
Formworks	4.4	44	117	6.4	9.8	10693.6	8158.45	90	1.16
Reinforcement	4.4	44	117	6.4	9.8	10693.6	8158.45	90	1.16
PC pouring for foundation	259.2	1559.84	3770.56	307.04	729.12	795823.2	14574.12	103	10.09
RC pouring for foundation	264.32	1493.09	3801.22	281.7	734.3	801501.79	14574.12	105	10.16
RC pouring for columns	133.2	801.66	1911.01	156.45	367.15	400747.78	14574.12	105	5.69
RC pouring for Beams & slabs	133.2	801.66	1911.01	156.45	367.15	400747.78	14574.12	105	5.69

As shown in Table 3, the maximum value of API is calculated from reinforced and plain concrete pouring for foundation to be 10.16 and 10.09, respectively as there are huge number of equipment utilized in this activities with high usage percentages. The reason that excavation generates considerable pollution (API 9.82) as it uses 2 loaders with high Hp (260 Hp) and high usage percentage (90%).In addition to that the soil has high percentage of silt that produce a lot of dust during execution of this activity.

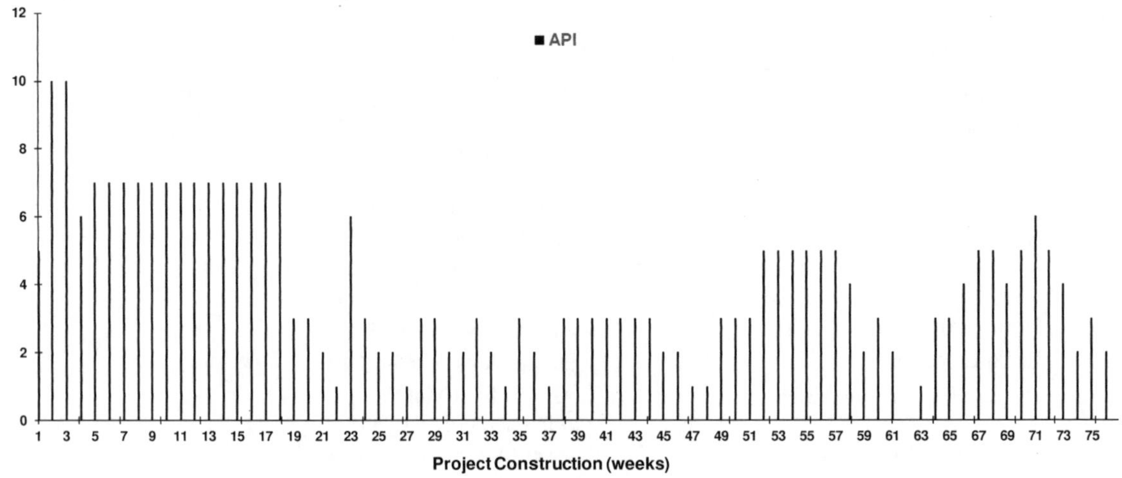

Figure 3. The resulted total pollution index distribution

Figure 3 shows the screen shot of the resulted total pollution index distribution throughout the project life that corresponding to the project planned scheduling. It is shown that total project duration is 76 weeks and the maximum pollution value is 69, which happened at 2nd and 3rd weeks (during excavation activity). In order to change the pollution distribution and minimize the peak points, we may change the construction methods, rescheduling project activities (by changing the relation amongst activities) or change crews 'configuration

5. Summary and Conclusions

Air pollution has been and continues to be viewed as a serious problem that affects human health. This paper presented a pollution module that aims at assessing the air pollution generated from buildings construction projects during the planning and execution phases. Emissions resulting from building construction projects was divided into three main categories: dust, harmful gases, and noise. The Activity Pollution Index accounts for the three main categories by calculating their impacts individually. Further it calculates the overall project pollution histogram corresponding to a certain scheduling and hence, the planner can know the project pollution every week and see if there is a need to modify their plan, for example, by rescheduling, by changing the construction methods or crew configurations to reduce the pollution resulted from the activities construction. Visual Basic for Application (VBA) is used to facilitate data entry and to perform interim calculations of the designed API. Construction authorities and environmental bodies can utilize this research to assure that the total pollution of building construction projects is within the permissible threshold. Also, construction contractors can utilize the model to follow the environmental regulations.

6. References

[1] Fan H 2017 A Critical Review and Analysis of Construction equipment emission factors *Procedia Eng.* 96 351–58

[2] Sandanayake M, Zhang G, Setunge S, Li C, and Fang J 2016 Models and method for estimation and comparison of direct emissions in building construction in Australia and a case study *Energy Build.* 126 128–38

[3] Li H , Wong T and Chen Z 2000 Environmental management of urban construction projects in China *J. Constr. Eng. Manag.*136 320–4

[4] Soderman M 2006 Environmental management in the construction industry - a comparative analysis of Skanska's environmental risk assessment *Dep. Econ. thesis*

[5] Chen Z, Li H, and Wong T 2005 Environmental planning : analytic network process model for environmentally conscious construction planning *J. Constr. Eng. Manag* 131 92–10

[6] Guggemos A, Asce A and Horvath A 2006 Decision-support tool for assessing the

environmental effects of constructing commercial buildings *J. Archit. Eng.* 12 187–95
[7] Hummer J, Asce F, Arocho I, ASCE A and Rasdorf W 2017 Approach to assessing tradeoffs between construction equipment fleet emissions and cost *J. Constr. Eng. Manag.* 143 1–10
[8] Arocho I, Rasdorf W, Hummer J, and Lewis P 2017 Time and cost characterisation of emissions from non-road diesel equipment for infrastructure projects *Int. J. Sustain. Eng.* 10 123–34
[9] Barati K and Shen X 2017 Optimal Driving Pattern of On-Road Construction Equipment for Emissions Reduction *Procedia Eng.* 180 1221–28
[10] USEPA (United states environmental protection agency) 2009a *Emiss. Factor Heavy Constr. Oper. US Environ. Prot. Agency, Washington, DC* available www.epa.gov/ttn/chief/ap42/ch13/final/c13s02-3. pdf (accessed 19 May 2017)
[11] USEPA (United states environmental protection agency)2009b Emiss. Factor Fugitive Dust Sources.*US Environ. Prot. Agency* available www.epa. gov/ttn/chief/ap42/ch13/final/c13s02.pdf (accessed 19 May 2017)
[12] USEPA (United states environmental protection agency) 2009c *EPA* Emiss. factor for unpaved Road available www.epa.gov/ttn/chief/ap42/ch13/final/c13s0202. pdf (accessed 19May 2017)
[13] USEPA (United states environmental protection agency) 2009d Emission factor for aggregate handling and storage piles *US Environmental Protection Agency, Washington, DC* available at www.epa.gov/ttn/chief/ ap42/ch13/final/c13s0204.pdf (accessed 19 May 2017)
[14] Egyptian Promulgating Law No.9 2009 Maximum threshold limit of pollutant according to environmental regulations in Egypt

Acknowledgments
The first author acknowledges the mission department of the ministry of higher education of Egypt for granting him scholarship to carry out his PhD study in Egypt –Japan University for Science and Technology and fully support this research work.

IOP Conf. Series: Earth and Environmental Science **151** (2018) 012019 doi:10.1088/1755-1315/151/1/012019

"Rio Preto Urgente": A Proposal for Conservation and Recovery

R de M Teles[1], A da P Rocha[2], G R do Vale, L V S Pinheiro, M R Magalhães[2] and C A Leão[2]

1 Doctor Professor, IFMA - Monte Castelo, Maranhão, BR
2 City Hall of São Benedito do Rio Preto, Maranhão, BR
E-mail: rogerioteles@ifma.edu.br

Abstract. The Association of Young People of Alto Rio Preto (JUAP) and the Association of Children and Friends of São Benedito, Maranhão, Brasil (AFASB) develop the "Avante Camaradas" (Forward Comrades) Project and the "Alonga Vida" (Lenghten Life) Operation, for the maintenance of the Preto River and the Mocambo River, its main tributary. From these actions arose the necessity of technical study of the Preto River. In this research we carried out expeditions in the river, we verified its bed, tributaries, vegetation, number of households and riverside. The study area was from Tutanguira village to ETA CAEMA (Water Treatment Station of the Environmental Sanitation Company of Maranhão). Flora, with identification of 89 species and fauna, with identification of 101 bird species, 28 species of herpetofauna (amphibian and reptiles) and 11 species of mammals were studied. Environmental impacts such as cutting and burning vegetation, extraction of wood for sawmill, coal and fishing, sand extraction and pebble stone are frequent in the river area. A socioeconomic diagnosis of the community of the study area was made, being verified that the majority is farmer and with low schooling. The river runs the risk of being no longer perennial and becoming intermittent due to the degradation caused by population growth and the monoculture of soybean and eucalyptus. One suggests the execution of an urgent maintenance and recovery project in the Preto River.

1. Introduction

Maranhão is a state located in the northeast of Brazil (01°01' to 10°21'S and 41°48' to 48°40'W), with a total area of 329,555.8 km², limited with the Atlantic Ocean, to the north, and with the states of Piauí to the east, Tocantins to the south and southwest, and Pará to the northwest. [1]

To the east of the state of Maranhão is located the municipality of São Benedito do Rio Preto (03°20'02"S and 43°31'40"W), with an area of 931,592 km^2, a population of 17,799 inhabitants and a human development index (HDI) of 0.541 [2], [3].

The city is cut by one of the main tributaries of the Munim River Basin, the Preto River, which rises south of the municipality of Buriti, Maranhão, Brazil (03°56'42"S and 43°02'3"W) about 230 km, bathing the municipalities of Buriti, Brejo, Milagres do Maranhão, Anapurus, Mata Roma, Urbano Santos, Chapadinha, São Benedito do Rio Preto and Nina Rodrigues. [4]

In many regions of the world, there have been declines in freshwater reserves, as was already recorded in the Human Development Report of the United Nations Program UNDP [5]. Some of these degradations are: deforestation, removal of riparian forests, practice of irrigated agriculture with the use of agrochemicals and livestock without technical support, allied to the expansion of the urban zone, domestic and industrial waste improperly released into the environment [6].

According to Braga and Rebouças (2017) [7], unsustainable activities have caused serious environmental impacts in Brazilian rivers, such as disappearance or diminution of riparian forests and habitats for aquatic fauna, elimination of the buffer zone between terrestrial ecosystems and aquatic environments, increase of particular suspended material in the water and significant changes in sediment composition of aquatic ecosystems. This also interferes in the quality and quantity of water in a river basin [8]. Like many other Brazilian rivers, the Preto River has suffered a lot of degradation due to the anthropogenic action, mainly in its sources.

Since 2007, two non-governmental non-profit organizations, the Association of Children and Friends of São Benedito do Rio Preto (AFASB) and the Association of Young People of Alto Rio Preto (JUAP) have been developing the "Avante Comaradas" (Forward Comrades) Project - a social program with an educational character with the objective of assisting in the moral, social, spiritual and civic formation of the young person, forming good citizens and conditioning them to a level of adequate preparation to better know how to face the difficulties. The Project is developed by volunteer instructors (teachers) and all financial investment is the result of spontaneous donations from people in the community.

It is a basic course of citizenship and humanization in which young people receive notions of Sociology, Anthropology, Health, Moral, Human Relations, Human Values, Citizenship / Politics, Environmental Education, Civics, Philosophy and Arts (Music, Theater, Dance and Plastic). Upon completion of the course, young people are able to live in society, feeling valued and valuing their neighbor, as a good citizen, knowing their rights and duties.

Undoubtedly, the main consequence of the citizen training promoted by the "Avante Camaradas" Project was the "Alonga Vida" (Lenghten Life) Social Action, a socio-environmental project that aims to revitalize the Preto River and its main tributary, the Mocambo River. It is a joint social action among the residents of the municipality, which has been taking place since 2010, promoting the maintenance, preservation and revitalization of the river bed, the banks and the riparian forest (ebb) of the Preto and Mocambo rivers, from the city of Urbano Santos to São Benedito do Rio Preto and bringing information to the riverside population about the importance of this preservation of river banks and their streams. Expeditions carried out by the "Alonga Vida" Social Action, showed several types of attack resulting from human action on the river [9].

In Brazil, the National Water Resources Policy [10] brings in its foundations that water is a public property and that its management should be decentralized and count on the participation of the Public Power, users and communities.

In the participatory management of water resources, people in the community must act consciously in decision making, and environmental education is essential for people's training and awareness [11]. Strategies should be based on the recovery and preservation of water bodies, especially springs [12], since predatory actions interfere in the natural cycles and the availability of good quality water, which compromises the sustainability of water resources [13].

The results of the Social Action "Alonga Vida" revealed the need for a more in-depth study of the real situation of the Preto River, which culminated with the research object of this article. This research aims to recognize the Preto River basin, to identify the source and the mouth of each of its tributaries, to diagnose the conditions of the river bed and forest and its tributaries, as well as to carry out a socioeconomic study of the communities that inhabit the area delimited in this study.

2. Methods

For the development of this work, the study area was initially recognized through expeditions throughout the Rio Preto basin from its source in the municipality of Buriti de Inácia Vaz to São Benedito do Rio Preto.

The area of study of this stage of the research belongs to the municipality of São Benedito do Rio Preto, upstream of the catchment point of the Environmental Sanitation Company of Maranhão - CAEMA, in an area comprised of the following coordinates: point 01: 3 ° 26'0 "S, 43 ° 27'0" W; lane 02: 3 ° 24'0 "S, 43 ° 33'0" W; lane 03: 3 ° 30.0 ° S, 43 ° 27.0 ° W; and point 04: 3 ° 24'0 "W, totalling 344.28 km².

In this area, a study of the hydrography and survey of the environmental protection areas was carried out. The flora, taxonomy and landscape were also investigated in two points, APA Fazenda Estiva and Marçal village, through parallel transects with plots of 10m x 10m [14]
The areas studied for the study of avifauna were Tutanguira, João Pedro, Bacuri, Gomes, Fautina, APA Fazenda Estiva and the Negro river riparian areas, through transepts and fixed points for listening observation. The herptofauna (amphibians and reptiles) [15]-[17] and mastofauna (mammals) [18]-[21] were also surveyed.
Environmental impacts were also investigated in the affected area, such as cutting, burning, logging, predatory fishing and mineral extraction.
In order to obtain socioeconomic data of the families living in the area of influence of the project, a survey was carried out with the population of 17 villages in the vicinity of the Preto River, where age, sex, schooling, economic activity and ties to class entities were verified.

3. Discussion and Results

3.1. Hydrography

Within the area delimited for this study, the Preto River crosses 43 km and receives 24 tributaries, being: Pau Mole stream (source: 3°26'13.19"S, 43°27'48.31"W, mouth: 3°25'59.66"S, 43°26'1.60"W), Bacuri stream (mouth 3°15'18.37"S, 43°15'54.75"W), Zé Raimundo stream (source: 3°24'55.84"S, 43°27'9.84"W, mouth: 3°25'33.39"S, 43°26'41.71"W), Santa Rosa stream (mouth: 3°22'13.08"S, 43°25'23.05"W), Gomes stream (source: 3°21'53.79"S, 43°26'15.68"W mouth: 3°21'27.54"S, 43°25'50.05"W), Olaria stream (source: 3°19'49.82"S, 43°29'47.16"S, mouth: 3°19'36.32"S, 43°29'47.16"W) Santa Luzia stream (source: 3°20'46.10"S, 43°26'38.70"W), Faustina stream (source: 3°21'29.15"S, 43°26'3.84"W mouth: 3°21'26.11"S, 43°25'49.97"W), Mangueira stream (source: 3°20'42.43"S, 43°25'46.34"W), Prata stream (source: 3°11'34.61"S, 43°15'41.81"W, mouth: 3°19'29.46"S, 43°26'16.14"W), Juçara stream (foz: 3°18'53.29"S, 43°26'58.88"W), Mocambo River (mouth: 3°18'52.26"S, 43°20'55.20"W), Pardo stream (source: 3°12'7.02"S, 43°16'19.66"W, mouth: 3°19'28.34"S, 43°27'11.80"W), Buriti stream (source: 3°12'7.02"S, 43°27'54.41"W, mouth: 3°20'39.31"S, 43°27'56.97"W), Cícero stream (source: 3°18'35.05"S, 43°29'18.33"W, mouth: 3°19'37.03"S, 43°29'18.33"W), Izabel stream (source: 3°20"17.92"S, 43°29'16.46"W, mouth: 3°18'35.05"S, 43°29'9.65"W), Jaime stream (mouth: 3°19'36.16"S, 43°26'51.12"W) Arlindo stream (mouth: 3°19'37.44"S, 43°29'41.78" W), Pedemeti stream (source: 3°19'44.98"S, 43°30'10.18"W, mouth: 3°19'21.38"S, 43°30'1.37"W), Marfisa stream (mouth: 3°19'9.31"S, 43°30'22.62"W), Barra Grande stream (source: 3°29'16.38"S, 43°31'6.17"W, mouth: 3°22'22.44"S, 43°25'27.47"W), Mapá stream (source: 3°18'16.07"S, 43°30'10.18"W, mouth: 3°19'11.94"S, 43°30'44.93"W), Cemitério stream (mouth: 3°19'11.94"S, 43°30'44.93"W), Genilson stream (mouth: 3°19'50.32"S, 43°31'5.05"W).
From these tributaries, only 3 resist to the drought period, being. The most voluminous, the Mocambo River, is born in the municipality of Urbano Santos. The fact that the river has most of its affluent inactive during a great part of the year has compromised the volume of water in its bed.
In the area studied in this research, there are two areas of environmental preservation (APA). The first, private, is Fazenda Estiva, located on the banks of Highway MA 224, in the municipality of São Benedito do Rio Preto. The second, created through a state decree, is the Environmental Protection Area of Upaon Açu / Miritiua / Alto Preguiças, which covers 17 municipalities, among them São Benedito do Rio Preto.

3.2. Flora

A total of 89 plant species belonging to 49 families were identified, with the most representative families being *Fabaceae* (11), *Arecaceae* (7), *Myrtaceae* (7), *Malvaceae* (5), *Anacardiaceae* (4), *Caesalpinioideae* (4) and *Mimosaceae* 4). In the transect on the banks of the Preto River, the dominant species is the "mofumbro" (*Viborgia polygaliformis*), pioneer species that extends its successional stage until the secondary one. On the banks of the Marfisa Stream, "becuiba" (*Virola bicuhyba*), "Açaí" (*Euterpe oleracea*), "Pau-pombo" (*Tapirira guianensis Aubl*) and "Buriti" (*Mauritia flexuosa*) are predominat species. These species are adapted to hydromorphic soil, having a fundamental role for the wetlands ecosystem [22].

3.3. Fauna

Bird Fauna

In the area of the Preto River studied in this research, 101 species of birds were found, distributed in 39 families, with no species threatened of extinction [16] and only two considered almost endangered [16], "Maracanã" (*Primolius maracana*) and "Aracuã-de-sobrancelhas" (*Ortalis superciliaris*). The APA Fazenda Estiva was the point that presented the greatest diversity of species, where 62 taxa of the 101 were located in all sampling areas

The observation points for Bacuri village (João and Bacuri streams, together, due to their geographical proximity) presented the second largest number of bird species, 58 taxa. Then the Gomes and Faustinas points, with 40 taxa and the Tutanguira point, with 28 species. These last three points present strong antropic influence and present great similarity in the composition of the species, which can be explained by low phytophysiological heterogeneity of the environment [23].

The survey on the banks of the Preto River did not show a significant number of species, 44 taxa, although it was the longest route. Certainly, this is verified by the high degree of anthropization of these margins, which has affected the most dependent species, without adaptations.

3.4. Herpetofauna and Mastofauna

Twenty-eight species that comprise the herpetofauna (amphibians and reptiles) belonging to 17 different families were found, and the status was of little concern (LC) for all, regarding the degree of threat of extinction [16]. On the other hand, it was found that the species of mammals (11 species) were distributed in 9 families. The mallard cat (*Puma yagouarondi*) was the only one threatened with extinction, with vulnerable degree, VU [16].

3.5. Environmental Impacts

Many are the impacts of anthropocentric origin on the Preto River basin, many of them with an intimate relation with the cultural historical context of natural resources management of the region.

The cutting and burning of vegetation on the banks of the river is often verified as a way of cleaning areas for family farming. It is not uncommon for fire to escape control and spread to areas beyond the interests of these farmers, which leads to major degradation [24]. Many are the damages brought by these fires, such as the release of carbon dioxide into the atmosphere, destruction of the soil micro-biota, destruction of retention forests and even riparian forests, which accelerates the flow of water, damaging the water supply of the groundwater, [25].

The extraction of wood for sawmill is already little verified in the forests of the studied area, since the trees are practically extinct and will take many decades to be recovered. It is also quite common to see the extraction of wood for the production of coal near the river and its tributaries, which is detrimental to the forests that are necessary for the survival of these springs.

Predatory fishing is also a cause of deforestation along the banks of the Preto River in at least three ways. One is the removal of small portions of riparian forests to have a clean place for fishing. This causes a lot of damage, since the removal of the ciliary forest accelerates the sedimentation process, with elevated accumulation of organic matter, besides the interference in the ecological restoration [26]. The second form is the removal of vegetation to create a physical barrier that prevents the passage of fish, facilitating their capture. This intervention compromises the reproduction of fish, putting species at risk even of extinction [22]. In the region there is also the "moita" or "coivara" fishery in which riparian trees are cut so that their branches and leaves are accumulated submerged in the river to serve as a fishery which will be captured after a few months. The latter, in addition to causing siltation and compromising reproduction, may still cause contamination of water with excess organic matter [27].

Another form of frequent degradation in the river is the extraction of sand and pebble for the civil construction. This withdrawal is done either manually or through machines that dredge material out of the river. Many are the damages caused by this form of extraction, such as the formation of craters, destruction of riparian vegetation, destruction of biological chains and contamination of water by fossil fuels [28].

3.6. Socioeconomic Situation of the Riverside Communities

In this research a socioeconomic diagnosis was carried out with the population located in the areas near the Preto River, which comprises 17 villages, totaling 1387 inhabitants. The indicators were age, schooling, sex, economic activity and link with class entities. The data are shown in Figure 1.

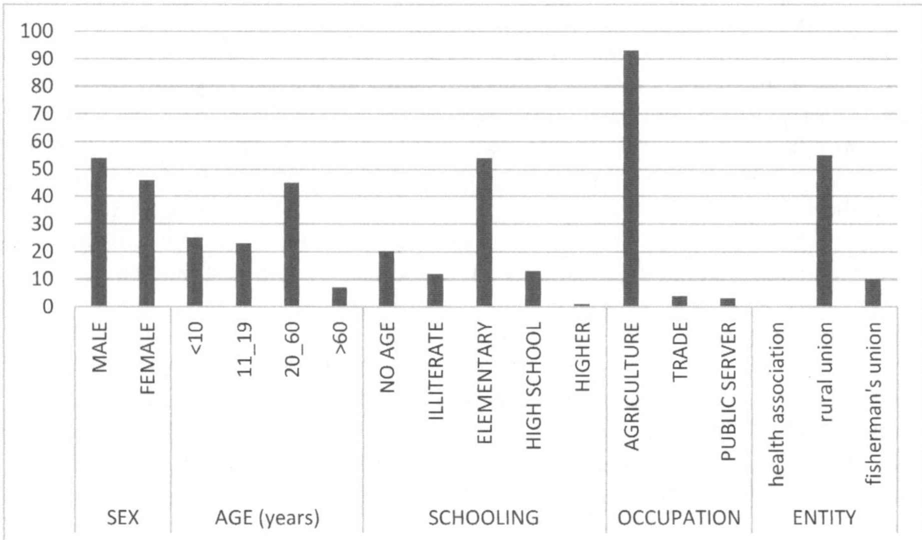

Figure 1. Socioeconomic diagnosis of the inhabitants of the study area

In these communities, there are residences built of bricks or of "taipa" (mud and straw), as some are covered of tiles, but others are of straws. Children have access to schools and most families participate in government income transfer programs. It is observed that the population has low or no schooling, most of them are farmers and most of them are associated with some class institution, most of them being in the Rural Workers' Union. In the visits to the communities, it was observed that agriculture is subsistence and are mainly cultivated rice, beans, maize and cassava.

4. Conclusions

Many environmental impacts due to population growth, monoculture and high number of fires threaten the availability of water resources in the Preto River basin, the water volume of which has been reduced each year. If actions are not taken to maintain and recover the river, it runs the risk of being no longer a perennial river and becoming intermittent, which would put at risk many activities that are currently performed by the community, since the river is responsible for the water supply of more than 90% of the population of São Benedito do Rio Preto.

From the analysis of the socioeconomic data collected, it is perceived that they are people lacking basic needs and who survive primarily from the resources and potentials of the place. This leads to the conclusion that there should be an environmental education plan so that these people can produce their food in a sustainable way, that is, without causing damage to the environment in which they live.

In this sense, this research shows the urgent need for a recovery plan for Preto River, passing through the immediate recovery of the water catchment forests to the water tables and the riparian forests. As it is estimated that many areas can not be recovered immediately, a specific study is recommended for the construction of dams to capture water in order to revitalize the sources of the river.

If there is no immediate intervention by local authorities, the damage to the environment and the economy of the region will be so significant that the financial resources currently committed will be insignificant in the amount that should be committed to repair future damages.

5. Future Perspectives

Execution of a project to monitor, preserve and recover the Preto River in the next ten years, where it is estimated that topographic stabilization and restoration of organic soil can be achieved, as well as revegetation of native species, where vegetation is not naturally restored.

For reforestation, seedling control and pest control will be required. Other needs include burn and deforestation controls, as well as the systematic evaluation of all stages of the process.

It is also recommended the immediate recreating of the Environment Secretariat as an executive body for environmental management of the municipality, including promoting projects at the state and federal levels.

6. References

[1] Brazilian Institute of Geography and Statistics (IBGE) 1997 Zoneamento Geoambiental do Estado do Maranhão. Rio de Janeiro, IBGE

[2] Brazilian Institute of Geography and Statistics – IBGE. (2010). *Censo 2010*. Available at http://www.censo2010.ibge.gov.br/sinopse/index.php?uf=21&dados=1

[3] United Nations Development Program - UNDP, Institute for Applied Economic Research – IPEA, & Fundação João Pinheiro - FJP. (2012). *Atlas do Desenvolvimento Humano no Brasil 2013*. Available at http://www.atlasbrasil.org.br/2013/o_atlas/ idhm

[4] Pressoti A 2008 *Avaliação de Impactos Ambientais da Sojicultura em um Ecossistema Aquático na Microrregião de Chapadinha - MA*. (Master's dissertation, Federal University of Maranhão, São Luís). Available at http://bdtd.ibict.br/vufind/Record/UFMA_84bcdb3c6f5fde317648cf051fb47f70/Details

[5] United Nations Development Program – UNDP 2006 Relatório de Desenvolvimento Humano (RDH-PNUD). Relatório de Desenvolvimento Humano

[6] Bomfim E, Gadelha C, Figueira H, Amorim J and Amorim D. 2015 Sustentabilidade Hidroambiental de Nascentes na Bacia Hidrográfica do Rio Gramame no Estado da Paraíba, Brasil. *Sociedade & Natureza*, 27 (3), 453-468.

[7] Braga B and Rebouças A 2017 *Capital Ecológico. Uso e Conservação*. Available at http://ambientes.ambientebrasil.com.br/agua/artigos_agua_doce/impactos,_recuperacao_e_conservacao_de_ecossistemas_aquaticos.html.

[8] Kobiyama M, Mota A and Corseuil C 2008 *Recursos hídricos e saneamento*. Curitiba: Organic Trading.

[9] Teles R and Rocha 2017 "Lenghten Life" Operation: Environmental Education, Citzenship and Preservation of Rio Preto *European Journal of Sustainable Development* 6, 4, 95-104

[10] BRASIL (1997). *Lei n. 9.433*, de 8 de janeiro de 1997. Available at http://www.planalto.gov.br/ccivil_03/leis/L9433.htm

[11] Oliveira M 2008 *O papel da Educação Ambiental na Gestão dos Recursos Hídricos: caso da Bacia do Descoberto/DF* (Master's dissertation, University of Brasília, Brasília-DF). Available at http://repositorio.unb.br/handle/10482/3693

[12] Rogerio P 2017 *Caracterização Revitalização da nascente da biquinha no bairro Bromélias*. Available at www.meuartigo.brasilescola.com.br.

[13] Tundisi J 2003 *Água no século XXI, enfrentando a escassez*. São Carlos, SP: RIMA.

[14] Nogueira M and Schiavini I 2003 Composicao floristica e estrutura da comunidade arborea de uma mata de galeria inundavel em Uberlandia, MG., Brasil. Bioscience Journal, v. 19, n. 2.

[15] Segalla M 2014 Brazilian amphibians–List of species. Sociedade Brasileira de Herpetologia. Volume 3 - Número 2 - Julho.2014. Available at: http://www.sbherpetologia.org.br

[16] Costa H and BÉRNILS R 2014 *Répteis brasileiros: Lista de espécies*. V 3 (3) Sociedade Brasileira de Herpetologia. Available at: http://www.sbherpetologia.org.br

[17] Freitas M 2015 *Herpetofauna no nordeste brasileiro: guia de campo* TECHNICAL BOOKS.

[18] IUCN. 2016 IUCN *Red List of Threatened Species*. Version 2016.2. Available at: www.iucnredlist.org

[19] Chico Mendes Biodiversity and Conservation Institute (ICMBIO) 2016 *Livro Vermelho da Fauna Brasileira Ameaçadas de Extinção*.

[20] Sigrist T 2012 *Mamíferos do Brasil: uma visão artística*. Avisbrasilis

[21] Becker M and Dalponte J 2015 *Rastros de mamíferos silvestres brasileiros: um guia de campo* TECHNICAL BOOKS.
[22] Begon M, Townsend C and Harper J 2009 *Ecologia: de indivíduos a ecossistemas.* Artmed
[23] Magnato L, Martins S and Pereira O 2011 Heterogeneidade florística das fitocenoses de restingas nos estados do Rio de Janeiro e Espírito Santo, Brasil *Revista Árvore*, v.35, n.2, p.245-254, 2011
[24] Porro R, Porro M and Sakiara N 2015 Identidade social, conhecimento local e manejo adaptativo de comunidades tradicionais em Babaçuais no Maranhão. *Ambiente & Sociedade*, v. 18, n. 1
[25] Vasconcelos M 2013Características químicas de solos sob cultivo tradicional no município de vargem grande-MA. *Agropecuária Científica no Semiárido*, v. 9, n. 3, p. 26-29
[26] Abdon M 2004 *Os impactos ambientais no meio físico: erosão e assoreamento na bacia hidrográfica do rio Taquari, MS, em decorrência da pecuária.* (Doctoral thesis) Universidade de São Paulo.
[27] Fearnside P 2009 Hidrelétricas como "fábricas de metano": O papel dos reservatórios em áreas de floresta tropical na emissão de gases de efeito estufa. *Oecologia Australis*, v. 12, n. 1, p. 100-115.
[28] Annibelli M and Souza Filho C 2007 Mineração de areia e seus impactos sócio-econômico-ambientais. In: Conpedi National Congress

Acknowledgments
The authors of this work thank JUAP and AFASB entities, as well as the Federal Institute of Education Science and Technology (IFMA), the City Hall of São Benedito do Rio Preto – MA and FAPEMA, for the institutional support

Road Traffic Noise Forecast (RTNF) in the Process of Creating Urban Space - Case Study of Poland

K Szopińska[1]

1 Department of Geomatics and Spatial Economy, UTP University of Science and Technology, Al. prof. S. Kaliskiego 7, 85-796 Bydgoszcz, Poland
E-mail: k.szopinska@utp.edu.pl

Abstract. Road traffic noise is one of the main nuisance factors in a city. This creates the necessity of determining and monitoring not only its level in relation to the existing urban structure, but also precisely determining and monitoring the level of noise over time, accounting for the designed spatial and functional solutions. In Poland, issues connected with noise are taken on in ecophysiographic reports and forecasts of its impact on the environment, which are a component of the local spatial development plan (LSDP) creation processes. Unfortunately, studies carried out on the noise environment are done in a selective manner, while environmental reports assume the consistency of nuisance phenomena over time. In connection with the above, the present article focuses on the problem of noise nuisance derived from a planned road solution while making a planning decision in the process of creating an LSDP. The specific aim is to present an original Road Traffic Noise Forecast (RTNF) understood as an element of the procedure of choosing an optimal spatial and functional solution for noise sensitive areas in terms of maintaining noise-level standards. The above will be illustrated on the example of a selected LSDP of the Polish city of Bydgoszcz.

1. Introduction

Spatial planning deals with the organization of planning space. This is an ongoing process based on adapting spatial components in regards to each other and aiming to restore an optimal design in the event of it being disturbed by the effect of developmental factors [1]. It is also an action aimed at creating such a state of the natural environment in which the functions it serves result from the necessity of fulfilling people's needs while accounting for the principles of ecological development. Noise is one of the main factors degrading the environment. It is commonly encountered in urbanized areas, causing discomfort among the people and having a negative influence on the management of the city's resources, as well as lowering the value of real estate within the vicinity of areas generating noise [2]. The influence of noise on the human body has been the topic of many scientific publications. A number of studies confirm the relationship between the level of environmental noise and the occurrence of many adverse health effects, which include: ischaemic heart diseases, arterial hypertension [3], diabetes [4], stroke [5], or sleep loss and problems with concentration [6]. Noise derived from road traffic is included in the most bothersome group of noise sources [7]. Inhabitants of big cities are particularly at risk of its negative effects. This is the result of, above all, the strong expansion of urbanized areas, ever-shrinking planning space, and ineffective noise policy [8]. The increasing urbanization of cities, which is strongly connected with industrialization, may contribute to generating noise nuisance in noise sensitive areas, the aim of which is to ensure proper noise conditions, where noise levels specified by the legislator are not exceeded. In accordance with Polish legislation [9], during the planning of urban space, matters of protection against noise ought to be considered. Authorities of Polish cities fulfill this requirement in ecophysiographic studies as well as a forecast of the impact on the environment, which constitute an element of the local spatial development plan (LSDP) creation process [10]. Unfortunately, the above studies of the noise environment are carried out selectively, while environmental reports assume the consistency of nuisance phenomena over time. As a result of this, all proposed actions regarding noise protection are the result of incomplete analyses. This often leads to the erroneous localization of noise sensitive areas and contributes to the emergence of social conflicts. Due to noise sensitive areas being located in the direct neighborhood of roads, the above especially concerns nuisance generated by planned transportation solutions.

Thus, the aim of this article is to offer an insight into the subject matter of protection against road traffic noise and draw attention to the problem of noise nuisance derived from a planned road solution while making the planning decision in the process of creating an LSDP. The detailed aim is to present an original Road Traffic Noise Forecast – RTNF, understood as an optimizing tool of the planning phase of the LSDP creation process, facilitating the minimization of noise pollution derived from the planned road solution in noise sensitive areas. The above will be illustrated on the example the Local Spatial Development Plan of the Akademickie-Wschod neighborhood in the Fordon District of the Polish city of Bydgoszcz [11]. The research will be used to verify the study hypothesis of whether the introduction of the planned road solution to the city space will affect the acoustic conditions prevailing in the vicinity of the planned noise sensitive areas. The research was realized using the following study methods: studies of literature and laws regarding the protection of the environment against road traffic noise, studies of source materials, including planning documents (http://mpu.bydgoszcz.pl/), information from land and building registers, the strategic noise map (SNM) of Bydgoszcz [12] and a base map of the study area, as well as data presented on the web portal – http://geoportal.mojregion.info/. Moreover, the following computer software was used for analyses: GeoMedia Professional (program for spatial analyses) and CadnaA (program for noise analyses).

2. Problem of Road Traffic Noise in Light of Binding Polish Legal Regulations

In Europe, the basic legal act regulating issues of protection against noise is the Environmental Noise Directive (END) [13]. The main aim of the directive is to unify, in the member states, actions involving the reduction of the level of noise in the environment. In Poland, the implementation of the above directive is reflected in the provisions of the Environmental Protection Law Act [10]. In Poland, protection against noise relies on ensuring the best possible noise state of the environment by maintaining the level of noise at values not exceeding those of two acceptable groups of indicators. The first of them is applied when determining and controlling the conditions of using the environment in relation to a single day (L_{AeqD} – equivalent level of noise A for daytime house, from 6^{00} to 22^{00}, and L_{AeqN} – equivalent level of noise A for the time of night, from 22^{00} to 6^{00}). The second group of indicators is applied in the long-term policy dealing with the protection of the environment against noise pollution and developing strategic noise maps. This group includes L_{DEN} – long-term average level of noise A expressed in decibels (dB) determined during all days and nights in a year, and L_N – long term average level of noise A expressed in decibels (dB) determined during all hours of the night in a year (from 22^{00} to 6^{00}). Out of concern for people's health, the acceptable levels of noise in the environment have been specified in each EU member state. Acceptable values of the individual indicators depend on the source of noise and intended land use. In Poland, depending on the type of land, sensitivity and type of noise source, the permissible level oscillates between 50-70 dB during the daytime hours and 45-65 dB during night-time hours. For example, the following acceptable levels of road traffic noise apply for residential areas [14]: for single family housing real estate: L_{DEN}=50dB, L_N=45dB, for multi-family housing real estate: L_{DEN}=68dB, L_N=59dB. The maximum values are laid out in the national provisions of EU member states, as there are no common legal regulations in this scope [15]. Each country has a different definition of a noise sensitive area, imposes different limits, and realizes noise policy aims in different ways. In the majority of countries, areas zoned for residential development, as well as hospitals and areas of healthcare facilities are subject to protection against noise. In certain countries, e.g. Belgium, Germany or the United States, permissible noise levels are set out for the current situation as well as the anticipated state.

The basic tool for tackling noise is an SNM, which specifies the areas of reach of harmful sounds in the environment. A SNM falls into all key areas of the noise protection policy. As a GIS system, it is included in the informative layer of city management [16]. The process of creating SNMs is a time-consuming analysis of demographic data as well as data regarding the means of land management and use. It is realized using advanced computer software intended for modeling noise in space. Noise calculations can be carried out using programs for noise analyses, e.g.: CadnaA, SoundPlan, ArcAkus, etc., based on the GIS database. Data and conclusions stemming from SNMs (in the form of emission maps, exceedance maps, and maps of sensitive areas) are the basis for preparing and updating the Program of Environmental Protection Against Noise - PEPAN. Directive [13] requires the member states to periodically (every five years) prepare SNMs for all agglomerations in their area, as well as relay data to the European Commission. As of currently, three phases of works on SNMs have been completed. Phase 1 covered the preparation of SNMs for cities of over 250 thousand inhabitants (completed in 2007). Phase 2 covered works on SNMs for all cities of over 100 thousand inhabitants, along with updating data from 2007 (completed in 2012). Phase 3 was completed in mid-2017. In Poland, in the year 2013, additional updates were carried out, resulting from changes in the permissible levels of noise in the environment.

3. Problem of Protection against Road Traffic Noise in the Process of Creating Polish Cities

In Poland, under the current legal situation, LSDPs play the main role in shaping urban space. Their main purpose is to establish the intended land use in such a way that meets the economic needs of various fields, and is known and predictable, as well as socially accepted, while abiding by the principles of sustainable development or environmental protection, including protection from environmental noise pollution. An LSDP realizes the above aim, above all, by indicating areas of various land use functions and providing regulations as to their development, including noise sensitive areas. In Poland, areas covered by noise protection include, among others, areas intended for residential and residential-commercial development, buildings connected with the permanent or temporary stay of children and youth (e.g. schools, kindergartens, preschools), or areas zoned for hospitals or nursing homes [14]. In an LSDP, in addition to indicating areas of various intended use, the borders of and means of developing areas subject to protection arising from separate regulations are provided; special conditions for their development are set out, along with limitations in their use. In the context of issues connected with noise, quiet areas (e.g. recreational and leisure areas in cities) or limited use zones surrounding airports may serve as examples. Their basic aim is to maintain a good quality of the noise environment as a result of introducing limitations in the use of real estate or the requirement of applying technological solutions enabling the noise comfort in a building to be maintained.

In the event of the occurrence of an excessive level of road traffic noise resulting from a conflict at the meeting point of residential and transportation land uses, corrective measures are severely limited. The complete elimination of noise nuisance generated by roads is not possible. Reducing road traffic noise at the stage of city planning is extremely difficult due to the occurrence of existing, strictly specified spatial structures which significantly limit planning activities. Keeping this in mind, in order to minimize noise, modernization- or investment-based forms of noise protection are applied, which, in connection with a proecological approach to city planning, make it possible to maintain a proper noise environment. Such actions include promoting the notion of sustainable city transport, which meets social and economic expectations and, at the same time, has a positive influence on the environment; decreasing the transportation intensity by the proper organization of urban space (e.g. the construction of ring roads around areas of residential housing development); creating zones restricting or calming road traffic by changes in the organization of traffic and decreasing speed limits; applying quiet road surfaces, a feature of which is reducing the noise level to even as little as 10dB; locating roads in deep trenches or between earthen mounds; constructing soundproof tunnels over a road; road traffic noise shielding using noise screens, buildings and structures, as well as green belt; installing sound-proof windows along with air-conditioning in areas exposed to excessive levels of noise.

All of the above actions must be based on objectively existing and disclosed evidence. Ecophysiographic studies as well as the forecast of environmental impact, which undertake the topic of protection against noise, especially in the context of designed spatial and functional solutions, are to be used for this purpose when creating an LSDP. In the above reports, information regarding the noise environment is derived, above all, from the Provincial Inspectorate of Environmental Protection (PIEP), contained in reports on the state of the environment. In these documents, analyses of road traffic noise are carried out selectively. Moreover, PIEP, when realizing the tasks of National Environmental Monitoring, carries out noise measurements using automatic and mobile stations only at selected points. The obtained results are, therefore, not representative of the entire city, and do not account for noise derived from all sections of the road network. Moreover, all of the carried out studies assume the consistency of parameters over time. As of currently, changes in the intensity and the structure of road traffic by type over time on the existing and planned road sections are not accounted for when creating LSDPs. Also not accounted for is road traffic noise generated from neighboring areas, which can influence how the noise environment of the planned space is shaped (including data regarding the parameters of road sections, intensity and structure of road traffic, or changes in the transportation needs of the inhabitants influencing the choice of a specific transportation system or road section), as well as the time needed to fully realize the provisions of the LSDP.

4. Road Traffic Noise Forecast (RTNF) as an Optimizing Tool for Planning Polish City Space

Looking for an optimal spatial and functional solution ought to include noise forecasts, which will allow for the level of noise generated by the planned transportation solutions to be determined in a precise manner. For this purpose, it is recommended that new tools such as Road Traffic Noise Forecasts (RTNF) be accounted for when planning the structure of a city. Their findings will be the basis for making planning decisions and will influence the selection of an optimal spatial and functional solution for noise sensitive areas in terms of meeting noise requirements resulting from binding legal regulations. The concept accounts for using SNM resources along with making use of geoinformation data. An RTNF requires a series of activities which, accordingly to Figure 1, have been assigned to the following points:

- Action 1 – identifying the spatial and functional structure of the area for the existing and planned state.
- Action 2 – carrying out RTNFs based on the GIS database along with a program of remedial actions in the form of variants of spatial and functional solutions for noise sensitive areas based on the results of the RTNFs.
- Action 3 – selecting the optimal spatial and functional solution for areas of sensitivity based on the results of RTNFs.

In Poland, it is recommended to account for RTNFs in the process of creating LSDPs using SNM resources by the design solutions shown in Figure 1. Data on the level of road traffic noise derived from SNM information resources of the city ought to be included in the input materials of the procedure for creating local spatial development plans (Stage 0). When preparing the Environmental Impact Forecast, the results of RTNFs should be accounted for (STATE 1, STATE 2, STATE 3) (Stage III). The new element has been presented in the Figure 1 in red. The results of the RNF have been presented in the Environmental Impact Forecast and ought to serve as the basis for once again analyzing the decisions made in the drafted local spatial development plan in terms of maintaining noise standards for noise sensitive areas (Stage III).

The spatial and functional structure of the land (Action 1) for the planned state should be realized in accordance with the findings of the spatial and functional solution proposed by the city planners. Imission maps for three states of road traffic noise should be generated within the framework of the presented RTNF (Action 2), accounting for various spatial solutions of the area. These states are:

- STATE 1 – existing road solution accounting for current traffic intensity.
- STATE 2 – planned road solution accounting for the current traffic intensity.
- STATE 3 – planned road solution accounting for future traffic intensity.

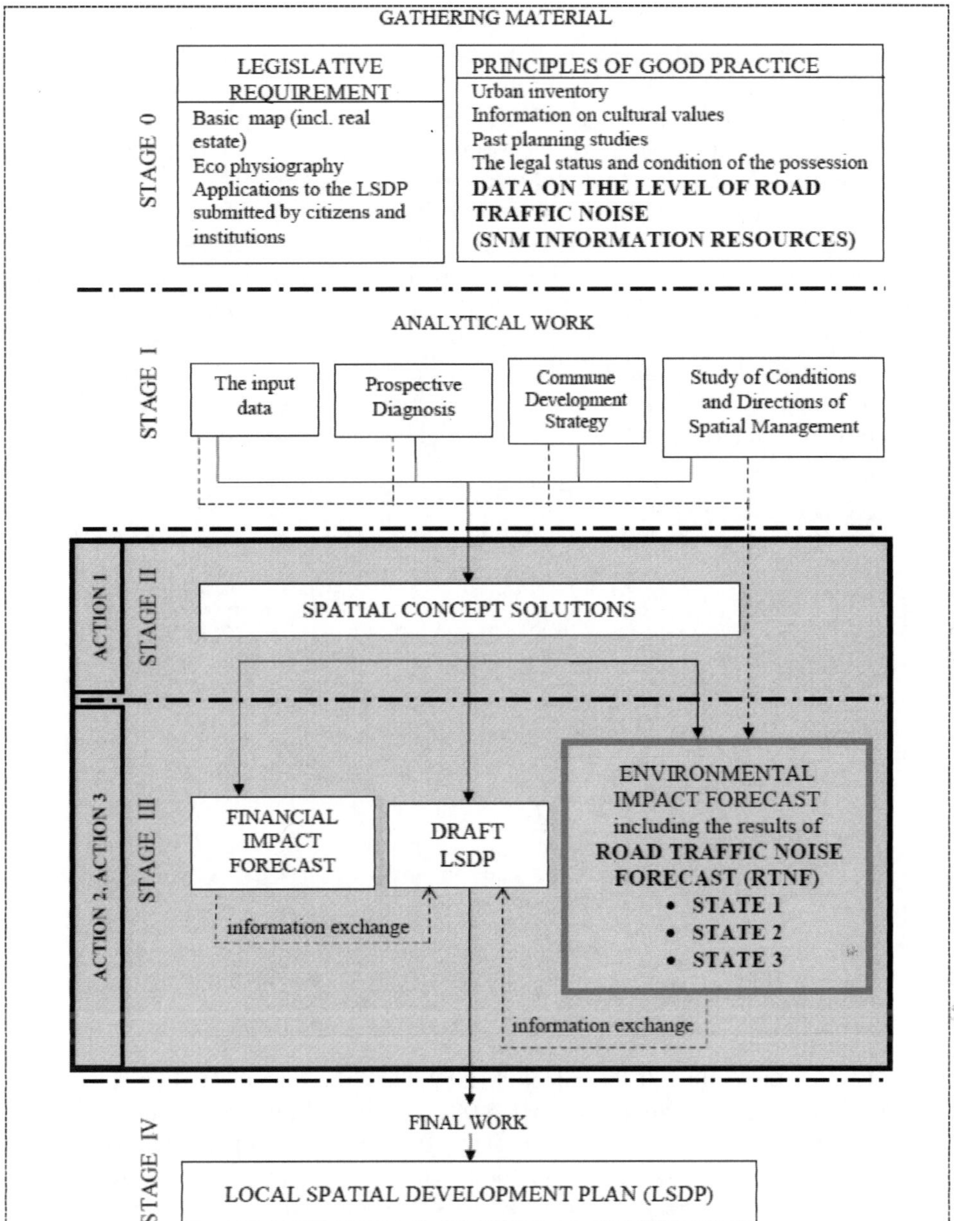

Figure 1. Scheme of creating a local spatial development plan accounting for the road traffic noise forecast.

In accordance with the presented algorithm for the spatial and functional solution proposed by city planners, the level of road traffic noise generated by existing road solutions ought to be assessed. Data regarding the current traffic intensity should be obtained in the year of making the planning decision (STATE 1). Next, the level of noise generated by the planned road solution ought to be determined. For this purpose, the planned stretches of road should firstly be considered for the existing traffic intensity for which the data were obtained in the year of making the planning decision (STATE 2). Due to the fact that the time necessary to carry out the spatial and functional solutions proposed by city planners should be accounted for in the studies, STATE 3 was introduced to the presented algorithm. It accounts for future traffic intensity determined in the assumed time frame, which enables the full realization of all activities in the urban space. The forecast period introduced to STATE 3 ought to account for the time necessary to carry out the proposed tasks while accounting for economic, technical, social and environmental aspects. In this way, a forecast showing the noise generated by a planned road solution accounting for the future traffic intensity will be created. Moreover, in the framework of Action 2 for STATE 1, STATE 2 and STATE 3, analyses of spatial and functional solutions for noise sensitive areas ought to be carried out, and the program of remedial actions, which will become the basis for selecting the optimal solution, presented (Action 3).

Below, we present the results of introducing a Road Traffic Noise Forecast to the process of drafting a selected LSDP. For the purposes of the research, the Local Spatial Development Plan of the Akademickie-Wschod neighborhood in the Fordon District of Bydgoszcz, dominated by multi-family residential buildings dating back to the 1980s, was selected. Bydgoszcz, located in the northern part of the country, is one of the largest centres of settlement in Poland. The city is populated by nearly 360,000 inhabitants and situated on the Brda River and Bydgoszcz Canal, with its eastern part bordering the Vistula River. In Poland, it is the tenth largest city in terms of area (176 km^2) and the eighth in terms of population (Fig. 2).

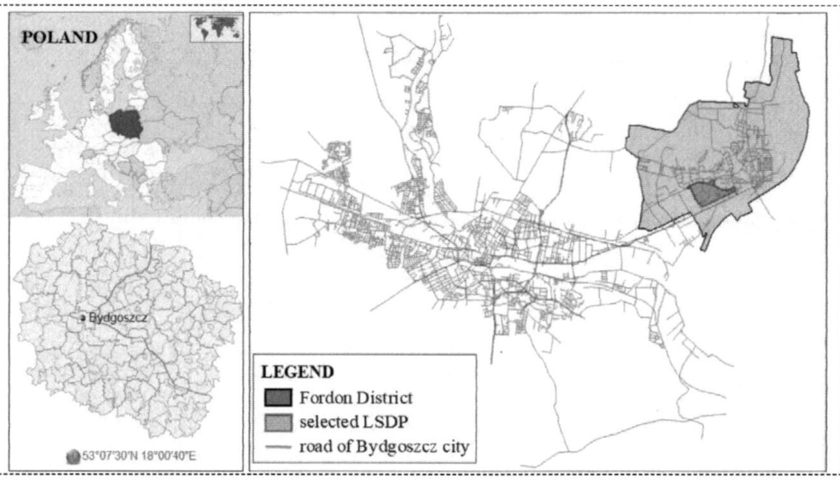

Figure 2. Road network of the city of Bydgoszcz marking the analyzed LSDP.

In accordance with the provisions of the selected LSDP [11], the investigated territory will have new noise sensitive areas, covering a total area of 15.09 ha (25% of the total area of the LSDP), with planned construction of 37 residential buildings (Fig. 3). In addition to this, there are plans to extend the existing road system. Such modernisation includes the construction of new roads and the extension of existing roads (Fig. 3b and Fig. 3c). In order to see whether the planned extension of the road system (referred to as the "projected road solution") would contribute to a change in the acoustic conditions within the area of the projected noise sensitive areas, the author created a GIS database, and then, using Cadna A, created a set of immission maps for the L_{DEN} indicator of road noise, taking into account various spatial solutions. Figure 3a shows a road traffic noise immission map for STATE 1, with the existing roads characterized by traffic loads as of 2011. Figure 3b shows an immission map for STATE 2, where existing roads have the traffic load predicted for the year 2035. Figure 3c presents an immission map for the noise generated by the projected road solution and the traffic predicted for 2035. Figures 3a,b,c indicate the noise level (L_{DEN} indicator) for each of the situations. As shown, including the road solution projected by the planners within the structure of the investigated area contributes to a change in the noise level within the area of the projected housing estate. These findings are confirmed by the L_{DEN} indicator, which changes according to the situation and is highest for STATE 3 (see Table 1).

Table 1. Road traffic noise level - L_{DEN} indicator for three states.

Measuring point	Road traffic noise level L_{DEN} [dB]		
	STATE 1	STATE 2	STATE 3
1	63.7	68.2	69.1
2	58.9	68.7	69.2
3	64.6	59.8	60.7
4	56.8	66.2	67.1
5	60.2	559	54.7
6	63.9	58.6	57.5
7	64.1	64.3	64.7
8	62.9	63.4	63.7
9	54.9	63.5	64.7
10	60.1	63.8	64.2

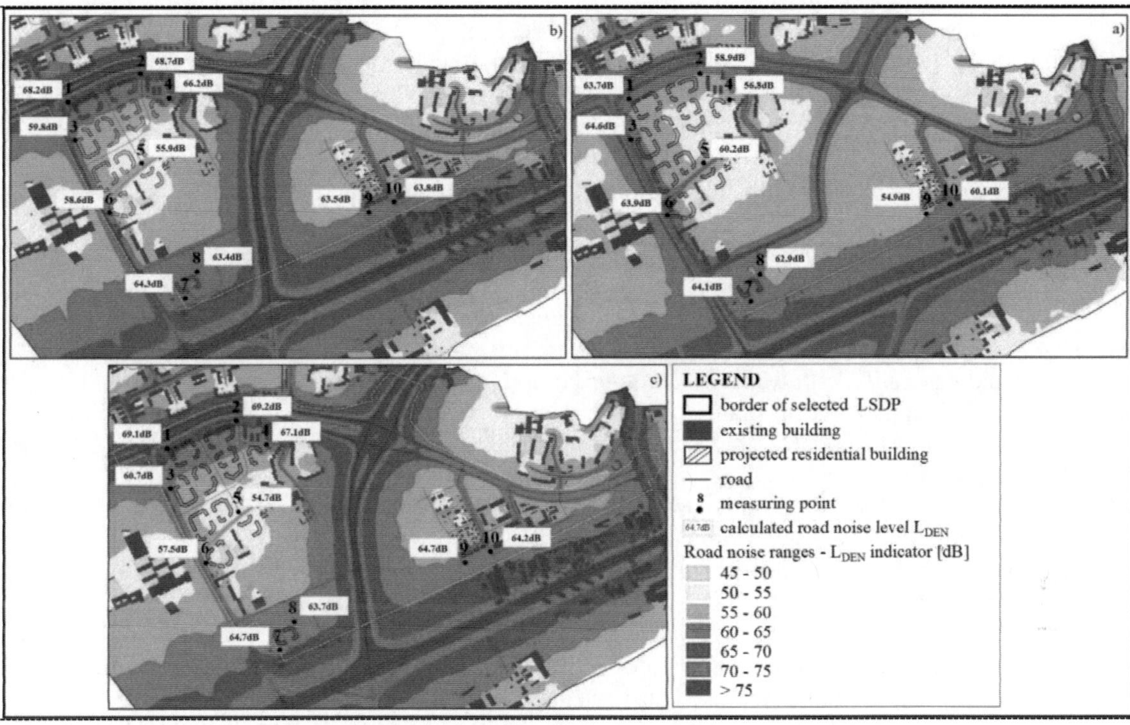

Figure 3. Analysis of the spatial and functional solutions in accordance with the investigated spatial development plan for a) STATE 1, b) STATE 2, c) STATE 3.

5. Conclusion

The fact that noise research of the forecasted state is not taken into account while creating local spatial development plans often leads to noise sensitive areas and areas of noise nuisance being located at distances which contribute to the emergence of spatial and social conflicts. The above data are confirmed by the research, which in turn support the initial hypothesis. Including the projected road solution in the investigated area contributed to a change in the noise levels in the environment, which in turn has an influence in the acoustic situation in the vicinity of the projected areas of noise sensitivity. Introducing the original Road Traffic Noise Forecast to the process of creating LSDPs using SNM resources will make it possible to determine the spatial and functional structure of noise sensitive areas while maintaining noise standards specified by the legislator. Accounting for the results of RTNFs will prevent the common practice of applying vague statements regarding the quality of the noise environment in studies accompanying the development of LSDPs. Reliable information in terms of changes in the environmental noise levels is essential to making appropriate investment decisions. Due to the fact that the presented RTNF refers to the means of adopting local plans, municipal self-governments are the main recipients. In this view, the presented tool may influence the shaping of urban space in accordance with the principles of sustainable development and spatial order.

6. References

[1] Krajewska M, Źróbek S and Šubic-Kovač M 2014 The role of spatial planning in the investment process in Poland and Slovenia *Real Estate Management and Valuation* 22(2) pp. 52–66

[2] Szopińska K and Krajewska M 2016 Sensitive property valuation taking into account noise aspects* *Geographic Information Systems Conference and Exhibition "GIS ODYSSEY 2016" Conference proceedings (Perugia Italy)* ed Bieda A, Bydłosz J and Kowalczyk A pp 246–252

[3] Burden of Disease from Environmental Noise. Quantification of Healthy Life Years Lost in Europe 2011 WHO Regional Office for Europe: Copenhagen, Danmark, pp. 1-106 (available online: URL http://www.euro.who.int/__data/assets/pdf_file/0008/136466/e94888.pdf)

[4] Sørensen M, Andersen Z J, Nordsborg R B, Becker T, Tjønneland A, Overvad K and Raaschou-Nielsen O 2013: Long-term exposure to road traffic noise and incident diabetes: a cohort study *Environmental Health Perspectives (Online)* 121(2) 217

[5] Halonen J I, Hansell A L, Gulliver J, Morley D, Blangiardo M, Fecht D, Toledano M B, Beevers S D, Anderson H R, Kelly F J et al. 2015 Road traffic noise is associated with increased cardiovascular morbidity and mortality and all-cause mortality in London *European heart journal* 36(39) pp. 2653–2661

[6] Perron S, Plante C, Ragettli M S, Kaiser D J, Goudreau S and Smargiassi A 2016 Sleep disturbance from road traffic, railways, airplanes and from total environmental noise levels in Montreal *International journal of environmental research and public health* 13(8) 809

[7] Salomons E M and Berghauser Pont M 2012 Urban traffic noise and the relation to urban density, form, and traffic elasticity *Landscape and Urban Planning* 108(1) pp. 2–16

[8] Szopińska K and Krajewska M 2016 Methods of Assessing Noise Nuisance of Real Estate Surroundings *Real Estate Management and Valuation* 24(1) pp. 19–30

[9] Act of 27 March 2003 on planning and spatial planning (Ustawa z dnia 27 marca 2003 r. o planowaniu i zagospodarowaniu przestrzennym, Tekst jedn., Dz.U. Nr 0/2017, poz. 1073 – in Polish)

[10] Act of April 27 2001 Environmental Protection Law (Ustawa z dnia 27 kwietnia 2001 r. Prawo Ochrony Środowiska. Tekst jedn. Dz. U. Nr 0/2017, poz. 519 – in Polish)

[11] Local Spatial Development Plan of Akademickie-Wschód neighborhood in the Fordon District of Bydgoszcz (Miejscowy Plan Zagospodarowania Przestrzennego osiedla Akademickie – Wschód w jednostce Fordon w Bydgoszczy zatwierdzony Uchwałą nr XLIV/1367/2001 Rady Miasta Bydgoszczy z dnia 28 listopada 2001 roku – in Polish) (available online: URL http://mpu.bydgoszcz.pl/) (accessed on 10 September 2017)

[12] Strategic Noise Map of Bydgoszcz City (available online: URL http://mapy.bydgoszcz.pl/) (accessed on 10 June 2017)

[13] Directive 2002/49/EC of the European Parliament and of the Council of 25 June 2002 relating to the assessment and management of environmental noise – Official Journal of the European Communities L 189 of 18 July 2002

[14] Regulation by the Minister of Environment of 14 June 2007 on permissible noise level in the environment (Rozporządzenie Ministra Środowiska z dnia 14 czerwca 2007 r. w sprawie dopuszczalnych poziomów hałasu w środowisku, Tekst jedn. Dz. U. Nr 0/2014, poz. 112. – in Polish)

[15] King E A and Murphy E 2016 Environmental noise–'Forgotten'or 'Ignored'pollutant? *Applied Acoustics* Vol. 112 pp. 211–215

[16] Kwiecień J and Szopińska K 2013 Implementation of the EU Noise Directive in Process of Urban Planning in Poland *29th Urban Data Management Symposium / International Archives of the Photogrammetry, Remote Sensing and Spatial Information Sciences* XL-4/W1 pp. 45–49

IOP Conf. Series: Earth and Environmental Science **151** (2018) 012021 doi:10.1088/1755-1315/151/1/012021

Spindly $BiFeO_3$ Nanoparticles for Photodegradation of Organic Pollutants Under a Compact Fluorescent Lamp

S M Lam[1], Z H Jaffari[1], J C Sin[2] and A R Mohamed[3]

1 Department of Environmental Engineering, Faculty of Engineering and Green Technology, Universiti Tunku Abdul Rahman, Jalan Universiti, Bandar Barat, 31900 Kampar, Perak, Malaysia
2 Department of Petrochemical Engineering, Faculty of Engineering and Green Technology, Universiti Tunku Abdul Rahman, Jalan Universiti, Bandar Barat, 31900 Kampar, Perak, Malaysia.
3 School of Chemical Engineering, Universiti Sains Malaysia, Engineering Campus, 14300 Nibong Tebal, Pulau Pinang, Malaysia.
E-mail: lamsm@utar.edu.my

Abstract. Spindly bismuth ferrite oxide ($BiFeO_3$, BFO) nanoparticles were fabricated by $Bi(NO_3)_3$, $Fe(NO_3)_3$, NaOH and urea solutions at 125°C. The BFO nanoparticles have a spindle-shaped of ~120 m in length and ~50 nm in width. Such nanoparticles could be used in the photocatalytic degradation of Malachite Green (MG) dye under irradiation by a 105 W compact fluorescent lamp. They were very stable and could be easily recovered by an applied magnetic field. There was an optimal photocatalyst amount of 1.0 g/L, at which the degradation efficiency of MG was improved to 72.6% under exposure of visible light for 240 min. The active hydroxyl (•OH) radicals formed during the photocatalytic process were tested using a photoluminescence–terephthalic acid (PL–TA) measurement, which were validate to be significantly affected by the photocatalyst amount. Other organic pollutants including phenol, bisphenol A and methylparaben could also be photodecomposed in the presence of similar conditions. These features demonstrated the BFO nanoparticles practical applications in environmental remediation.

1. Introduction

Since the photochemical utilization of solar energy concept was introduced, much scientific works have been performed to establish photocatalytic materials for splitting H_2O to generate H_2 energy, reducing CO_2 to fuel molecules and destruction of harmful organic pollutants. In order to harvest the visible light energy (~43%) of the total solar energy efficiently, one approach is to create photocatalytic materials that have appropriate band gaps to absorb the visible light. Of major interest in the photocatalytic materials, bismuth ferrite ($BiFeO_3$, noted as BFO) is recognized as one of the room temperature multiferroic materials with rhombohedral phase having the polar R_3c space group [1]. The recent researches in BFO materials as potential photocatalysts arise from their small band gaps (2.0–2.8 eV), chemical stability and low cost [2,3]. The small band gap of BFO materials enables them response in the visible light spectrum. For example, Gao et al. [4] revealed that BFO nanophotocatalysts displayed photoactivity for degradation of methyl orange under exposure of visible light. Humayun et al. [5] also tested that the aqueous phenol solution can be degraded by the TiO_2/BFO nanocomposites in the presence of visible light. Nevertheless, it should be mentioned that

IOP Conf. Series: Earth and Environmental Science **151** (2018) 012021 doi:10.1088/1755-1315/151/1/012021

BFO alone typically indicates a poor visible light photoactivity than those of other bismuth-based photocatalyts including Bi_2WO_6 and $BiVO_4$ [6,7].

In order to surmount these lapses and enhance the BFO photocatalytic performance, ion doping, coupling with semiconductor metal oxides and synthesis of perovskite oxides with larger surface area are highly desirable. It is well known that the microstructure of photocatalyst played a pivotal role on the catalytic performance [8,9]. Nanoscale photocatalyst is anticipated to improve photocatalytic performance because of a larger specific surface area and more surface active sites for the reaction. A myriad of wet chemical technique including co-precipitation, sol-gel, hydrothermal techniques and some wet-chemical solution processes have been extensively utilized to fabricate the BFO nanoparticles. Among these, the urea-assisted hydrothermal method [10,11] is relatively facile, cheap, surfactant-free and effective. Additionally, it can be conducted at short periods, normal pressure and at low temperature (<200°C). Despite this method has been utilized to synthesize the BFO microparticles, its ability to form different structures at adequately huge amounts for practical applications including photocatalysis still requires to be established.

In the present study, a novel spindly BFO nanoparticles has been prepared by an urea-assisted hydrothermal technique at low temperature (~125°C). Such a new structure has not been reported elsewhere. The prepared BFO was then characterized by various characterization techniques such as X-ray diffraction (XRD), field emission-scanning electron microscopy (FESEM), energy dispersive X-ray spectroscopy (EDX), Fourier transform infrared spectroscopy (FTIR) and UV–vis diffuse reflectance spectroscopy (DRS) measurements. The •OH radicals concentration for different BFO amounts were evaluated upon the photocatalytic processes. More importantly, the prepared BFO displayed high photocatalytic activities in the degradation of organic pollutants such as malachite green (MG), phenol, bisphenol A (BPA) and methylparaben (MP) under irradiation by a low-power (105 W) compact fluorescent lamp.

2. Experimental

$Bi(NO_3)_3 \cdot 5H_2O$ (0.005 mol), $Fe(NO_3)_3 \cdot 9H_2O$ (0.005 mol), urea (0.1 mol) and nitric acid (2 mL) was first prepared in 20 mL distilled water and stirring continuously for 20 min. The mixture was slowly added into 10 M NaOH solution under magnetic stirring. Subsequently, the mixed suspensions were placed into a 100 mL stainless-steel Teflon-lined autoclave for the hydrothermal treatment. After being tightly closed, the autoclave was heated up to 125°C for 60 min and then cooled to room temperature naturally. The as-formed products were filtered, washed with distilled water and ethanol and dried at 70°C for 3 h.

The samples were then analyzed using X-ray diffraction (XRD) pattern recorded using a Philips PW1820 diffractometer with Cu Kα radiation at a scanning rate of 2° min^{-1} in the range of 20°–80°. The morphology of the samples was scanned using a Quanta FEG 450 field emission-scanning electron microscope (FESEM) with an energy dispersive X-ray spectroscopy (EDX) analysis. UV–vis diffuse reflectance spectroscopy (DRS) spectrum was recorded on a Perkin Elmer Lambda 35 UV-vis spectrophotometer with $BaSO_4$ as a reference. The infrared spectrum was tested using a Thermo scientific Nicolet IS10 Fourier Transform Infrared (FTIR) spectrometer at scanning range from 500 to 4000 cm^{-1}. Photoluminescence (PL) spectroscopy was conducted at room temperature using a Perkin Elmer Lambda S55 spectrofluorometer using a Xe lamp with an excitation wavelength of 315 nm.

In a typical experiment, the photocatalytic experiments were performed under a low-power (105 W) compact fluorescent lamp. The average light intensity striking the surface of the reaction solution was about 6,500 lux, as determined by a digital luxmeter. This lamp was positioned about 12 cm above the reaction solution. The catalyst (100 mg) was added in 100 mL malachite green (MG) solution (10 mg/L) in a beaker. The mixture was subjected to magnetic stirring for 1 h in the dark prior to the exposure of light irradiation. At specific time intervals, 5 mL of the sample was removed from the system, centrifuged and the MG concentration at different time intervals was evaluated by an UV–visible spectrophotometer (HACH DR6000). For mineralization test, 2 mL of filtered supernatant was added into COD vial and placed inside the Hach DRB 200 COD reactor at 150°C for 2 h. The COD vial was then measured using an UV-Vis spectrophotometer to determine mineralization efficiency of MG dyes. The photodegradation efficiencies of other organic pollutants including phenol, BPA and MP in the presence of prepared samples were monitored based on our previous report [12].

IOP Conf. Series: Earth and Environmental Science **151** (2018) 012021 doi:10.1088/1755-1315/151/1/012021

3. Results and Discussion

Figure 1a displays the XRD pattern of the as-prepared BFO samples. All of the diffraction peaks were indexed to rhombohedral perovskite structure of BFO (JCPDS Card file No. 20–0169) [4,5]. The strong and sharp XRD peaks in the figure indicated that the as-prepared BFO crystals were highly crystalline. Further evidence of formation of BFO samples came from the EDX analysis. The presence of Bi, Fe and O peaks can be easily found in Figure 1b. The UV–vis DRS spectrum of the BFO product (Figure 1c) depicted a steep absorption edge which lay between 500 and 580 nm. The observed region was a characteristic region for the absorption band of the rhombohedral structure of BFO [10,11]. The band gap energy of the BFO sample can be acquired from the plot of $(F(R)hv)0.5$ versus photon energy (E). The band gap energy was measured to be 2.5 eV (Figure 1d), which was concordance with the reported BFO catalyst [10,11].

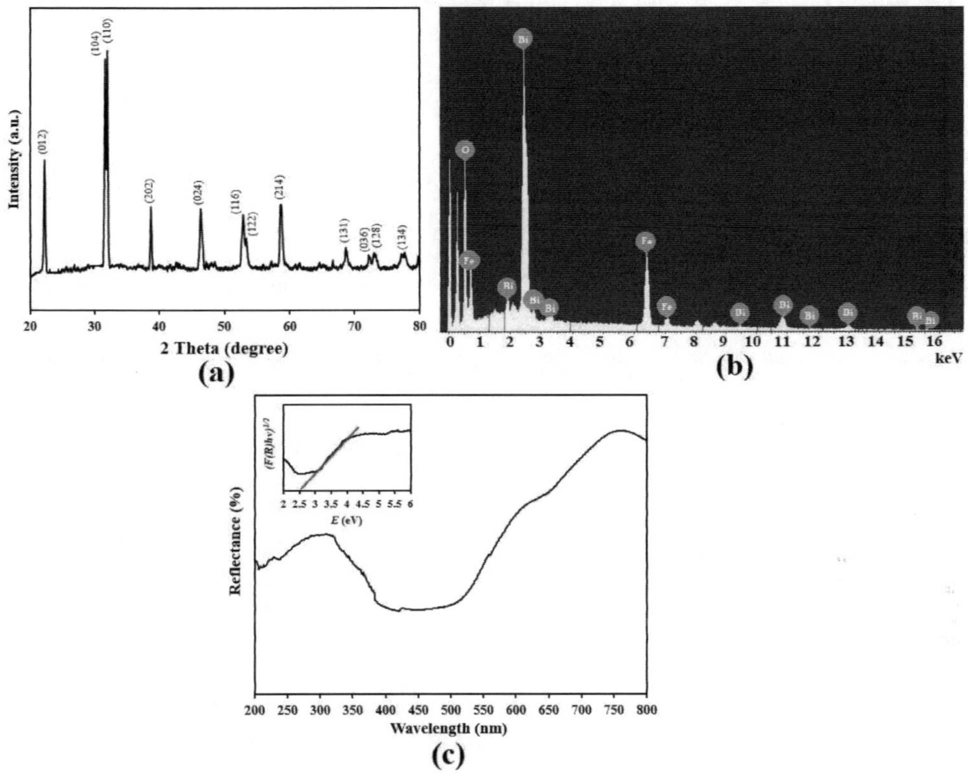

Figure 1. (a) XRD spectrum, (b) EDX spectrum, (c) UV-vis DRS spectrum and (d) plot of $(F(R)hv)^{0.5}$ vs E (hv) of spindly BFO nanoparticles.

FESEM analysis of as-prepared BFO samples was carried out under different magnifications. Figure 2a shows the spindle-like BFO particles were synthesized in a large scale. This image indicated also that the obtained BFO spindle-like nanoparticles presented good dispersion and uniformity. The magnified image in Figure 2b indicated that average length of these spindles were about 120 nm and average width of about 50 nm.

(a) (b)

Figure 2. FESEM images of spindly BFO nanoparticles at different magnifications (a) 20,000x and (b) 100,000x.

Figure 3 presents FTIR spectrum of spindly BFO nanoparticles. Two strong absorption peaks observed around 450 cm^{-1} and 556 cm^{-1} in the FTIR spectrum could be assigned to O–Fe–O bending vibrations and the Fe–O stretching of FeO_6 groups in FeO_6 octahedral unit in perovskite compounds [13,14]. The absorption peak that appeared at 1626 cm^{-1} could be attributed to the asymmetric stretching of carboxylate ligands. Moreover, the broad band around 3448 cm^{-1} could be arisen from the anti-symmetric and asymmetric stretching of absorption H_2O and OH^- groups. This FTIR result confirmed that spindly BFO nanoparticles possessed perovskite structures, which was in agreement with the result of XRD (Figure 1a).

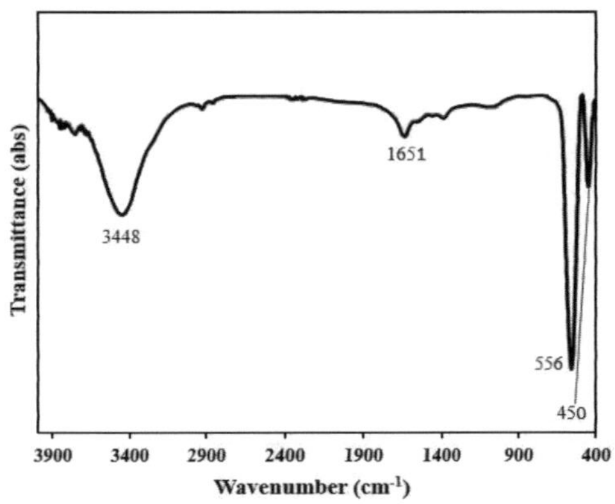

Figure 3. FTIR spectrum of spindly BFO nanoparticles.

The photoactivity of as-prepared BFO samples was examined by degrading MG under visible light irradiation. Figure 4a displays the absorption spectrum of MG solution in the presence of BFO nanospindles at different durations. The absorption peak at 625 nm conforming the MG molecule and decreased steadily as the irradiation time prolonged. Figure 4b shows that the MG solution was degraded about 72.6% by BFO nanospindles under 240 min. Further experiments were conducted to examine the photoactivity of photocatalysts. When the solution with the nanospindles was kept in the dark or when the solution was irradiated without the BFO nanospindles, only small amounts of MG were adsorbed or degraded. The degree of MG degradation of the two experiments attained 2.1% and 18.8%, respectively under identical conditions. It inferred that the as-prepared BFO samples were active photocatalysts.

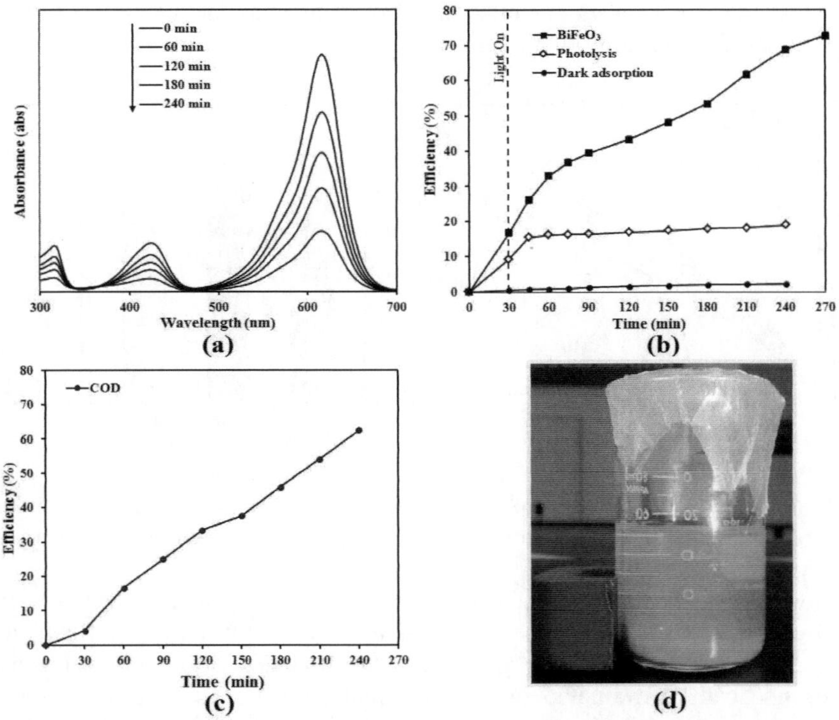

Figure 4. Photoactivity of spindly BFO nanoparticles in the MG degradation. (a) UV-vis absorption spectra of MG solution at different time intervals, (b) MG degradation dependence on irradiation time at different conditions, (c) Changes of COD during the photodegradation of MG and (d) Isolation of the spindly BFO nanoparticles using an external magnetic field.

The COD test (Figure 4c) was conducted to examine the performance of spindly BFO nanoparticles on the MG mineralization under irradiation of a 105 W compact fluorescent lamp. The initial COD value was 24.0 mg/L and then this value gradually weakened upon the irradiation in 240 min. This verified the mineralization extent of MG resulting from an efficient degradation of benzene-ring products. The COD efficiency of MG attained at 62.5% at the point that the solution was degraded, which showed several organic intermediates were still remained in solution. Furthermore, the recovering process of photocatalysts from the reaction medium is another important factor for their practical applications. Figure 4d shows that spindly BFO nanoparticles found to be relatively easier to filter from the mixtures owing to magnetically attractive property.

The influence of the photocatalyst amount on the degradation of MG was investigated at different BFO nanospindles under visible-light irradiation (as recorded in Figure 5a). When the amount of photocatalyst at 1.0 g/L, the degradation efficiency was 72.6%, which was higher than that at 0.25, 0.5 and 2.0 g/L. Higher amount of photocatalyst indicated more availability of total active sites or surface area of the photocatalyst. A decline in degradation efficiency was typically found at photocatalyst overloading because of optical opaque and agglomeration of the excess photocatalysts in the solution. This indicated that an optimal photocatalyst amount was 1.0 g/L.

The photodegradation processes of photocatalyst amount displayed a pseudo-first-order kinetic model against to the irradiation time. The linear relationship is expressed in equation (1) [12]:

$$\ln (C_o/C) = kt \quad (1)$$

where C_o is the initial concentration of a pollutant, C is the concentration of pollutant at time t, and k is the reaction constant of first-order reaction. The results are illustrated in Figure 5b. The results clearly demonstrated the optimal photocatalyst amounts 1.0 g/L with the maximal degradation rate constant of 0.0059 min^{-1} as compared to those of other photocatalyst amounts.

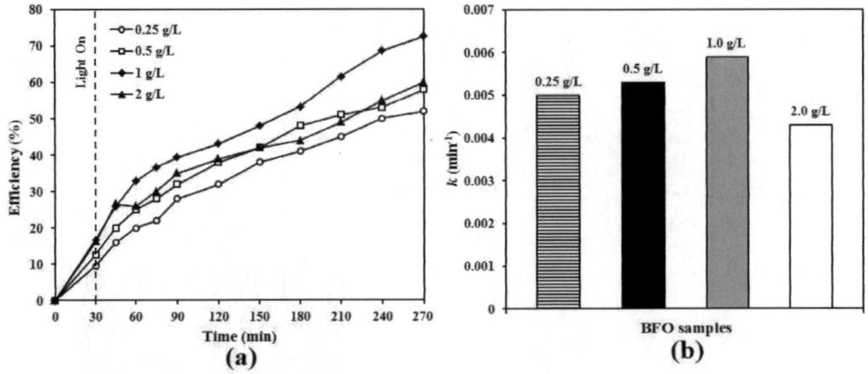

Figure 5. (a) Photocatalytic degradation of 10 mg/L MG at different photocatalyst amounts and (b) kinetic contants for degradation of MG under a compact fluorescent lamp.

Photocatalyst-based catalytic mechanism included a portion of photogenerated charge carriers reacted with adsorbed O_2 and H_2O to form active species such as •OH radicals that further partook in the degradation of organic pollutant [1]. The generated •OH radicals over BFO nanospindles in the photocatalytic reaction can be evaluated by PL technique with terephthalic acid (TA) as a probe molecule. The generated •OH radicals in the photocatalytic reaction could readily react with TA and produce the highly fluorescent compound, 2-hydroxyterephthalic acid (2HTA) (equation (2) [1].

$$\bullet OH + TA \rightarrow HO{-}TA\ (2HTA) \quad (2)$$

Figure 6 displays the PL spectra changes observed during the irradiation of the as-prepared BFO samples in the aqueous basic solution of TA. Generally, PL intensity at about 425 nm was accordance to the amount of generated •OH radicals. It can be clearly seen that the amount of generated •OH radicals on 1.0 g/L BFO nanospindles were larger than that of other samples. With increasing the BFO nanospindles amount, the excess •OH radicals would react each other (equation (3)) [15] rather involve the degradation process of organic pollutants. This result was also consistent with the results of photocatalytic test in Figure 5a.

$$\bullet OH + \bullet OH \rightarrow H_2O_2 \quad (3)$$

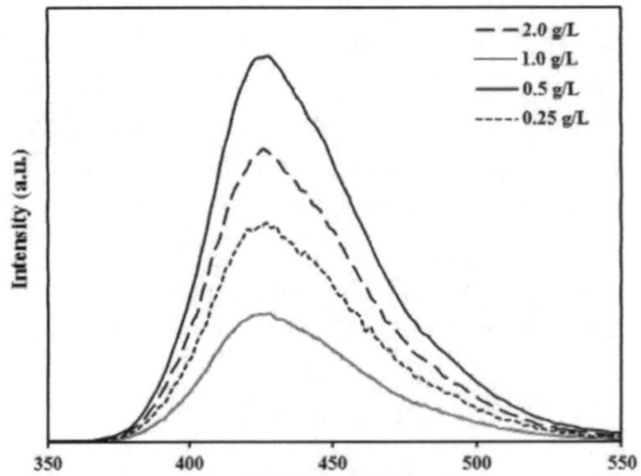

Figure 6. PL spectra of BFO nanospindles at different photocatalyst amounts in basic TA solution for 120 min of irradiation.

In addition to MG degradation, the spindly BFO nanoparticles could be utilized in photodegradation of other organic pollutants as shown in Figure 7. Phenol and bisphenol A (BPA) were effectively decomposed under the same experimental conditions as those in the degradation of MG. The results indicated that the spindly BFO not only can degrade the dye colorants but also can degrade the non-absorbing visible organic compounds. When BFO nanoparticles were added into the phenol and BPA solutions, the degradation efficiencies were found to be 59.1% and 55.4%, respectively. For methylparaben (MP), the BFO nanospindles increased the degradation efficiency from 3.1% (for photolysis alone) to 42.6% in 240 min. Obviously, BFO nanoparticles indicated different photoactivities in degradation of various organic pollutants and however, it could improve the degradations of all pollutants aforementioned owing to their photocatalytic ability.

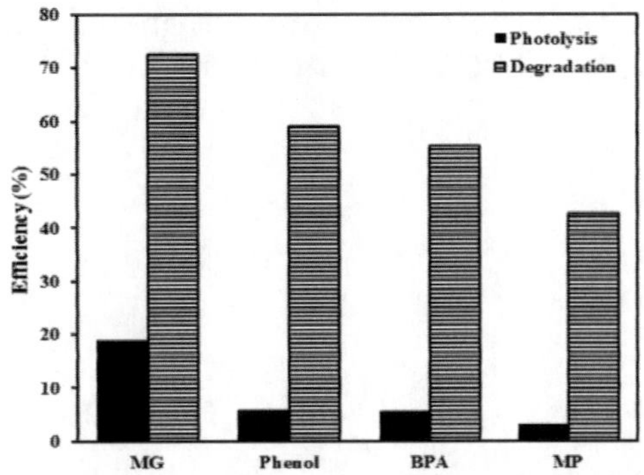

Figure 7. Photodegradation of (a) MG, (b) phenol, (c) BPA and (d) MP under irradiation by a 105 W compact fluorescent lamp (photocatalyt amount: 1.0 g/L; volume of pollutant: 100 mL; concentration of pollutant: 10 mg/L; treatment time: 240 min).

4. Conclusion

Spindly BFO nanoparticles were successfully synthesized via an urea-assisted hydrothermal route. The proposed method was facile and suitable to yield a large production of BFO nanospindles. The structural, morphological and optical properties of the as-prepared BFO nanospindles were confirmed by XRD, UV–vis DRS, FESEM, EDX, TEM and FTIR analyses. The BFO nanospindles have showed good photocatalytic performance for the MG degradation under visible light irradiation. The amount of BFO nanospindles of about 1.0 g/L has resulted in the superior degradation rate of MG solution at which the generated •OH radicals showed their maximum as shown in the PL–TA test. These nanospindles could be utilized in photodegradation of organic pollutants such as MG, phenol, BPA and MP. The light source and spindly BFO nanospindles have no harmful effects and readily available. These provide the photocatalytic system practicality in degrading various organic pollutants.

5. References

[1] Lam S M, Sin J C and Mohamed A R 2017 *Mater. Res. Bull.* 90 15

[2] Ihlefeld J F, Podraza N J, Liu Z K, Rai R C, Xu X, Heeg T, Chen Y B, Li J, Collins R W, Musfeldt J L, Pan X Q, Schubert J, Ramesh R and Schlom D G 2008 *Appl. Phys. Lett.* 92 142908

[3] Basu S R, Martin L W, Chu Y H, Gajek M, Ramesh R, Rai R C, Xu X. and Musfeldt J L 2008 *Appl. Phys. Lett.* 92 091905

[4] Gao T, Chen Z, Niu F, Zhou D T, Huang Q L, Zhu Y X, Qin L S, Sun X G and Huang Y X 2015 *J. Alloy Compd* 648 564

[5] Humayun M, Zada A, Li Z J, Xie M Z, Zhang X L, Qu Y, Raziq F and Jing L Q 2016 *Appl. Catal. B: Environ.* 180 219

[6] Zhang L S, Wang W Z, Chen Z G, Zhou L, Xu H L and Zhu W 2007 *J. Mater. Chem.* 17 2526

[7] Wang W, Huang X, Wu S, Zhou Y, Wang L, Shi H, Liang Y. and Zou B 2013 *Appl. Catal. B: Environ.* 134–135 293

[8] Lam S M, Sin J C, Abdullah A Z and Mohamed A R 2013 *Mater. Lett.* 93 423–426.

[9] Tian Y L, Chang B B, Lu J.L, Fu J, Xi F N and Dong X P 2013 *ACS Appl. Mater. Interfaces* 5 7079
[10] Wei J, Zhang C and Xu Z.,2012 *Mater. Res. Bull.* 47 3513
[11] Li P, Li L, Xu M J, Chen Q and He Y. B 2017 *Appl. Surf. Sci.* 396 879
[12] Lam S M, Sin J C, Abdullah A Z and Mohamed A R 2013 *Ceram. Int.* 39 2434
[13] Yin J, Liao G Z, Zhou J L, Huang C M, Ling Y, Lu P and Li L S 2016 *Sep. Purif. Technol.* 168 134
[14] Ramezanalizadeh H and Manteghi F 2017 *J. Photochem. Photobiol. A: Chem.* 346 89
[15] Tong T, Zhang H, Chen J G, Jin D R and Cheng J R 2016 *Catal. Commun.* 87 23

Acknowledgements
This work was supported by the Universiti Tunku Abdul Rahman (UTARRF/2016–C2/S03 and UTARRF/2017–C1/L02) and Ministry of Higher Education of Malaysia (FRGS/1/2015/TK02/UTAR/02/2 and FRGS/1/2016/TK02/UTAR/02/1).

Photocatalytic degradation of organic pollutants using surfactant-free hydrothermally prepared flower-like BiOBr hierarchical structures under visible light irradiation

J C Sin[1], C A Lim[1] and S M Lam[2]

[1] Department of Petrochemical Engineering, Faculty of Engineering and Green Technology, Universiti Tunku Abdul Rahman, Jalan Universiti, Bandar Barat, 31900 Kampar, Perak, Malaysia.
[2] Department of Environmental Engineering, Faculty of Engineering and Green Technology, Universiti Tunku Abdul Rahman, Jalan Universiti, Bandar Barat, 31900 Kampar, Perak, Malaysia.
Email: sinjc@utar.edu.my

Abstract. Flower-like BiOBr hierarchical structures were successfully prepared via a facile hydrothermal route in the absence of any surfactants. Various characterization tests were carried out to analyze the as-prepared BiOBr samples. By utilizing the flower-like BiOBr as photocatalysts, enhanced photocatalytic performance on the sunset yellow degradation was observed in comparison with the BiOBr nanosheets and commercial TiO_2. The photocatalytic enhancement was ascribed to the unique hierarchical porous surface structure with good crystallinity and optical properties, which could improve the electron-hole pairs separation and led to high yield of reactive oxygen species for photocatalytic reaction. Other organic pollutants such as bisphenol A and methylparaben were also successfully degraded over flower-like BiOBr under similar experimental conditions.

1. Introduction

Bismuth oxybromide (BiOBr), as an important V-VI-VII ternary semiconductor with exceptional physical and chemical properties has recently drawn much attention for their great potential applications [1] [2]. BiOBr has a tetragonal structure consisting of $[Bi_2O_2]$ positively charged slabs and interleaved by double slabs of Br atoms to form [Br-Bi-O-Bi-Br], which enable the formation of internal static electric field. Moreover, the highly dispersive band structure of BiOBr also led to high mobility of the photogenerated charge carriers. BiOBr has exhibited exceptional performance in view of its application in the field of biosensors, dye-sensitized solar cells, pigments and heterogeneous catalysis [1]-[5]. It has been demonstrated that the ultimate performance and application of BiOBr materials are considerably dependent on the morphology and dimensionality in micro- and nano-scale systems [6] [7]. Compounds with the same chemical compositions but different morphologies and dimensionalities can lead to substantial variations in performance [7]. Hitherto, BiOBr crystals with various morphologies have been developed including nanobelts, nanoparticles, nanoplates, flower shaped, hierarchical and complex BiOBr microarchitectures [1],[7]-[10]. In particular, three-dimensional hierarchical structures combined the features of low-dimensional nanoscale building blocks not only showed unique properties different from those of the mono-morphological structures, but also provided large number of surface activation sites to realize region-dependent surface reactivity [11].

The increasing environmental deterioration and the continuing energy crisis have stimulated intense interest on the heterogeneous photocatalytic technologies. As a versatile semiconductor, BiOBr has recently received enormous scientific attention as significant photocatalytic material for its excellent performance for the environmental pollutants degradation under visible light irradiation [1],[12]. For example, Mera et al. [12] reported the exceptional photocatalytic activity of hierarchical BiOBr microspheres on the anion dye degradation under solar irradiation. A study of Li et al. [1] also showed the efficient degradation of rhodamine B and methylene blue over belt-like BiOBr hierarchical nanostructures irradiated under visible light. In this study, 3D BiOBr hierarchical structure with flower-like morphology was synthesized via a simple and surfactant-free hydrothermal route. The as-prepared samples were characterized by different techniques and used for sunset yellow (SSY), bisphenol A (BPA) as well as methylparaben (MP) degradation under visible light irradiation. An improvement of photocatalytic activity was observed when as-prepared samples were used as photocatalysts compared to two-dimensional (2D) BiOBr nanosheets and commercial TiO_2 photocatalysts.

2. Experimental

The flower-like BiOBr hierarchical structures were prepared by a facile hydrothermal method. In a typical experiment, $Bi(NO_3)_3{\bullet}5H_2O$ (4 mmol) was first dissolved in 80 mL acetic acid under magnetic stirring. Then, 40 mL KBr solution (4 mmol) was added with constant stirring. The resulting reaction mixture was sealed in a Teflon-lined stainless steel autoclave and maintained at 120°C for 12 h. After cooling the autoclave naturally, the as-formed precipitates were collected by filtration and repeated washing with deionized water and ethanol for several times. The resulting products were dried in oven at 80°C, ground into fine powder and finally calcined at 400°C for 3 h.

The obtained samples were characterized by field-emission scanning electron microscopy (FESEM, JEOL JSM-6701F), energy dispersion X-ray (EDX, Oxford X-max), X-ray diffraction (XRD, Philips PW1820 diffractometer), fourier transform infrared spectroscopy (FTIR, Spectrum RX 1) and UV-vis diffuse reflectance spectroscopy (UV-vis DRS, Perkin Elmer Lambda 35).

The photocatalytic experiment was as follows: the photocatalyst (200 mg) was suspended in the organics aqueous solution (10 mg/L, 200 mL) and stirred in the dark for 30 min to establish the adsorption-desorption equilibrium. The degradation was carried out at room temperature under irradiation of a 55 W compact fluorescent lamp with average light intensity of 14,500 lx. Air bubbling at a flow rate of 2 mL/min was added during the reaction process. 4 mL of sample suspensions were withdrawn at given time intervals and centrifuged. The SSY concentration was analyzed by UV-vis spectrophotometer at a wavelength of 480 nm while the concentrations variation of BPA and methylparaben were obtained using a high performance liquid chromatography. The HPLC separations were performed with a mobile phase mixture of acetonitrile and water at wavelength of 225 and 254 nm for bisphenol A and methylparaben, respectively. The organic pollutants degradation was expressed as C/C_o versus irradiation time, where C_o is the equilibrium concentration of organic pollutants and C is the concentration at time t. Meanwhile, the comparison studies with commercial TiO_2-P25 and 2D BiOBr nanosheets prepared by the solvothermal method according to Jiang et al. [13] were also conducted.

The photoluminescence (PL) technique using terephthalic acid as a probe molecule was performed to detect the hydroxyl radicals (•OH). Experimental procedure was similar to that of photocatalytic process. 5×10^{-4} M terephthalic acid aqueous solution in 2×10^{-3} M NaOH solution replaced the organics aqueous solution. The PL spectra of generated 2-hydroxyterephthalic acid at 425 nm were measured on a Perkin-Elmer Lambda S55 spectrofluorometer using excitation wavelength of 315 nm. The PL intensity of 2-hydroxyterephthalic acid was proportional to the amount of •OH produced in the degradation system.

3. Results and Discussion

The general morphologies of the as-prepared BiOBr samples were characterized by FESEM which indicated the synthesis of flower-shaped BiOBr. The flower-like BiOBr were synthesized in large quantity with diameters ranging from several hundred nanometers to micrometers (Figure 1a). The microflowers also exhibited hierarchical structure composed of many nanosheets, which were interlaced with each other and have formed porous flowery BiOBr through oriented aggregation

IOP Conf. Series: Earth and Environmental Science **151** (2018) 012022 doi:10.1088/1755-1315/151/1/012022

(Figure 1b). The flower-like assemblies of nanosheets were made of Bi, O and Br elements as confirmed via EDX analysis. A signal of C was also detected, which attributed to the adhesive carbon tape used in the EDX analysis (Figure 1c). No other crystalline phase related to impurity was identified in the XRD spectrum which further verified the formation of pure BiOBr. The XRD diffraction peaks can be corresponded to the tetragonal structure of BiOBr (JCPDS no. 09-0393) (Figure 1d). In addition, the intense and sharp diffraction peaks indicated the high crystallinity of the as-prepared BiOBr samples.

Figure 1. (a) Low and (b) High magnification FESEM images, (c) EDX spectrum and (d) XRD pattern of as-prepared flower-like BiOBr hierarchical structures.

The FTIR spectrum was recorded in the range 4000-400 cm^{-1} (Figure 2a). The main absorption bands at 500-1200 cm^{-1} were ascribed to stretching vibrations of Bi-O bonds. Particularly, the peak at 520 cm^{-1} was assigned to the stretching vibration bonds in crystal tetragonal BiOBr [14]. A strong absorption at wavelengths lower than 430 nm was observed in UV-vis DRS spectrum, implying that the as-prepared samples can be promising in solar photocatalysis (Figure 2b). The band gap energy was determined using E_g (eV) = 1240/λ (nm), where E_g is the band gap energy and λ is the wavelength (nm) of absorption onset. The E_g of the samples was measured to be 2.88 eV, which was close to the values reported in literatures [12],[13].

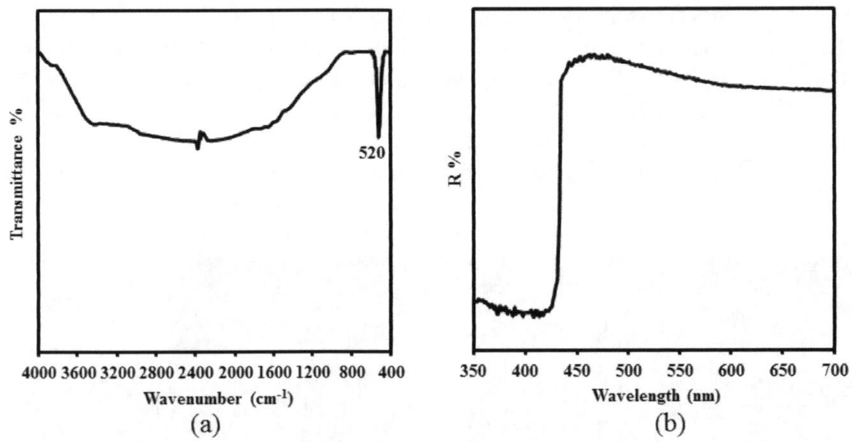

Figure 2. (a) FTIR spectrum and (b) UV-vis DRS spectrum of as-prepared flower-like BiOBr hierarchical structures.

The photocatalytic activities of the as-prepared BiOBr samples were evaluated by the SSY degradation under visible light irradiation (Figure 3). SSY is an important synthetic chemical which extensively used in food, pharmaceutical and cosmetic industries. Several recent cases of SSY contamination to food products have been reported in several countries [15]. Indeed, the complex molecular structure of SSY was difficult to be degraded by microorganisms, thus threating the water ecology. Various in vivo tests also showed that SSY can cause chromosomal aberration in the liver and germinative cells, DNA fragmentation and increase of morphological abnormalities in spermatozoids of mice [15],[16]. Blank experiment of SSY photolysis indicated that the self-degradation of SSY in the absence of catalysts can be neglected. However, when the catalysts were used, the degradation of SSY obviously accelerated with increasing of the irradiation time. It was apparent that the flower-like BiOBr showed higher photocatalytic activity than those of 2D BiOBr nanosheets and commercially available TiO_2 photocatalysts. Within the reaction time, the degradation efficiency of SSY achieved 53.5%, however, for BiOBr nanosheets and TiO_2 the degradation efficiencies were only 46.2% and 41.4%, respectively.

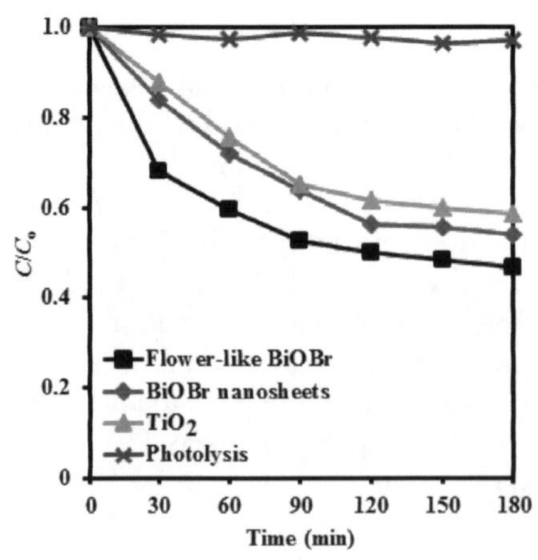

Figure 3. C/C_0 versus time curves of photodegradation of SSY.

The enhanced photocatalytic performance over flower-like BiOBr can be ascribed to the unique hierarchical porous surface structure with good crystallinity and optical properties, which could lead to efficient separation of electron-hole pairs and generation of large number of reactive oxygen species.

Herein, BiOBr first absorbed the light and transferred the electron from valence band to the conduction band, leading to the generation of the separated electron-hole pairs. Then, the reactive oxygen species such as •OH, superoxide oxide anion radical ($O_2•^-$) and hydroperoxyl radical ($HO_2•$) can be generated, which significantly degraded the organic pollutants adsorbed on the catalyst surface. To reveal the •OH involved in the photocatalytic degradation of SSY, PL spectroscopy technique with terephthalic acid as a probing molecule was employed (Figure 4). It was apparent that the •OH formation rate in the presence of flower-like BiOBr was higher than the BiOBr nanosheets and TiO_2, revealing an improved photocatalytic activity of the as-prepared flower-like BiOBr.

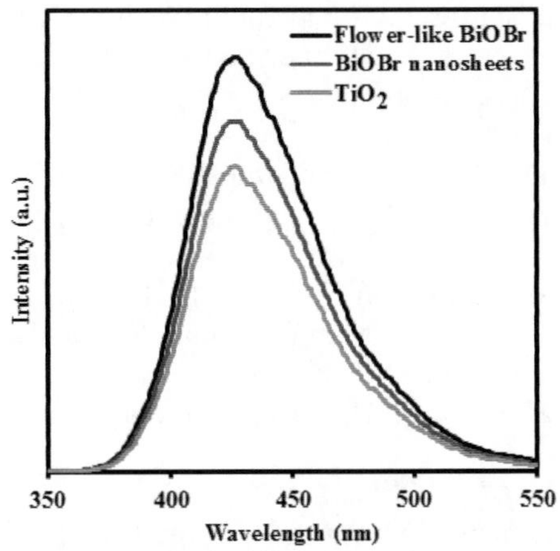

Figure 4. PL spectra of different photocatalysts in an aqueous basic solution of terephthalic acid after 180 min irradiation.

Furthermore, the degradation of colourless BPA and MP were performed to further ascertain the visible light photocatalytic activity of as-prepared BiOBr samples (Figure 5). Like SSY, BPA and MP are toxic and resistant to biodegradation [17]. Although the activities were poor, the flower-like BiOBr hierarchical structures still showed their visible light photocatalytic activity without the photosensitization effect. These results also suggested that the as-prepared BiOBr samples showed different activities in photocatalytic degradation of different organic pollutants, nevertheless, they could enhance the degradation of pollutants mentioned above owing to the catalytic ability.

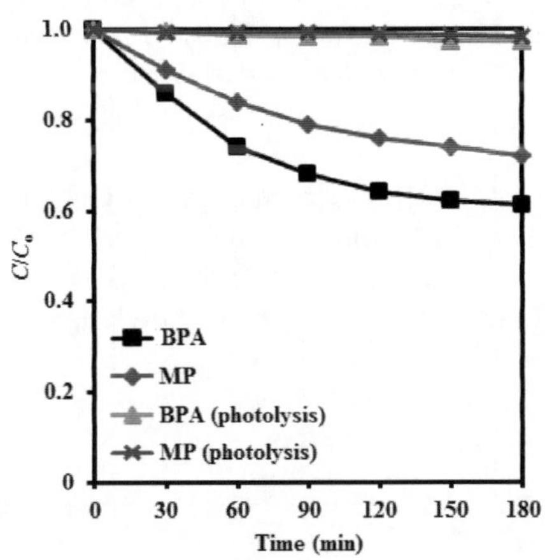

Figure 5. Photocatalytic degradation of several other organic pollutants by flower-like BiOBr.

4. Conclusions

In conclusion, 3D flower-like BiOBr hierarchical structures were successfully prepared in this study. The prepared BiOBr samples were characterized by FESEM, EDX, XRD, FTIR and UV-vis DRS measurements. Under visible light irradiation, the flower-like BiOBr demonstrated enhanced photocatalytic activity in comparison with BiOBr nanosheets and commercial TiO_2. The characterization results and theoretical analysis suggested the flower-like BiOBr as efficient photocatalysts can be rationally ascribed to the unique hierarchical porous surface structure with good crystallinity and optical properties which could enhance the electron-hole pairs separation and generated large number of reactive oxygen species. In addition to the facile preparation method and enhancement photocatalytic performance, preliminary results also demonstrated the prepared flower-like BiOBr were promising catalysts for environmental remediation. Further studies on these catalysts will be helpful to understand the exact growth mechanism of the flower-like BiOBr and to optimize their catalytic performance, which can be an interesting topic to address in future studies.

5. References

[1] Li H P, Liu J Y, Hu T X, Du N, Song S and Hou W G 2016 *Mater. Res. Bull.* 77 171
[2] Fan D W, Wang H Y, Khan M S, Bao C Z, Wang H, Wu D, Wei Q and Du B 2017 *Biosens. Bioelectron.* 97 253
[3] Luz A, Conradt J, Wolff M, Kalt H and Feldmann C 2013 *Solid State Sci.* 19 172
[4] Wu S S, Dong Q, Wang J, Jia Q M, Sun Y L, Shan S Y and Wang Y M 2015 *J. Nanomater.* 2015 489029
[5] Ma H C, Zhao M, Xing H M, Fu Y H, Zhang X F and Dong X L 2015 *J. Mater. Sci. Mater. Electron.* 26 10002
[6] Wei Z D and Wang R 2016 *Chin. Chem. Lett.* 27 769
[7] Li R, Gao X Y, Fan C M, Zhang X C, Wang Y W and Wang Y F 2015 *Appl. Surf. Sci.* 355 1075
[8] Ai Z H, Wang J L and Zhang L Z 2015 *Chin. J. Catal.* 36 2145
[9] Wu X Y, Zhang K K, Zhang G K and Yin S 2017 *Chem. Eng. J.* 325 59
[10] Xue C, Xia J L, Wang T, Zhao S S, Yang G D, Yang B L, Dai Y Z and Yang G 2015 *Mater. Lett.* 133 274
[11] Sin J C, Lam S M, Satoshi I, Lee K T and Mohamed A R 2014 *Appl. Catal. B: Environ.* 148-149 258
[12] Mera A C, Váldes H, Jamett F J and Meléndrez M F 2017 *Solid State Sci.* 65 15
[13] Jiang Z, Yang F, Yang G D, Kong L, Jones M O, Xiao T C and Edwards P P 2010 *J. Photochem. Photobio. A: Chem.* 212 8
[14] Liu Z S, Liu J L, Wang H Y, Cao G and Niu J N 2016 *J. Colloid Interface Sci.* 463 324
[15] Rovina K, Prabakaran P P, Siddiquee S and Shaarani S M 2016 *Trends Anal. Chem.* 85 47
[16] Sayed H M, Fouad D, Ataya F S, Hassan N H A and Fahmy M A 2012 *Mutat. Res.* 744 145
[17] Sin J C, Lam S M, Mohamed A R and Lee K T 2012 *Int. J. Photoenergy* 2012 185159

Acknowledgments
This research was supported by UTARRF schemes (UTARRF/2016-C2/S03 and UTARRF/2017-C1/L02) from Universiti Tunku Abdul Rahman. The support by Ministry of Higher Education of Malaysia through Fundamental Research Grant Schemes (FRGS/1/2015/TK02/UTAR/02/2 and FRGS/1/2016/TK02/UTAR/02/1) was also gratefully acknowledged.

Utilization of Na-bentonite to Improve pH-buffering Capacity of Acid Sulfate Soils in Natural Gas Transmission Pipeline Rights-of-Way, Thailand

N. Chittamart[1*], S. Tawornpruek[1], D. Ketrot[1], S. Aramrak[1], K. Chittanukul[2], R. Sattapun[2]

1 Department of Soil Science, Faculty of Agriculture, Kasetsart University, Bangkok, 10900 Thailand
2 Project Environmental Management Department, PTT Public Company Limited, Bangkok, Thailand
E-mail: fagrnpc@ku.ac.th

Abstract. Na-bentonite, non-hazardous material usually uses as a drilling mud in Horizontal Directional Drilling (HDD) technique. Its alkalinity may reduce severe acidic conditions of acid sulfate soils (ASS) in natural gas transmission pipeline Rights-of-Way in Thailand (RoW). This study focused on the effects of Na-bentonite on pH buffering capacity (pHBC) of ASS. Five major acid sulfate soils in RoW were collected and incubated with incremental amounts of 0.4 M $Ca(OH)_2$ and 0.4 M HCl to determine their pHBC. Hydroxyl equivalent alkalinity derived from pHBC curves was used to define the appropriate rate of Na-bentonite. The various soil:Na-bentonite ratios of 1:0.1, 1:0.15, 1:0.2, 1:0.3, 1:0.4, 1:0.5, 1:1 were prepared to examine their pH change. The results showed that active acid sulfate soils especially in subsoil had very high pHBC and Na-bentonite can increase soil pH to the optimal pH of 6.0, implying that Na-bentonite was beneficial for acid sulfate soils.

1. Introduction

PTT Public Company Limited is a national oil and gas company that has played an important role in strengthening national energy security of the Kingdom of Thailand by transporting natural gas via pipeline system to customers, industrial and commercial sectors since 1981. PTT has been constructing and operating natural gas pipeline system network for more than 4,500 kilometers, including both onshore and offshore areas. Recently, there is the new project, named the 5th Transmission Natural Gas Pipeline Project (5TP) with approximately 415 kilometers length from eastern to central part of Thailand laid through various kinds of soils and land uses.

Acid sulfate soil is one of major soil types in the Rights-of-Way (RoW) of 5th Transmission Natural Gas Pipeline Project. Most of them occurred in the lower central plain of Thailand [1]-[3]. Generally, acid sulfate soils can be divided into i) potential acid sulfate soils which contain sulfide minerals near the soil surface, ii) active acid sulfate soils, where sulfuric acid is produced by oxidation of sulfide minerals and their pH become ultralow (pH < 4.0), and iii) post active acid sulfate soils, where sulfuric acid is neutralized and soil pH has risen above 4.0 by liming or drainage [3]. In addition, wetland rice is a major crop grown on these soils [4]. So, the disturbance of these soils by excavation causes exposure of severe acid and aluminium toxicity [5] to surface soils.

In general, the construction of natural gas transmission pipeline in some area needs Horizontal Directional Drilling (HDD) technique and Na-bentonite, an alkaline material, is usually used as a

IOP Conf. Series: Earth and Environmental Science **151** (2018) 012023 doi:10.1088/1755-1315/151/1/012023

drilling mud. This alkalinity may be useful for reducing severe acidic conditions of major agricultural acid sulfate soils in natural gas transmission pipeline RoW. However, there are only few scientific reports demonstrating the utilization and impact of Na-bentonite on agricultural soils [6]-[7], and there are no any scientific reports on utilizing Na-bentonite to improve acid sulfate soils in Thailand. Therefore, the objective of this study mainly focused on the effects of Na-bentonite on pH buffering capacity (pHBC) of acid sulfate soils in natural gas transmission pipeline RoW of Thailand. Our research is therefore novel to investigate an effective estimation of the appropriate rate of Na-bentonite and management of acid sulfate soils on the pipeline RoW.

2. Materials and Methods

2.1. Acid Sulfate Soil Samples and Na-bentonite

Selected topsoil and subsoil samples from three active acid sulfate soils namely Sena series (Se), Ongkharak series (Ok) and Rangsit series (Rs), and 2 potential acid soils namely Chachoengsao series (Cc) and Bang Nam Priao series (Bp) were collected in the natural gas transmission pipeline Rights-of-Way (Figure 1). Some physicochemical properties of the studied soils were analyzed according to the standard methods of National Soil Survey Center [8] and are given in Table 1. Commercial grade Na-bentonite was used in this study. Basic properties of Na-bentonite including pH, calcium carbonate equivalent (CCE) and alkalinity were analyzed and given in Table 2. Na-bentonite used in this study had a pH of 9.26 and very low CCE, indicating that Na-bentonite was not a carbonate liming material, but its alkalinity mostly derived from hydroxyl ion [OH^-] from its structure. Commercially, bentonite is a rock term used to designate a naturally occurring, very fine grained, light weighted, and opacity material, largely composed of clay minerals, i.e. montmorillonite with a small proportion of feldspar, calcite, silica and gypsum. Chemically, montmorillonite is a hydrous aluminosilicate containing small amounts of alkali and alkaline earth metals. Montmorillonite has very high cation exchange capacity (CEC) and water holding capacity with large expansion [9]-[10].

Table 1. Physicochemical properties of acid sulfate soils

Soil series	Soil horizon	Depth (cm)	pH 1:1 Soil:H_2O	EC 1:5 dS m^{-1}	Clay content (-----------%-----------)	Organic carbon	CEC ($cmol_c$ kg^{-1})
active acid sulfate soils							
Se	Ap	15	4.12	0.6	58.8	1.8	33.2
	Cg	100	2.76	2.49	32.8	5.0	36.5
Rs	Apg	7.5	4.99	0.21	48.0	2.5	29.8
	Cg	85	2.58	2.85	47.2	4.4	34.9
Ok	Apg	18	4.41	0.54	50.4	3.1	29.7
	Cg	85	2.68	3.29	33.2	4.7	36.1
Potential acid sulfate soils							
Bp	Apg	8	5.10	0.35	59.2	2.0	31.6
	Cg	75	6.75	0.56	63.6	0.5	29.6
Cc	Apg	15	6.79	0.14	47.2	2.4	28.8
	Cg	100	5.06	0.48	58.8	1.1	31.5

Remarks: pH 1:1 by pH meter in soil:water of 1:1, EC 1:5 by electrical conductivity meter in soil:water of 1:5, clay content by pipette method, organic carbon by Walkley and Black titration method, Cation exchange capacity (CEC) by 1 M NH_4OAc pH 7.0 method.

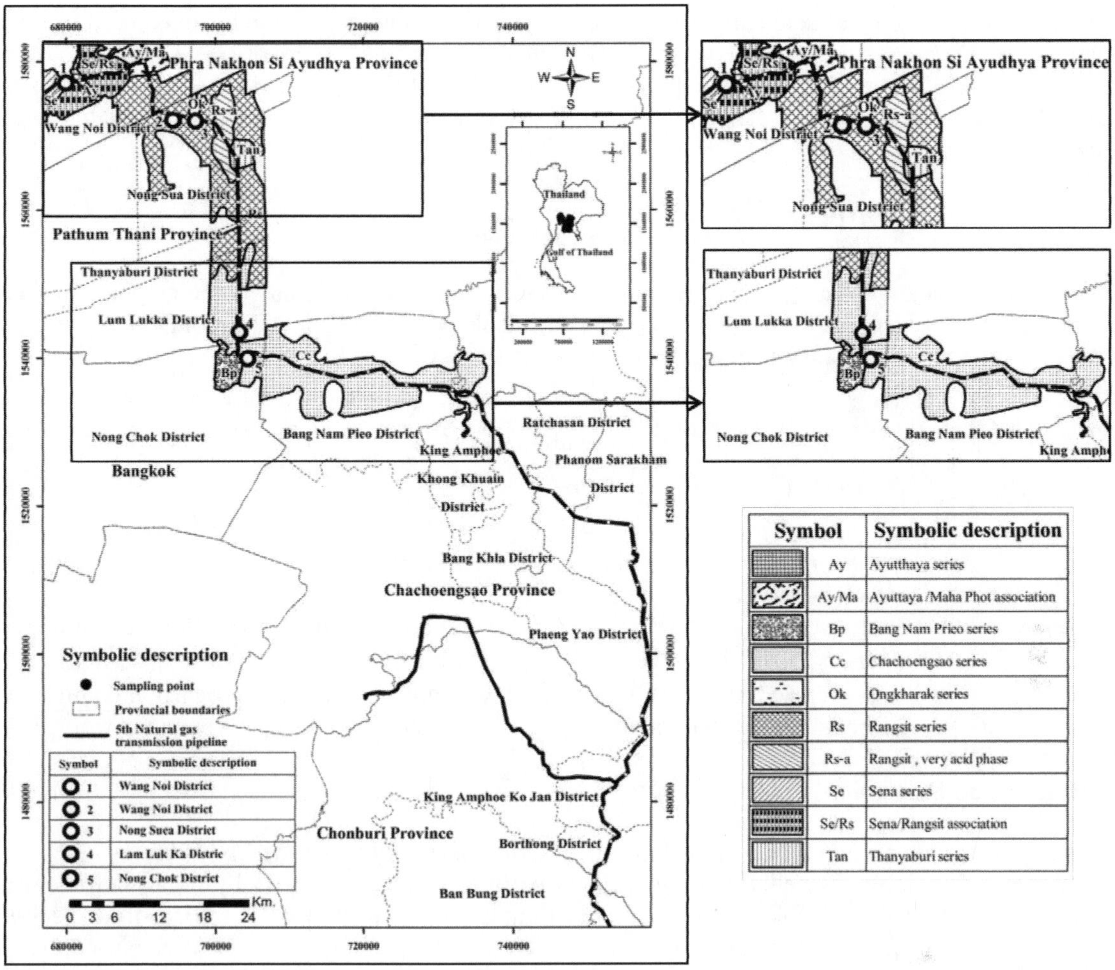

Figure 1. Soil sampling sites in the 5th transmission natural gas pipeline Rights of Way.

Table 2. Some selected properties of Na-bentonite

Properties	Analytical values
pH	9.26
CCE (%)	22
Alkalinity (mmol OH kg^{-1})	962

Remarks: pH by pH meter, CCE (calcium carbonate equivalent) and alkalinity determined by HCl dissolution and black titration.

2.2. pH Buffering Capacity Measurement

Soil pH buffering capacity (pHBC) of soil samples in Table 1 were determined by the modified method of Aitken and Moody [11] by adjusting concentration of alkaline ($Ca(OH)_2$) and acid (HCl) from 0.04 M to 0.4 M. The adjustment was made to meet target pH 6.0. Four grams of soil samples were incubated with the incremental amounts of 0.4 M $Ca(OH)_2$ and 0.4 M HCl of 0, 0.1, 0.25, 0.5, 1, 2.0, 4.0, 8.0 ml or being equally to 0 , 20, 50, 100, 200, 400, 800, 1600 mmol OH^- kg^{-1} soil, and 0, 10, 25, 50, 100, 200, 400, 800 mmol H^+ kg^{-1} soil, respectively. The incubation was conducted for 6 days prior to measure their pH.

Soil pHBC was calculated by fitting with sigmoid function according to the method described by Nelson and Su [12] and Luo *et al.*[13] as the following equation

$$pH = pH_{min} + \frac{a}{1+e^{\frac{-(A-A_{mid})}{b}}} \qquad \text{(Eq. 1)}$$

Where A is the amounts of acid added (inserted as a negative value) and alkaline added (and inserted as a positive value) A_{mid}, a, b and pH_{min} are fitted constants, was fitted to the pH measurement data by iteration using Solver in Microsoft Excel 2013. To calculate pHBC for the studied soils fitted with the sigmoid function, Eq. 1 was rearranged to obtain A as a function of pH, and then differentiated to the following Eq.2.

$$pHBC = \frac{dA}{dpH} = \frac{ab}{(a+pH_{min}-pH)(pH-pH_{min})} \qquad \text{(Eq. 2)}$$

The fitting parameters from sigmoid function (Eq.1) were then used to calculate the amounts of OH^{-1} (mmol kg^{-1}) to raise soil pH to pH 6.0 by Eq. 3. The calculated amounts of OH^- from Eq.3 was used to further calculate as the bentonite equivalent.

$$A = A_{mid} - b \times Log\left(\frac{a}{\text{target pH}-pH_{min}} - 1\right) \qquad \text{(Eq. 3)}$$

2.3. Determination of the Effects of Na-bentonite on pH Change of Acid Sulfate Soils in Natural Gas Transmission Pipeline Rights-of-Way

The selected soil samples were incubated with dried Na-bentonite at the rates of 0, 100, 200, 300, 400, 500, 600, 700, 800, 900 g kg^{-1} soil for 6 days at 25 C° then pH of each soil mixture was determined. The linear regression model was used to fit between logarithmic pH (Log[pH]) and bentonite rates (Log[bentonite]). The bentonite rates derived from the regression model were compared with those derived from the pHBC model.

3. Results and Discussion

3.1. Soil pHBC of Acid Sulfate Soils in Natural Gas Transmission Pipeline Rights-of-Way

The sigmoid function was well fitted with pH curve in this study indicated by high coefficient of determination (R^2) values (Table 3). The sigmoid fitting parameters and pHBC at pH_{mid} of acid sulfate soils in Table 3 revealed that all parameters are in diverse values among studied soils. However, the active soil series (Se, Rs and Ok soil series) had high pHBC values ranged from 51-159.3 mmol kg^{-1} pH $unit^{-1}$ and higher than potential acid sulfate soils (Bp and Cc series) (40.4-78.5 mmol kg^{-1} pH $unit^{-1}$) indicated that these soils needs more alkaline or acid to change one pH unit at pH_{mid} or infection point of slope in the sigmoid curve. High pHBC values of these soils are affected by acidity from oxidation of pyrite indicating by very low pH value especially in subsoil samples (Table 1). On the other hand, the pHBC of soils is depended on mineral phase, organic matter content and the concentration of Al^{3+} in soil solution, taking into account the increase in pHBC at low pH [14][15]. In addition, the studied acid sulfate soils have similar high clay content, high cation exchange capacity but quite different in organic content and pH values. So, the variations in organic carbon and soil pH made these soils different in their pHBC. This is due to organic carbon and Al^{3+} directly related to effective cation exchange capacity which was closely correlated with pHBC [12] [15].

Moreover, the pHBC of studied acid sulfate soils varied considerably with pH (Figure 2) and markedly increased at pH = 3.0 and pH = 10.0. These characteristics are due to their nature of permanent charges of soil clay which markedly differ from the study of Nelson and Su [12] who reported the lower pHBC of variable charge acid soils from Australia and Papua New Guinea than the acid sulfate soils in our study. However, there are considerably different in the magnitudes of pHBC between the studied acid sulfate soils (Figure 2), that is due to very low pH and higher organic contents especially in the subsoil horizons of Se, Rs and Ok series.

Table 3. Sigmoid fitting parameters and pHBC of acid sulphate soils

Soil series-genetic horizon	a	b	A_{mid}	pH_{min}	Adj. R^2	pH_{mid}	pHBC at pH_{mid} (mmol kg^{-1} pH $unit^{-1}$)
Se-Apg	9.97	127.1	79.4	1.57	0.999	6.56	51.0
Se-Cg	10.53	350.3	565.7	1.00	0.991	6.27	133.0
Rs-Apg	10.53	167.6	73.4	1.25	0.933	6.51	63.7
Rs-Cg	11.04	323.9	617.7	1.26	0.874	6.78	117.4
Ok-Apg	10.37	151.1	142.3	1.47	0.988	6.66	58.3
Ok-Cg	10.24	407.5	631.3	0.98	0.986	6.10	159.3
Bp-Apg	10.56	207.3	122.8	1.08	0.870	6.36	78.5
Bp-Cg	9.98	100.8	7.9	1.30	0.977	6.29	40.4
Cc-Apg	10.54	130.9	16.3	1.39	0.955	6.66	49.7
Cc-Cg	9.39	145.6	59.5	1.44	1.000	6.14	62.0

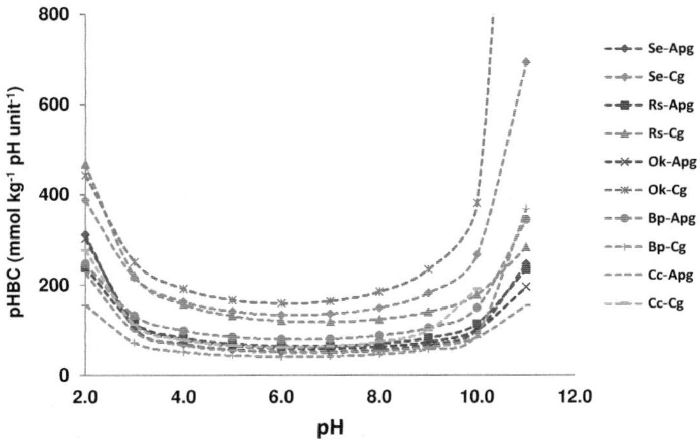

Figure 2. The relationship between pH buffer capacity (pHBC) and pH for acid sulfate soils.

3.2. Effects of Na-bentonite on pH Change of Acid Sulfate Soils in Natural Gas Transmission Pipeline Rights-of-Way

The well fitted linear regression (R^2 range of 0.90-0.99) between the logarithmic of pH and bentonite added rate for acid sulphate soils (Figure 3 and Table 4) indicated that bentonite significantly increased pH of acid sulfate soils especially in the subsoil layers of Se, Rs and Ok series which have higher slope than the topsoils, and the slopes of potential acid sulfate soils (Bp and Cc series). These results demonstrated that alkalinity from Na-bentonite (approximately 962 mmol OH^{-1} kg^{-1}) had a capacity to raise pH up to 6.0 which is in normal rage. This pH value is recommended for liming requirement that restricts to dissolution of aluminium which is toxic to plant root. The predicted bentonite rate is strikingly different among soil samples. The higher application rate was in the subsoils of Se, Rs and Ok series due to these soils had very high pHBC values.

The bentonite rates derived from pHBC (Eq. 3) and those from the incubation method were both able to increase pH to 6.0 of each acid sulfate soil sample in the similar order of magnitude (Table 5). Our results imply that the parameters derived from sigmoid equation can be used to predict the optimal bentonite rate for reducing severe acidic pH, allowing to apply into each acid sulfate soil series in the natural gas transmission pipeline Rights-of-Way.

Table 4. Linear model for prediction of Na-bentonite to adjust pH up to 6.0

Soil series	linear regression model	R^2	Log (bentonite) at pH 6.0	Bentonite rate at pH 6.0 ($g\ kg^{-1}$)
Se-Ap	Log pH= 0.0871Log(bentonite)+ 0.6435	0.9871	1.55	35.15
Se-Cg	Log pH= 0.2674Log(bentonite)+ 0.0004	0.9717	2.91	810.16
Rs-Apg	Log pH= 0.0187Log(bentonite)+ 0.8093	0.9072	-1.67	0.02
Rs-Cg	Log pH= 0.2558Log(bentonite)+ 0.0591	0.9643	2.81	647.13
Ok-Apg	Log pH= 0.0737Log(bentonite)+ 0.6414	0.9813	1.86	71.70
OK-Cg	Log pH= 0.251Log(bentonite)+ 0.041	0.9829	2.94	864.68
Bp-Apg	Log pH= 0.0719Log(bentonite)+ 0.736	0.9933	0.59	3.86
Bp-Cg	Log pH= 0.0595Log(bentonite)+ 0.7285	0.9863	0.83	6.83
Cc-Apg	Log pH= 0.0597Log(bentonite)+ 0.7701	0.9588	0.13	1.36
Cc-Cg	Log pH= 0.0749Log(bentonite)+ 0.6667	0.9934	1.49	30.76

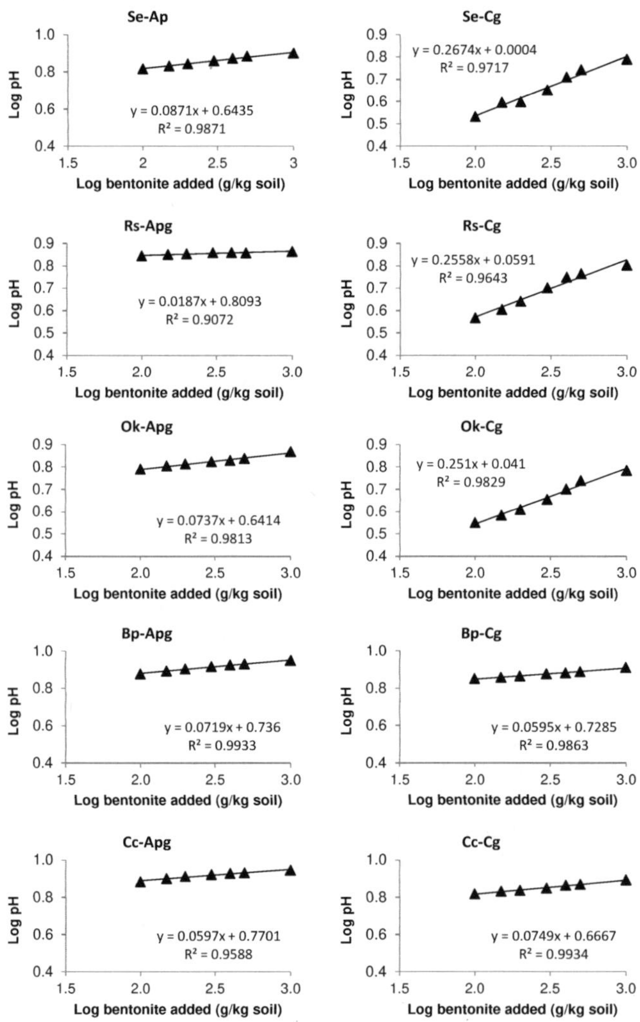

Figure 3. Linear regression of logarithmic pH [Log pH] and bentonite rates [Log bentonite].

Table 5. Comparison on Na-bentonite rate derived from incubation and pHBC prediction

Soil series	Bentonite rates (g kg^{-1}) needed for increasing soil pH to 6.0	
	incubation	pHBC
Se-Ap	36	70
Se-Cg	811	572
Rs-Apg	0.02	62
Rs-Cg	648	601
Ok-Apg	72	130
OK-Cg	865	649
Bp-Apg	4	115
Bp-Cg	7	3
Cc-Apg	2	2
Cc-Cg	31	58

4. Conclusions

The active acid sulfate soils with sulfuric materials (Rs, Ok, Se series) in the natural gas transmission pipeline RoW of Thailand have very high pHBC that require more OH^-/H^+ to change their pH, and relatively higher than the potential acid sulfate soils i.e. Bp and Cc soil series. The sigmoid function can be used to precisely predict soil pHBC of acid sulfate soils. Alkalinity derived from Na-bentonite could improve pHBC and raise soil pH up to the desired pH range, and also mitigate severe acidic conditions of these soils. Our findings imply that drilling mud Na-bentonite was beneficial to acid sulfate soils especially in the active acid sulfate soils at the natural gas transmission pipeline RoW, Thailand.

5. References

[1] Sinsakul S 2000 *J. Asian Earth Sci.* 18 415–426
[2] Janjirawuttikul N, Umitsu M and Vijarnsorn P 2010 *Res. J. Environ. Sci.* 4 336-358
[3] Sukitprapanon T, Suddhiprakarn A, Kheoruenromne I, Anusontpornperm S and Gilkes R J 2016 *Geoderma Regional* 7(3) 346–356
[4] Attanandana T and Vacharotayan S 1986 *Southeast Asian Stud.* 24 154-180
[5] van Breemen N 1973 *Soil Sci Soc Am J.* 37(5) 694–697
[6] Hongprayoon C, Osotsapa Y, Srichan P, Kamlang A, Wongmaneeroj A, Poompetch J, Piatanen P and Ratanachoo K 2006 *Final Report of Effect of Sodium Bentonite on Environment and Bentonite Drilling Mud Management (Unpublised work in Thai)* (Nakhon Pathom: Department of Soil Science, Kasetsart University Kamphaeng Saen Campus) p 104
[7] Thongpae S, Tawornpruek S, Suphapol S, Kritapirom S 2006 *Final Report of Quality and Soil Potential on Plantation after the Construction of the Natural Gas Pipeline (Unpublised work in Thai)* (Bangkok: Department of Soil Science, Kasetsart University) p 252
[8] National Soil Survey Center 1996 *Soil survey laboratory methods manual. Soil Survey Investigations Report No. 42, Version 3.0.* (Washington D.C: Natural Resources Conservation Service, USDA)
[9] Environmental Engineering Association of Thailand 2016 *Encyclopaedia of Bentonite for Engineering and Environment (in Thai)* (Bangkok: Environmental Engineering Association of Thailand) p 38
[10] Clem A G, Doehler R W 1961 *Proc. of the Tenth National Conference on Clays and Clay Minerals* (Austin, October 14-18) ed Swineford A (Texas: Pergamon Press) p 509
[11] Aitken R L and Moody P W 1994 *Aust. J. Soil Res.* 32(5) 975–984
[12] Nelson P N and Su N 2010 *Aust. J. Soil Res.* 48 (3) 201–207.
[13] Luo W T, Nelson P N, Li M H, Cai J P, Zhang Y Y, Zhang Y G, Yang S, Wang R Z, Wang Z W, Wu Y N, Han X G and Jiang Y 2015 *Biogeosciences* 12 7047–7056

[14] Verburg K, Braschkat J, Hochman Z, Moore A D, Helyar K R, Probert M E, Hargreaves J N G, Simpson R J 2003 *Handbook of soil acidity* ed Z Rengel (Marcel Dekker: New York) pp. 135–187

[15] Helyar K R and Porter W M 1989 *Acidity and Plant Growth* ed Robson A D (Academic Press: Sydney) pp 61–101

Study on Particle-Size Control of Hydrocyclone for Slurry Recycles

Zhou Cuihong*[1], Wang Shihan[1,2], Yang Changshun[3], Xu Ling[4]
1 College of Mechanical Engineering, Beijing Institute of Petrochemical Technology, Beijing 102617, PR China
2 College of Mechanical and Electrical Engineering, Beijing University of Chemical Technology, Beijing 100029, PR China
3 Beijing Municipal Road & Bridge Group Co., Ltd., Beijing 100045, PR China
4 Beijing MTR Construction Administration Co., Ltd., Beijing 100068, PR China
* Corresponding author: E-mail: zhoucuihong@bipt.edu.cn

Abstract: Particle size control is very important in the shield slurry separation. The structural parameters and separation efficiency of desilter separation efficiency were studied using numerical simulations. In this study, slurry treatment process was designed firstly, and then the influence including the diameter of the overflow port, the size of the cone, the length of cylinder and the diameter of the bottom flow on the separation efficiency were studied. In the hydrocyclone, the minimum tangential velocity occurs at the center of the cyclone, which makes it possible to separate materials more efficiently. The pressure distribution is consistent with the pressure characteristics of combined vortex motion. After obtaining the optimal design, the separation ability of the hydrocyclone was greatly improved. The 75 μm and 100 μm particles recycle efficiency increase 8% and 11% respectively. The results of this study provide a numerical basis for the optimization of hydrocyclone system.

1. Introduction

Slurry shield machines are used in subways, municipal engineering, and other underground tunnels in the construction of a wide range of applications. Shield slurry treatment equipment reveals that hydrocyclones are the most widely used desanding equipment [1]–[3]. The basic structure of a hydrocyclone, which is used for the separation of solid–liquid mixtures, includes an overflow port, underflow port, and feed port. Its basic principle is density–drag mechanical separation: the mixture is injected into the flow cavity where it forms a vortex on the interior surface; as the flow travels downwards, the chamber constricts, and an upward-travelling inner vortex forms [4]. The centrifugal inertia of particles drives them to the outer vortex, which exits through the underflow port. However, the centrifugal inertia will fail to keep the particles in the outer vortex if the drag experienced by the particles is sufficiently large; in this case, they will enter the inner vortex and exit through the overflow port [5],[6]. The main function of the cyclone is to remove mud particles with a particle size greater than 75 μm in the mud and retain particles with a particle size less than 75 μm.

Both theoretical and numerical methods have been adopted to study the theory of cyclones; however, owing to the complex flow field within a cyclone, theoretical research alone is not sufficient. With the common use of computational fluid dynamics (CFD), current theoretical research on hydrocyclones is mainly performed using numerical simulations. Boysan et al. [7] used the standard k-ε turbulence model to numerically simulate the internal flow field of a hydrocyclone; however, because k-ε is not suitable for the simulation of strongly swirling flow, the conclusion was that a strong vortex will produce additional viscosity. Cullivan et al. [8], Schuetz et al. [9], and Wang et al. [10] used Reynolds stress models in subsequent studies to simulate the flow field of a hydrocyclone with high accuracy. Delgadillo and Rajamani [11], Wang et al. [12], and Vakamalla and Mangadoddy [13] conducted in-depth studies on the separation and flow patterns of hydrocyclones based on the RSM and LES models. Using CFD, Banerjee et al. [14] developed a new complex mechanical dynamics model based on vortex flow in a limited environment for water splitting in hydrocyclones. Some researchers have studied the separation performance of hydrocyclones, taking into account factors such as the size of the cyclone, the diameter of the spigot, the inlet flow rate, curvature radius, and the inlet solid concentration [15]–[18]. Hwang and Dubey et al. studied the vortex finder with extra conical shapes and diameters of hydrocyclones through experiments and numerical simulations [19], [20]. This group used FLUENT to simulate the three-dimensional flow field of centrifuge and the dewatering efficiency of sludge pretreatment.

In China, most shield slurry separation processes are based on past experience or direct use of foreign processing equipment—i.e., there is no domestic targeted research. In this paper, we mainly studied particle size control using hydrocyclones in the shield slurry control process via simulations. By studying the velocity field and the pressure field distribution of the cyclone, the single-factor structural parameters of the hydrocyclone are optimized. And the separation performance under different inlet flow was studied. These provide guidance for ensuring the quality of the treated shield slurry.

2. Methods

2.1. The Process Flow

The larger diameter soil particles are first removed from the slurry by sieving, and the slurry enters the desander hydrocyclone after preliminary precipitation. The overflow from the desander, which contains the smaller solid particles, is piped to the desilter (Fig. 1) for secondary swirling, and a horizontal screw centrifuge concentrates the underflow of the desander and desilter into a solid phase. The water produced during the concentration process re-enters the sedimentation tank. The overflow sludge from the slime treatment, which contains fine particles, enters the regulating pot, thereby realizing the reuse of the slurry. In the final stage of the shield slurry treatment, the pretreated slurry is graded using a two-phase cyclone. The treated underflow stream is used as waste slurry, and the overflow is used as shield slurry for reuse. The shield slurry separation system is shown in Fig. 1.

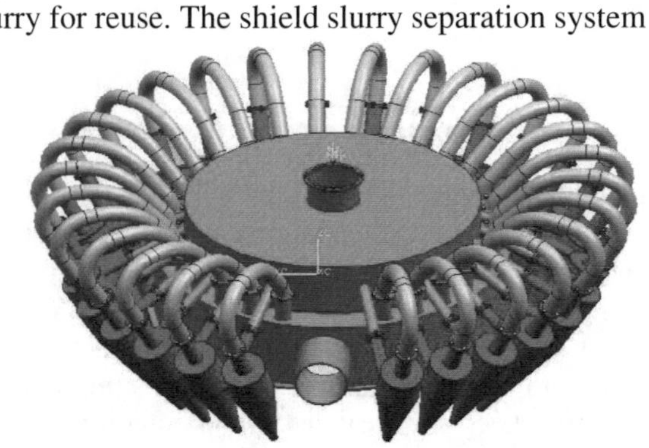

Figure 1. The desilter

The cyclone separation performance at this stage is the key factor to the quality of the slurry in the process; improper handling will cause the slurry to deteriorate. If the hydrocyclone separation efficiency is too low, large soil particles will be re-mixed into the cycle, resulting in a slurry density that is too large. Specifically, soil particles less than 75 μm will be lost from the slurry. In this study, we are particularly concerned with the separation of this grade of the particles because 75-μm particles ensure the operation of the shield.

2.2. Model and Mesh

The overall height of the hydrocyclone is 440 mm. The basic dimensions of the cyclone used for simulation are shown in Table 1. The rectangular inlet is tangent to the cylindrical circle with a length of 80 mm. Hexahedral meshing is applied to the model, and the entire calculation domain is divided into 13 sub-regions via horizontal and vertical cutting. Firstly, the Gambit pave and map commands are used to develop the grid at the interface of the inlet and the cylinder, and then they are used to mesh the entire cyclone cylinder. The meshes are all hexahedral meshes with uniform grid size. In this study, the velocity distribution of an interface was selected, and the distribution values did not change under different density grids. This proves that grid independence is true. After grid-independence verification, the segment was divided into 392166 individual cells.

Table 1. Basic dimensions cyclone

Project		Size
Main diameter of the cyclone (mm)		100
Overflow pipe	Diameter (mm)	30
	Insertion depth(mm)	50
Inlet (mm × mm)		16m×30mm
Cylinder length (mm)		120
Underflow	Diameter (mm)	15
	Length(mm)	20
Angle of cone(°)		20

3. Simulation Results and Analysis

The solid particle size is 75 μm, the mass fraction of the solid phase is 20%, and the inlet speed is 6m/s. The dynamic viscosity of the two-phase flow is 7 MPa·s. FLUENT was used for numerical simulation. The simulation results of the velocity distribution, pressure distribution, and separation performance are presented below.

3.1. Velocity Distribution

Fig. 2 illustrates the two-phase flow through the inlet and into the cyclone inside the hydrocyclone. The fluid inside the hydrocyclone forms two swirling flows in opposite directions of motion from the wall to the center: the fluid at the center flows upwardly and exits through the overflow port, whereas the fluid near the wall of the cyclone flows downwardly and exits through the underflow port. The overflow and underflow continue rotating, where the internal cyclone fluid swirl effect is better, which is consistent with the expected results.

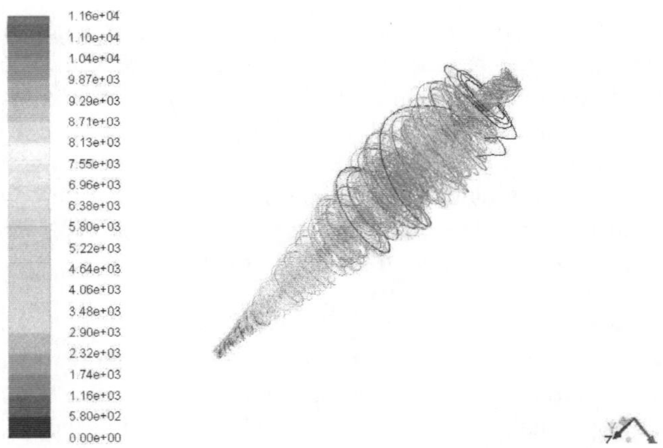

Figure 2. Flow path of the vortex chamber

The tangential velocity distribution curves of the flow within the cylinder were obtained at Fig. 3(a). Because of the no-slip conditions at the wall, the velocities at the walls of the hydrocyclone (radial position of 50 mm) and overflow tube (radial position of 15 mm at Z = 40 mm and Z = 50 mm) are zero. It can be seen that a cyclone column has formed within a stable tangential swirl—the flow state is consistent with combined vortex flow. The flow field in the cylinder is symmetrical, and the minimum tangential velocity occurs at the center of the cyclone, which is conducive to the generation of centrifugal inertia, which makes it possible to separate materials more efficiently. In the Z = 120 mm section, owing to the interface between the cylinder and the cone, the tangential velocity fluctuates. It can be seen from the simulation that the swirl distribution is generally consistent in different sections of the cylinder: the swirl is divided into two swirls—inner and outer—but the rotation direction is consistent. However, the outer swirl is significantly stronger than the internal swirl. The maximum-magnitude tangential velocity appears near the overflow port radius, forming a quasi-forced vortex zone.

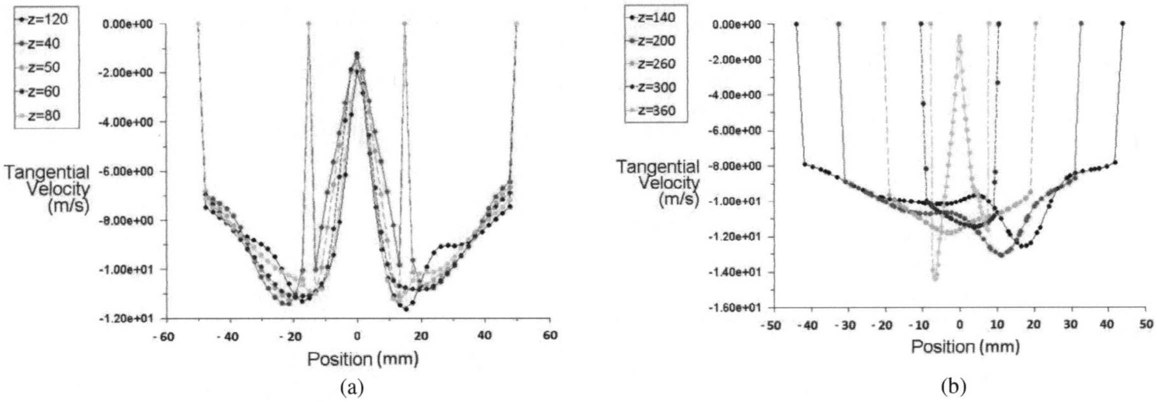

Figure 3. Tangential velocity distribution in the hydrocyclone: (a) cylinder region, (b) cone region.

Next, we consider the flow at five axial positions within the cone, namely, 140 mm, 200 mm, 260 mm, 300 mm, and 360 mm. Fig. 3(b) shows that the fluid within the cone generally consists of the higher swirl state, as the tangential flow velocity at different positions is generally consistent. This shows that the cone maintains the velocity of the swirl and accelerates the overall flow. However, the swirl is asymmetric, and the position of the maximum-magnitude tangential velocity varies in different sections. This position gradually shifts to the middle as cone constricts. This occurs because the inner vortex of the cylinder becomes less influential on the flow field of the cone. Within the Z = 360 mm section, the flow is travelling through the outlet, so the tangential velocity near the central position rapidly vanishes.

3.2. *Pressure Distribution*

Because the liquid enters the hydrocyclone at a fixed speed and develops a tangentially graded pressure distribution, the liquid in the cavity completes the flow state transition. The pressure distribution is consistent with the pressure characteristics of combined vortex motion, as shown in Fig. 4. The static pressure distribution shows that the static pressure in the cylindrical part of the hydrocyclone is minimized along the radial distance at the center of the hydrocyclone. This occurs because the vortices dissipate energy as they accelerate the flow through the overflow and underflow ports. Throughout the hydrocyclone, the static pressure isobars are approximately parallel to the cylinder wall, and the pressure decreases toward the center of the cyclone.

3.3. *Distribution of Solid Phase*

When the inlet velocity is 6m/s, mud particle diameter is 75 μm, slurry phase is 20%, simulated distribution of solid phase. Slurry enters the internal hydrocyclone with high-speed by centrifugal force, high density particles gradually deposited on the outer wall, and then continue to downwards flow out via the underflow. Water mainly gathered in the center to exclude from the overflow port. The two-phase separation is realized in the swirl cylinder, and the separation effect is more obvious in the cone section. It can also be seen that the oscillation of the spiral flow at the center of the cone is obvious in different sections (Fig. 5).

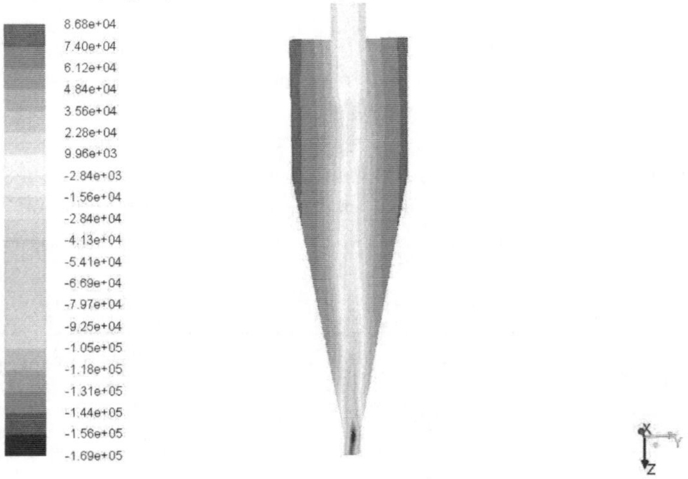

Figure 4. Static pressure distribution (X = 0 mm)

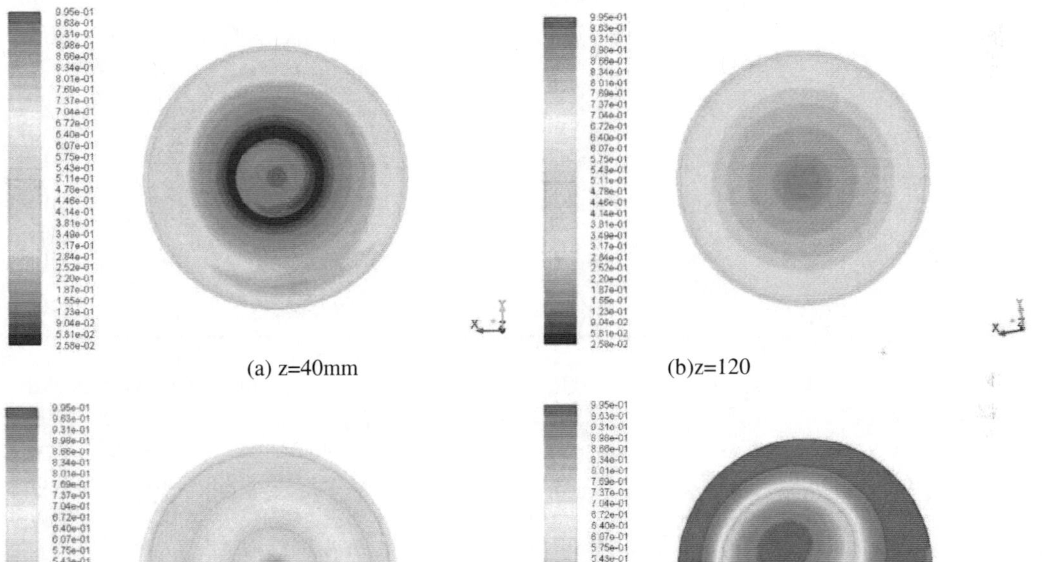

(a) z=40mm (b)z=120

(c) z=200 (d) z=340

Figure 5. Distribution of solid phase in different position of cyclone

3.4. Influence of Structural Parameters on the Separation Performance
Four diameters of soil particles 30 μm, 50 μm, 75 μm and 100 μm were selected to study the separation efficiency.

3.4.1. Overflow pipe diameter. Four hydrocyclones, with overflow pipe diameters of 26 mm, 30 mm, 34 mm, and 38 mm, were numerically simulated to evaluate the separation performance. The distribution of soil particles when the diameter of the overflow pipe is 30 mm. As the particle size of the soil becomes larger, the concentration of particles leaving through the overflow port decreases, thereby increases the separation efficiency, which is in accordance with the demands of the hydrocyclone.

The post-processing functions available in FLUENT provide direct access to the overflow and underflow two-phase and mixed-phase mass flow rates and the cyclone pressure drop. The separation efficiency and pressure drop data are shown in Table 2. For an overflow port diameter of 38 mm, separation efficiency of the larger particles decreases significantly. The results also show that the pressure drop decreases as the overflow pipe diameter increases, so there is less energy dissipation inside the cyclone. When the diameter of the overflow pipe is 34 mm, the differences between the separation efficiencies of particles of different diameters are largest. Thus, the best overflow pipe diameter is 34 mm.

Table 2. Separation efficiency and pressure drop under different overflow pipe diameters.

Overflow pipe diameter [mm]	Solid particle diameter [μm]	Separation efficiency [%]	Pressure drop [Pa]	Overflow pipe diameter [mm]	Solid particle diameter [μm]	Separation efficiency [%]	Pressure drop [Pa]
26	30	32	46653	34	30	23	33365
	50	42	43256		50	37	32486
	75	67	44738		75	64	32663
	100	82	48652		100	81	34102
30	30	27	38566	38	30	21	30306
	50	40	39628		50	36	29558
	75	66	38128		75	64	29036
	100	81	37521		100	75	29604

3.4.2. Cone taper angle. Selecting an overflow pipe diameter of 34 mm, five hydrocyclones, with taper angles of 14°, 16°, 18°, 20°, and 22°, are numerically simulated to evaluate the separation performance. When the cone angle is reduced to 14°, the small particles are excessively removed, and the separation efficiency of 100 μm particles is greatly reduced. When the taper angle is 18°, the separation efficiency of 30-μm particles is 24%, and the separation efficiency of 100 μm particles is 92%. Based on the pressure drop and separation efficiency, the optimal cone angle is 18°.

3.4.3. Underflow pipe diameter. Selecting an overflow pipe diameter of 34 mm and a taper angle of 18°, four underflow pipe diameters of 12 mm, 15 mm, 18 mm, and 21mm are numerically simulated. When the underflow diameter is 21 mm, the separation efficiency for 30 μm particles exceeds 30%, whereas the separation efficiency of 100 μm particles reaches 94%. This wastes soil particles beneficial to the pulp. When the underflow pipe diameter is 18 mm, the split ratio of the underflow is only 25%. The underflow diameter has little effect on the pressure drop of the cyclone, so the under flow diameter remains unchanged at 15 mm.

3.4.4. Cylinder height. Based on the above optimized parameters, the four cylinder lengths of 80 mm, 100 mm, 120 mm, and 140 mm are numerically simulated. The results of the simulations show that the length of the cylinder has little effect on the pressure drop of the cyclone, so the length of the cylinder remains unchanged at 100 mm.

4. Conclusion

In this paper, the influence of the internal flow field and hydrocyclone structure on the separation performance of the core device in a slurry separation system was studied using numerical simulations. The hydrocyclone separation performance experiment, simulation results, and experimental conclusions are summarized below:
(1) The flow field in the cylinder is symmetrical, and the minimum tangential velocity occurs at the center of the cyclone, which is conducive to the generation of centrifugal inertia, which makes it possible to separate materials more efficiently.
(2) The overflow pipe diameter of the hydrocyclone has a strong influence on the small particle separation efficiency in the mixed solution. A larger overflow pipe diameter facilitates recovery of small particles. The underflow diameter and cylinder length of the cyclone have negligible influence on the separation efficiency and pressure drop of the cyclone, much less than that of the overflow pipe and the cone angle.
(3) The inlet flow rate has a strong influence on the separation performance of the cyclone. Initially, increasing the flow rate improves the separation efficiency, but too much flow reduces in the slurry residence time, resulting in decreased separation efficiency. From the simulation analysis, the separation efficiency was highest (72.6%) at an inlet speed of 12 m/s.

5. References

[1] P. Fritz. "Additives for Slurry Shields in Highly Permeable Ground," *Rock Mechanics and Rock Engineering*, Vol.40 (1), pp.81-95, 2007.
[2] Y. Li, F. Emeriault, R. Kastner, et al. "Stability analysis of large slurry shield-driven tunnel in soft clay" *Tunnelling and underground space technology*, Vol. 24(4), pp.472-481, 2009.
[3] S.T. Azali, M. Ghafoori, G.R. Lashkaripour, et al. "Engineering geological investigations of mechanized tunneling in soft ground: A case study, East–West lot of line 7, Tehran Metro, Iran," *Engineering Geology*, Vol.166, pp.170-185, 2013.
[4] F.J. Souza, L. G. M. Vieira, J. J. R. Damasceno, et al. "Analysis of the influence of the filtering medium on the behaviour of the filtering hydrocyclone," *Powder Technology,* Vol. 107, pp.259-267, 2000.
[5] Peng D. Q., Lu Y. B.. "Review of hydrocyclones," *Coal preparation technology*, Vol. 5, pp.13-18, 2006.
[6] Jank A., Müller W., Waldhuber S., et al. "Hydrocyclones for the separation of impurities in pretreated biowaste," *Waste Management*, Vol. 64, pp.12-19, 2017.
[7] Boysan F., Ayers W. H., Swithenbank J. "A Fundamental mathematical modelling approach to cyclone design," *Chemical Engineering Research and Design,* Vol. 60, pp.222-230, 1982.
[8] J. C. Cullivan, R. A. Williams, C. R. Cross. "New insights into hydrocyclone operation," *Particulate Science & Technology*, Vol. 21(1), pp.83-103, 2003.
[9] Schuetz S., Mayer G., Bierdel M., et al. "Investigations on the flow and separation behaviour of hydrocyclones using computational fluid dynamics," *International Journal of Mineral Processing*, Vol. 73(s 2-4), pp.229-237, 2004.
[10] Wang B., Xu D. L., Chu K. W., et al. "Numerical study of gas–solid flow in a cyclone separator," *Applied Mathematical Modelling*, Vol. 30(11), pp.1326-1342, 2006.
[11] Delgadillo J. A., Rajamani R. K.. "Hydrocyclone modelling: Large eddy simulation CFD approach," Minerals & Metallurgical Processing, Vol. 22(4), pp.225-232, 2005.
[12] Wang B., K. W. C., Yu A. B.. "Numerical study of particle–fluid flow in a hydrocyclone," *Industrial & Engineering Chemistry Research*, Vol. 46(13), pp.4695-4705, 2007.
[13] T. R. Vakamalla, N. Mangadoddy. "Numerical simulation of industrial hydrocyclones performance: Role of turbulence modeling" *Separation and Purification Technology*, Vol. 176, pp.23-39, 2017.
[14] C. Banerjee, E. Climent, A. K. Majumder. "Mechanistic modelling of water partitioning behaviour in hydrocyclone," *Chemical Engineering Science,* Vol. 152, pp.724-735, 2016.
[15] R. Sabbagh, Michael G. Lipsett, Charles R. Koch, et al. "An experimental investigation on hydrocyclone underflow pumping," *Powder Technology*, Vol. 305, pp.98-107, 2016.

[16] Li J., Kuang S B, Zheng Q., et al. "Computational analysis and optimization of hydrocyclone size to mitigate adverse effect of particle density," *Separation and Purification Technology*, Vol. 174, pp.251-263, 2017.
[17] Zhao L. X., Jiang M. H., Xu B. R., et al "Development of a new type high-efficient inner-cone hydrocyclone" *Chemical Engineering Research and Design*, Vol. 90(12), pp. 2129-2134, 2012.
[18] Ni L., Tian J. Y., Zhao J. N.. "Experimental study of the effect of underflowpipe diameter onseparation performance of a novel de-foulant hydrocyclone with continuous underflow and reflux function" Separation and Purification Technology, Vol. 171, pp.270-279, 2016.
[19] Kuo-Jen Hwang, Shih-Ping Chou. "Designing vortex finder structure for improving the particle separation efficiency of a hydrocyclone," *Separation and Purification Technology*, Vol. 172, pp.76-84, 2017.
[20] R. K. Dubey, Eric Climent, C. Banerjee, et al. "Majumder. Performance monitoring of a hydrocyclone based on underflowdischarge angle," *International Journal of Mineral Processing*, Vol. 154, pp. 41-52, 2016.

Technology of Heating Accumulating with High Voltage and High Power Furnace Instead of Coal

Xiuhua Wu and Ruizhao Shao

Professor, Information and Electrical College of Shenyang Agricultural University, Shenyang, China
E-mail: wuxih70@126.com

Abstract. This paper discussed a kind of heating storage technology with high voltage and high power furnace. This technique used solid heat storage media, which changed electric power of 10 kV voltage level into thermal energy, to supply heat to the residents. In north of China, it could partly solve the pollution problem of traditional heating system. If the electric power was from the wind energy, solar energy or the valley electric power, the value and effect were much greater. This paper discussed the status, problems and efficiency of this technology and gave some practical and significant suggestion

1. Introduction

The social, economic and environmental problems caused by the traditional coal heating system were becoming more and more seriously in north of China. The coal was consumed largely, the heating cost of the residents could not be ignored and the environment pollution was more obviously. All this questions stimulated many researchers to solve or promote the situations. Today, the supply of electric energy was more and more sufficiently, which may come from wind, solar energy and many other clean sources. So it was a feasible method that using electric energy to replace coal to supply heating to the residents. Now it had been used in some enterprises, companies, residential quarters and some departments. Part of traditional coal-fired boilers were replaced, the pollution was decreased. The article was gonging to discuss this problem, analyze the technology, efficiency and social questions, in order to provide some benefit suggestions and advices.

2. Electric Energy Heating Technology

In China, the heating systems with electric energy could be classified intensive type and distributed type according to their scales. The intensive type included common boiler, heat accumulating boiler and heat pump. The distributed type[1] included electric room heater, electrothermal film and electric heating floor and so on. In this case, the common boiler possessed higher cost because it could not store heating. The heat accumulating boiler has the characteristic of higher equipment investment but lower operating cost because it could use the valley electric energy in those areas of the peak and valley electric prices differently. Now, the technology was of mature technique and steady heating output. When the boiler operated longer time the cost becomes relatively lower. So it was the best choice of the electric heating device. (See Table 1)

Table 1. Comparison among three common heating ways[2]

Items	Coal heating	Electric heating	Heat accumulating with electric
Heating as（m²）	72	95	95

IOP Conf. Series: Earth and Environmental Science **151** (2018) 012025 doi:10.1088/1755-1315/151/1/012025

Equipment investment (¥)	4500	5000	8500
Heating days per year（D）	155	155	155
Energy per day（kW•h）	31	46	48
Unit prices per day (¥)	0.70	0.48	0.20
Indoor average temperature（°C）	17	19	19
Energy cost per day (¥)	21.70	22.08	9.6
Energy cost per year (¥)	3363	3422.4	1488.00
Unit price/m2/year (¥)	46.70	36.03	15.66

3. Electric Energy Heat Accumulating System

The electric energy heat accumulating system was a new technique invent. It changed electric energy of high voltage and high power into heating with a solid material. It introduced separating water and electric technology, high voltage controlling technology, heat accumulating and preservation technology to transform electric energy from wind energy, solar energy or valley power energy into heat energy and reserve it into some solid media. When it was needed, the stored heating would be changed into hot water, hot wind or steam to be used in residents heating system. In some special cases, the heat energy also could be fed back to the power grid in order to release electric supply stress.

3.1. Structure

The high voltage high power electric energy storage furnace could be used to fulfill functions as the former. Its structure was shown as Fig.1.

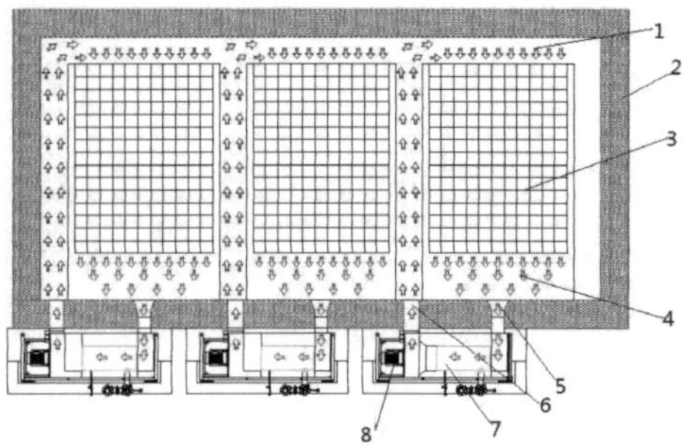

1.Low temperature air. 2. Heat keeping level
3.Heat accumulating part. 4. High temperature air.
5.Outlet of heat interchange
6.Inlet of heat interchange 7. heat interchange system
8. Air blower

Figure 1. Plane graph of high voltage high power electric energy storage furnace[3]

It included several parts. They were high voltage heating part, high temperature heat storage part, heat interchange part, heat output controller, air blower, heat keeping level, low temperature air circulation way, high temperature air circulation way, high voltage power controller, and so on.

The high voltage heating part[4] could directly transform the 10kV or higher voltage level electric energy into thermal energy without the transformer. The heating materials may be heating silks, heating slices or heating tubes. The high temperature heat storage part was some high density-high specific heat capacity heating accumulating material, which was of keeping temperature, fireproof, insulated。It could endure 800℃ temperature. The heat interchange part changed the heat energy stored by the accumulating material into hot water. The method was that the heat energy was transformed through high temperature air flow in the closed way between the accumulating body and the heat exchanger. The blowers drived the hot air flowing in the closed way. The revolution could be adjusted according to the heat supply demands. The air velocity could be changed by the hot water temperature. The high voltage power controller included time controller that controlled the start of the high voltage power switch and temperature controller that controlled the temperature of the heat accumulating body.

The high voltage high power electric[5] energy storage furnace was kind of new heat resources. It saved valley electric and discard wind energy into some solid heat stored materials. It didn't need transformers and could directly work under the 10kV or 66kV voltage level. The input power of one unit could be reach 100 MW and the heat accumulating capacity could be reach 800 MWh. At the same time, the equipment could consume the discard wind energy, supplied the heat 24 hours without interruption in large or great large areas, so it could replace the traditional heating system partly or completely. It not only reduced the consuming of unrenewable energy but also cut down the generation of the waste.

3.2. Principle

The equipment of this paper was working at the valley time of the power grid. It connected the high voltage switches with the auto-control system. When the switch was closed, the high voltage power grid could supply electric to the heat accumulating body continuously, then the electric energy was changed to heat energy here. The temperature of heat accumulating body was raised up by the absorbing of the heating. When the temperature reached a upper limit, the switch was open and at the same time the grid would stop providing electric energy to the body. At last, the heat energy stored in the heat accumulating body would be changed to the other type of heat energy and export through the output controller and heat exchanger. Because the heat transmission between the body and the exchanger was fulfilled by the hot air circle flowing and heat the hot water in the heating system, many technique questions that used the high voltage power grid to supply heating directly were all solved.

4. Efficiency Analysis

4.1. The Cost Analysis

In northern China, the discard wind power was commonly generated in winter while the heating supply period was also at that time. So, the cost of coal heating system and the discard wind power heating system could be compared.

Unit price of coal heating system could be calculated by 860 Kcal times transfer efficiency of electric boiler and price of coal, then divided by low order calorific value of coal and transfer efficiency of coal-fired boiler. The transfer efficiency of electric boiler was generally 95%, the price of coal was 600 Y/ton, the low order calorific value of coal was 5000Kcal/ton. The transfer efficiency of coal-fired boiler was 70%. At last the Unit price of coal heating system was about 0.14￥. If the price of the discard wind power was 0.14 ￥ per kWh, the cost of these two types was considerable.

The cost of the electric boiler was related with the unit power consumption and its heat efficiency. The unit power consumption of converting electric to hot water, steam or supply heating were as follow:

Supply heating: 1GJ = 277.8 kWh, hot water: 1Ton hot water of 50℃= 46.7 kWh, steam: 1ton saturated steam of 170℃= 760 kWh.

$$\text{Cost} = \text{Unit power consumption} \times \text{Heat efficiency} \times \text{Power price} \qquad (1)$$

In according to the current residential architecture of energy conservation design standard of Liaoning Province, Rule DB21/T1476-2006, the design heating load index of the architecture of saving 65% energy was 32.8W/㎡. In southern areas of Yingkou and Huludao City, the index was 25.7W/㎡。

The heat consumption of the residential architecture was related with design heating load, supply heating time per day and the supply heating period. It could be calculated by the formula as below.

$$Q_h = Q_q \times T \times (1+F) \times Z \times K \qquad (2)$$

Where Q_h, average heating consumption (kW/㎡); Q_q, design heating load index(kW/㎡); F, loss of the pipe networks and equipments, geting10%; T, supply heating time per day (h); Z, supply heating period (day); K, weighted average heating load correction factor, getting 0.5. (See Table 2)

Table 2. The heating consumption of some cities in Liaoning Province during the heating period[6]

Cities	Supply	Heating consumption Q_h

	days (D)	(kWh) 30 W/m²	40W/m²	50W/m²	60W/m²
Shenyang	152	60.19	80.25	100.00	120.38
Anshan	144	57.02	76.03	95.04	114.05
Fuxin	156	61.78	82.37	102.96	123.55
Faku	161	63.76	85.01	106.26	127.51
Dalian	131	51.88	69.17	86.46	107.53
Yingkou	145	57.42	76.56	95.70	114.84

4.2. The Investment Analysis

Equipment investment was calculated by this formula:

$$C_e = Q_q \times T / T_C \times (1+F) \times D \text{ (¥/m²)} \quad (3)$$

Where Tc, time of discard wind power supply（h）; D,unit price of devices（electric boiler was 850 ¥/ kW）(See Table 3)

Table 3. Analysis of investment in equipment[7]

Time of wind power supply	Investment of equipments（¥/m²）			
	30W/m²	40W/m²	50W/m²	60W/m²
7 hours	96	128	160	192
15 hours	45	60	75	90

The investment cost was: $Co = Q_q \times T \times (1+F) \times Z \times K \times P$ (¥/m²) （4）

Where , P is electric price （¥/kWh）。

For example, a house in Shenyang city, its heat load index was 40 W/m², its discard wind power unit price was 0.14¥/kWh, the supply heating period was 152 days, so its operation cost was

$$Co = Q_q \times T \times (1+F) \times Z \times K \times P = 0.04 \times 24 \times 1.1 \times 152 \times 0.5 \times 0.14 = 11.24¥/m^2 \quad (5)$$

The other costs were depreciation and maintenance of equipment, the pipe maintenance and personal management and so on. In those, the designed life of the electric heating furnace was 20 years. Its heating stored body could have 40% residual value. The pipes maintenance fee between the stove house and resident house was about 1.5¥/m² according to the experience. The repair charge of the stove was about 1～2¥/m². The personal management fee was salary times number times heating area per month. For example, the people of management were often 3 people when the area smaller than 100,000 m², 6 people when larger than 100,000 m². The average per month was 3500 each person. The personal management fee was 3500×3×6 / 100000/m²= 0.63¥/m².

4.3. Investment Payback Period

The investment payback period could be calculated use this formula[8]:

$$T_p = \frac{Q_q \times T / T_C \times (1+F) \times D \times j}{(\text{heatingfee} - a - s - g - r - j/20)} (year) \quad (6)$$

Where j, depreciation of equipment; a ,operational cost; s ,maintenance of equipment; g , pipe maintenance; r, personal management fee.

For example, a certain dwelling district in Shenyang city, the heating area was 100,000 ㎡, resident heating fee was 28￥/㎡, heating load index was 40W/㎡, heat exchange station management people were 3.The investment payback period was

$$T_p = \frac{0.04\times 24/7\times 1.1\times 850\times 0.6}{(28-11.24-1.5-1-0.63-3.6)} = 7.7(year) \qquad (7)$$

5. Instance Analysis

This essay take the [9]Longyuan supply heating project with area 200,000 ㎡ in Faku Town using discard wind power as an example. The indexes of this project were as below.

The general heating index of Faku Town was about 72W/㎡. The designed heating index of saving energy 65% architecture was 32.8W/㎡ according to the Liaoning province building energy conservation design rules. The index of this project adopted 50W/㎡ by measuring and calculating.

The total heating load was[10]:

$$Q_L = Q_q \times Ao\times(1+F) = 50\times 200{,}000\times(1+10\%) = 11MW \qquad (8)$$

Where, Ao, central heating area, getting 200,000 ㎡.

The heat consumption of the whole year of this project was:

$$Q = Q_h / q\times Q_q \times T\times(1+F)\times Z\times Ao \qquad (9)$$

=0.015/0.0328×0.05×24×(1+10%)×161×200000

=19,437,805kWh

According to the meteorological condition of Faku Town, during the supply heating period, the average outdoor temperature was -6.2℃, the average outdoor relative humidity was 58%, the temperature difference between indoor and outdoor was 22.2℃. So the q =32.8W/㎡; Q_h =15W/㎡ according to the DB21/T1476-2006 Rules of Liaoning province.

Because the total heating load was 11MW, the heating mode was designed 7 hours accumulating and 24 hours releasing. It adopted 4 high voltage high power electric stored heating furnaces. The total stored heating power was 40MW and the largest stored capacity was 280,000kWh, the output heating power was 12 MW, it could satisfy the supply demand.

The furnaces could be adapted voltage scale was 10kV~66kV. Its output method was hot water of under 85℃ temperature with ordinary pressure. Its design power was calculated as follow:

$$Pe = Q_q \times T/Tc\times(1+F+Fc)\times Ao = 0.05\times 24/7\times 1.15\times 2\times 10^5 = 37.71MW \qquad (10)$$

Where Pe, power of the furnace(MW); , Fc, heat loss of the furnace, getting 5%.

The project could select wind power enterprise or grid company to complete a nearby electric transmission lines to introduce the furnaces. Once the project was fulfilled, it could utilize the discard wind power 20,000MWh, reduce coal consumption 7,000 Tons, decrease the carbon emission 16,000 tons. Its benefits were the sale proceeds of the discard wind power, selling electric proceeds and supply heating proceeds.

Case one. The transmission lines were built or remade by the wind power enterprise and the fee was about ￥5,000,000.

The profit of the wind power enterprise was income of selling electric (20,000MWh*0.5214￥/kWh=￥10,428,000)plus allowance (20,000MWh *（0.45-0.14）=￥6,200,000), plus additional tax(added value tax*6%= ￥55,900),plus financial cost(￥5,000,000*7%= ￥350,000). The value was ￥3,822,100 and the payback period was 1.3 years.

The profit of the heating supply enterprise was income of selling heating energy (￥32*200,000 ㎡= ￥6,400,000)plus electric purchase fee(20,000MWh *0.14= ￥2,800,000), plus management fee(￥150,000 per year),plus pipes maintenance fee (￥300,000),plus equipment maintenance fee(￥200,000),plus depreciation cost(￥960,000 for 20 years). The value was ￥1,990,000 and the payback period was 8.5 years.

Case two, The transmission lines were built or remade by the gird company. The profit of the wind power enterprise was ￥4,170,000 and the profit of the heating supply enterprise was ￥1,990,000. The payback period was not changed.

6. Conclusion

According to the initial economic analysis, when the heating supply enterprise buyed the valley electricity, the price was under 0.45￥/kWh or the discard wind power was 200MW, the operational indicator of each side was satisfactory.

7. References

[1] GE Yanfeng, Technical Plan for Electric Heat Storage and Heating by Wind Energy Curtailment Based on Joint Dispatching of Heat and Electricity[J]. Smart power grids, 2015, 3(10):901-905

[2] LV Quan, Comparison of coal-saving effect and national economic indices of three feasible curtailed wind power accommodating strategies [J]. Automation of Electric Power Systems, 39(7):75-83

[3] Jiang Hao.Research on accommodating curtailed wind power by CHP installed electric boilers.(M), Dalian University of Technology.2013

[4] Li Yinling, The analysis of energy consumption for the urban district heating and its energy saving and emission reduction(M) ,North China Electric Power University,2012

[5] Fragaki A, Andersenb A N, Tokec D. Exploration of economical sizing of gas engine and thermal store for combined heat and power plants in the UK[J] . Energy, 2008. 33(11):1659-1670

[6] Incer ibrahim. On thermal energy storage systems and applications in building [J] .Energy and buildings,2002,24:377-388

[7] LI Peifeng, Model for Energy Consumption of Heating System and Its Application，Proceedings of the CSEE，Vol.33 No.23 Aug.15, 2013：19-27

[8] CHEN Lei1, XU Fei1 ,Implementation and Effect of Thermal Storage in Improving Wind Power Accommodation, Proceedings of the CSEE, Vol.35 No.17 Sep. 5, 2015:4283-4290

[9] HU Si-ke， XING Jiao-jiao, Solid Heat Storage Device Storage and Heat Release Characteristics of the Numerical Simulation Analysis, Fluid Machinery, 2014. Vol.42(8):78-82

[10] XU Yong, KE Xiufang, Feasibility analyses of a high temperature heat storage device using electrical heating, Energy Storage Science and Technology, 2015.11.Vol.4(6):627-631

IOP Conf. Series: Earth and Environmental Science **151** (2018) 012026 doi:10.1088/1755-1315/151/1/012026

Using Remote Sensing Techniques for Estimating Water Stress Index for Central of Nile Delta

Ayat Elnmer[1,2], Mosaad Khadr[2], Ahmed Tawfik[1,3]

1 Environmental Engineering Department, School of EECE, Egypt-Japan University of Science and Technology, E-JUST, P.O.Box 179, New Borg Al-Arab City, Postal Code 21934, Alexandria, Egypt.
2 Irrigation and Hydraulic Engineering Department, Faculty of Engineering, Tanta University, Egypt
3 Water Pollution Research Department, National Research Centre, Giza 12622, Egypt
E-mail: ayat.abdelwahab@ejust.edu.eg

Abstract. Water scarcity is one of the main challenges facing water management in Egypt. This in turn will have direct impacts on the agricultural sector which is a key sector for the socio-economic development in Egypt, and plays a significant role in the Egyptian national economy. In Egypt, water resources are limited to the Nile River, rainfall, deep groundwater, and potential desalination of sea. Climate change, rapid population growth, and economic development will significantly affect the future availability of water resources for agriculture sector, which consumes about 85% of total water resources in Egypt. Therefore, continuously monitoring of crop statutes and crop water consumption plays a vital role in water resources management in developing countries such as Egypt. Recently Remote sensing techniques provide decision makers with spatial information about crop statutes and water stress at region scale. Remote sensing techniques also have the ability to monitor large areas with saving time and cost. The main objective of this study is investigating the capabilities of satellite data in monitoring of water stress and crop statues in the central portion of Nile delta by using the water stress index (WSI). The water stress index was used to identify locations of poor irrigation in order to maximize the crop yield. The proposed model was calibrated and validated against the measured data by using 20 points of ground measurements for actual evapotranspiration (ET_C) and wet evapotranspiration (ET_{Wet}). The performance of the model was measured using various evaluation criteria. Validation results showed that satellite data are capable of estimating WSI since the comparison between WSI_{obs} and WSI_{est} resulted in R^2=0.5003. Therefore WSI can be considered as a quick, costless and moderate tool to provide farmers and decision makers with spatial information about crop statues and water stress.

1. Introduction

Water resources management in Egypt is considered the major obstacle that faces its future development. Egypt has very limited water resources where the Nile participates every year with fixed share (55 billion cubic meters). The agriculture sector consumes 85% of total water supplies while the crop consumption in the center of Nile delta consumes about 45% of total agriculture water[1]. The water demands in all sectors are increasing rapidly due to the population growth and the increment in urban activities. Hence the stresses on the agriculture sector are growing every year.

Detection the water stress in cultivated lands and evaluating its impacts on crop yield plays a vital role in monitoring the crop production and enables farmers and decision makers to optimize the irrigation system [2]. Water stress indexes could be useful tool in detecting the crop water stress [3]-[5]. Water stress could be derived from optical remote sensing or thermal remote sensing. In water stress indexes based on optical remote sensing, the used bands are short wave infrared band (SWRI) and near infrared band (NIR). SWRI and NIR are used to detect the drought effects on the water content and the water stress on vegetation [6]-[9]. SWRI is more effective in detection water stress of crops as it is very sensitive to the water thickness in plant leaf. Moreover it is less affected by ozone, water vapour and aerosol content [10]. There are several indexes based on SWIR and NIR such as land surface wetness index (LSWI) which measures the interaction of the liquid molecules in the crop [11]. Normalized difference water index (NDWI) is similar to LSWI and derived by Chen et al., 2005[12]. There are several methods to derived information about crops statues based on the relation between vegetation indexes (VI) and the surface temperature [9]. These methods have the ability to cover very large areas with little information; in addition they can deal with the surface variations. Price 1990 [13] was the first to use the relation between NDVI-Ts scatter to derive spatial maps for latent heat fluxes. Latent heat flux is very important in monitoring crop productivity and crop water consumption to manage the limited water resources [14], [15]. Maron et al., (1994) assumed that the water deficit could be derived from the trapezoidal shape IV-(T_S-T_a) where t_a is the air temperature and VI was the vegetation index [16]. The selected IV was SAVI to cope with the spectral changes in soil background. Hence the water deficit index (WDI) was depending on the ratio between the actual evapotranspiration ET_c and the potential evapotranspiration ET_o. Jiang-Islam (2001) modified the WDI by replacing the evapotranspiration with Priestley and Taylor equation [17]. Then potential evapotranspiration was replaced with the wet evapotranspiration and the relation between Normalized Difference Vegetation Index (NDVI) and the land surface temperature (LST) was derived in order to assess the water stress index (WSI) [18].

This study aims to assess the ability of WSI in detecting the water stress locations in the center of the Nile delta and validate WSI with old approach WDI.

2. Methodology

2.1. Study Area

The study area located over the center of the Nile delta with longitudes of 30°54′42.15″ - 31°1′37.35″ E and latitude of 30°43′43.75″ - 30°57′40″ N[19]. The center of the Nile delta is a flat region with homogenous clay- silt soil. The study area has a very rich irrigation system with two main canals Al-Qasid and Tanta Al-Melahia canal as shown in Figure 1. The main crops are wheat and clover in winter season while in summer, are maize and rice.

2.2. Data

The study area is located in zone N36 with UTM coordinate system and WG84 datum. Two free clouds images of Landsat 8 with 30 meter ground resolution were corrected radiometrically in order to calculate WSI and WDI through estimating the land surface temperature (LST), actual evapotranspiration ET_c and NDVI. Table (1) summaries Landsat 8 images dates and their Path/Row.

2.3. Water Stress Indexes

Figure (2) shows the methodology for estimating the water stress index for the study area. Two free clouds were radiometric corrected in order to study the water stress locations in the study area. Then Normalized Difference Vegetation Index (NDVI) was derived by using the following equation,

$$NDVI = \frac{\rho_{NIR} - \rho_{Red}}{\rho_{NIR} + \rho_{Red}} \qquad 1$$

Where; the red band with width with (0.67 – 0.90 μm) in Landsat 8 is called ρ_{Red} while near infra-red band with width (0.63 – 0.67 μm) is called ρ_{NIR} [20].

In order to estimate the water stress statues, the surface temperature should be estimated in accurately and efficiently way as it is considered the most driving factor in crop stress. Hence temperature could be estimated from Landsat 8 data using the algorithm of split window [15] with the equation,

$$LST = TB_{10} + C_1 (TB_{10} - TB_{11}) + C_2 (TB_{10} - TB_{11})^2 + C_0 + (C_3 + C_4 W)(1 - m) + (C_5 + C_6 W)\Delta m \qquad 2$$

Where, c_1 –c_6 split- window coefficient values [15], *m* is the mean of land surface emissivity (LSE), the difference of LSE is Δm, TB_{10}-TB_{11} are the brightness temperature of 10 and 11 Bands and W is the atmospheric water-vapour content.

The relationship between LST and NDVI was derived in order to estimate the maximum and minimum temperature through the study area using PYTHON language.

The actual evapotranspiration was derived by Surface energy balance algorithm for land (SEBAL) model over the Nile delta for winter season 2015-2016 in previous studies by the authors.

One of the famous method to estimate the Reference evapotranspiration (ET_o) is FAO- Penman Monteith formula which depends on several meteorological data such as temperature, humidity and wind speed proceed [21]. Water deficit index (WDI) was first introduced by [16] and related with the ratio between actual evapotranspiration and potential evapotranspiration and could be estimated by the following equation,

$$WDI = 1- \frac{ET_c}{ET_o} \quad 3$$

On the other hand, WSI is a modification of WDI where ET_o was replaced with E_w which replaced with Priestley-Taylor equation [22] and $ET_c$was replaced with Jiang and Islam method, hence the new index of WSI could be obtained by the following equation [18],

$$WSI = \frac{T_i - T_{min}}{T_{max} - T_{min}} \quad 4$$

Where, T_i is the temperature of each pixel, T_{max} is the maximum temperature of vegetation class with NDVI equal 0.0 which could be obtained from Ts-NDVI scatter plot and T_{min} is the minimum temperature of vegetation and presented by the average surface temperature of free water pixels.

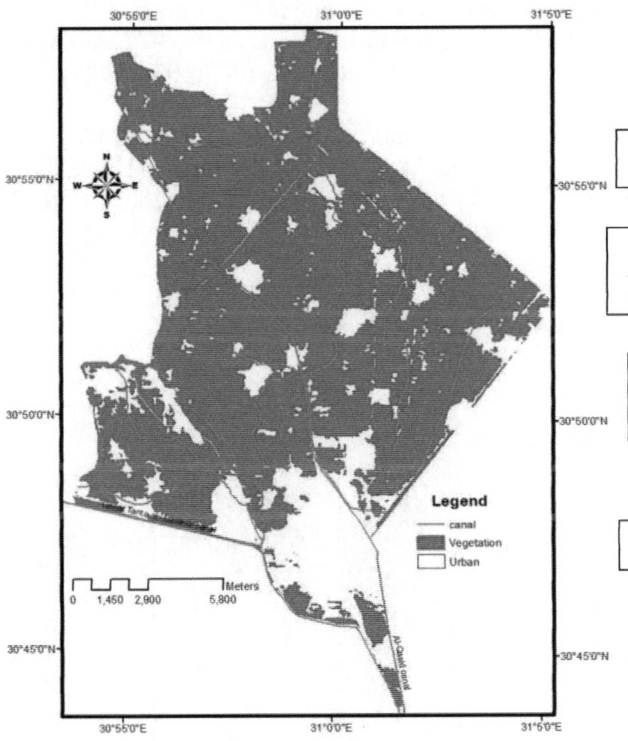

Figure 1. Location of the study area [19]

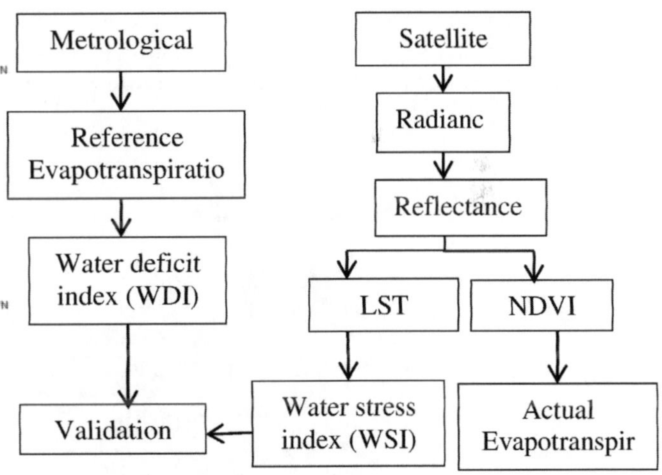

Figure 2. flow chart for the methodology

Table 1. Landsat 8 data for the study area

Acquired date	Path/Row	Pixel Size (m)
11th March 2016	39/177	30

IOP Conf. Series: Earth and Environmental Science **151** (2018) 012026 doi:10.1088/1755-1315/151/1/012026

30th May 2016	39/177	30

3. Results and Discussion

Well knowledge about the Water stress statues is the backbone in maximizing the crops productivity and minimizing the irrigation water consumption for the decision makers in order to manage the limited water resources efficiently. NDVI for the study area in March 2016 ranged from 0.223 to 0.511 with mean 0.3438 and standard deviation (SD) 0.0986 as it is considered the mid-season where the vegetation cover is full however it ranged from 0.0 to 0.271 with mean 0.1238 and standard deviation 0.0617 in May 2016 where the absence of the vegetation cover took place (harvesting season) as shown in Figure 3. The land surface temperature was derived for March using split window algorithm and ranged from 296.0 to 302.0 K with mean 299.67 K and SD 1.6 as shown in Figure 4 while it was 304.0 to 318.0 K with mean 313 and SD 2.723 in May. The relationship between NDVI and LST was drawn to get the maximum and minimum temperatures to determine the water stress locations as shown in Figure 5. It can be extracted from figure 5 that the Maximum temperature with NDVI 0.0 was 295 K in March while it was 309K in May.

This relation between NDVI and LST was the base to estimate the thermal index of water stress through estimating Tmax and Tmin. Figure 6 shows the spatial distribution of WSI over the center of the Nile delta in March and May 2016. During the growing season represented by March, lower values have been occurred and ranged from 0.0 to 0.75. On the other hand, higher values of WSI were obtained in May 2016 due to the harvesting season where they were 0.2 to 0.8as shown in Figure 6. The water stress was lower in March than May due to dense vegetation cover in growing months (March) compared to the harvesting season (May). Similar trends were obtained by Dangwal (2014) [23] where WSI was 0.1 to 0.2 in February 2009-2010 (growing season) and was 0.35 to 0.2 in April 2009-2010 (harvesting season). While WDI in March was 0.4 to 0.7 and it was 0.7 to 0.9 in May as shown in Figure 7.

Comparison between WSI and WDI was proceeded in order to validate the applicability of WSI in determination water stress locations. The comparison between WSI and WDI illustrated that the correlation R2 =62% as shown in Figure 8. This error between WSI and WDI could be related to the difference between ETo and E_w.

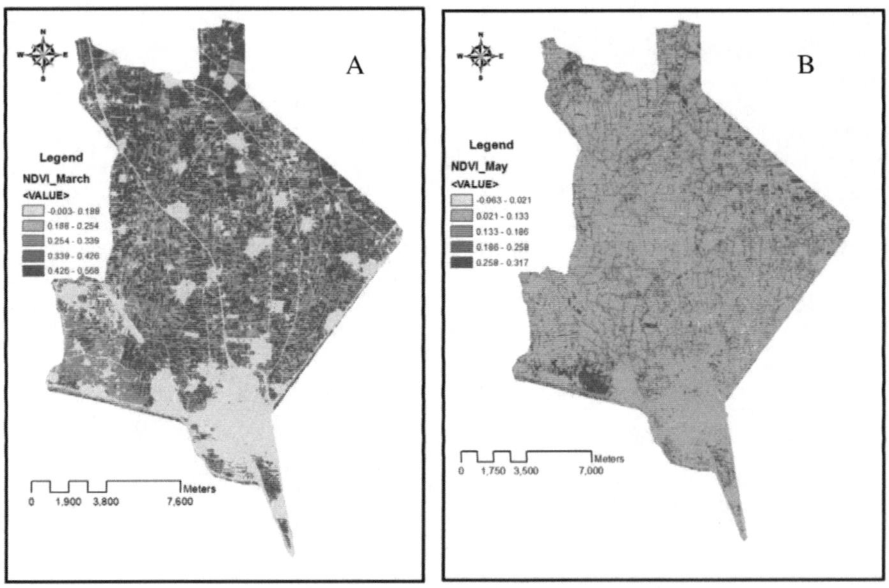

Figure 3. Distribution of NDVI over the center of the Nile delta, A) in March 2016, B) in May 2016

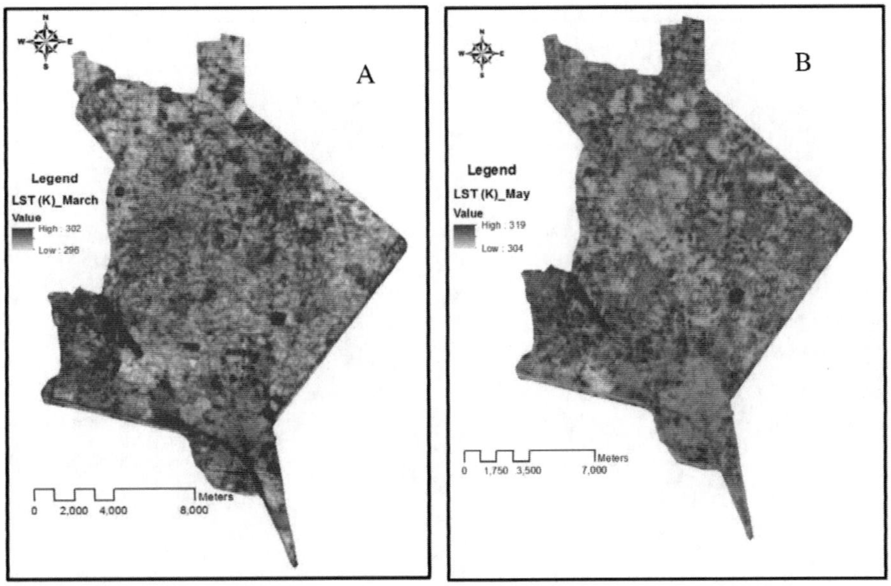

Figure 4. land surface temperature (LST) over the center of the Nile delta, A) in March 2016, B) in May 2016

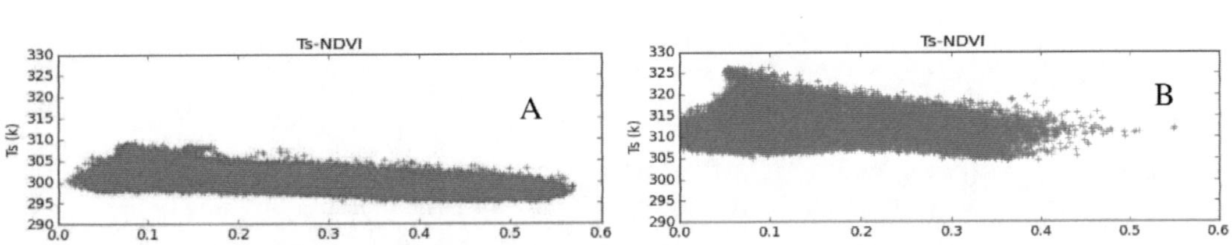

Figure 5. NDVI-TS scatter plot over the center of the Nile delta, A) in March 2016, B) in May 2016

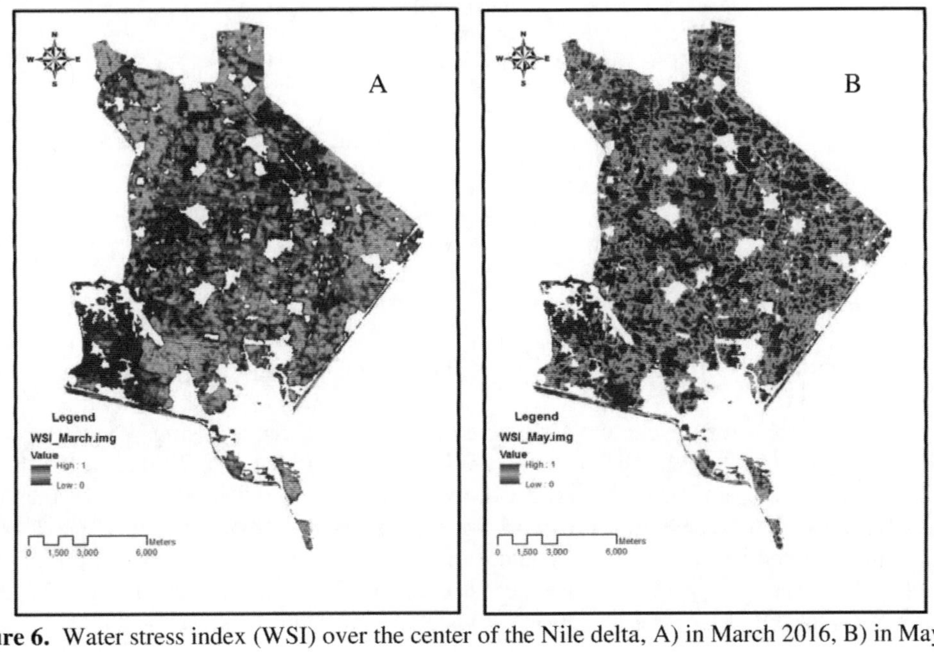

Figure 6. Water stress index (WSI) over the center of the Nile delta, A) in March 2016, B) in May 2016

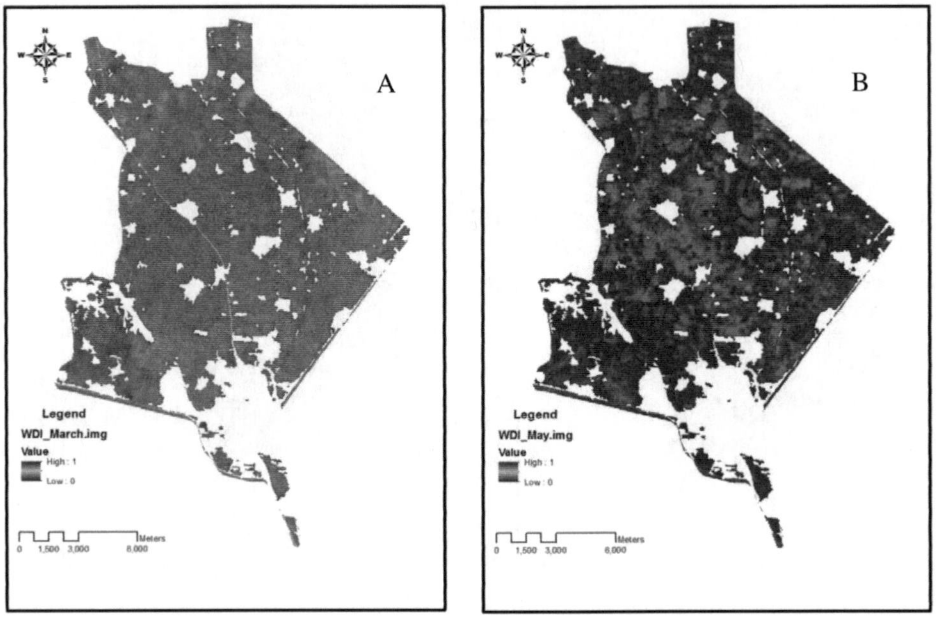

Figure 7. Water deficit index (WDI) over the center of the Nile delta, A) in March 2016, B) in May 2016

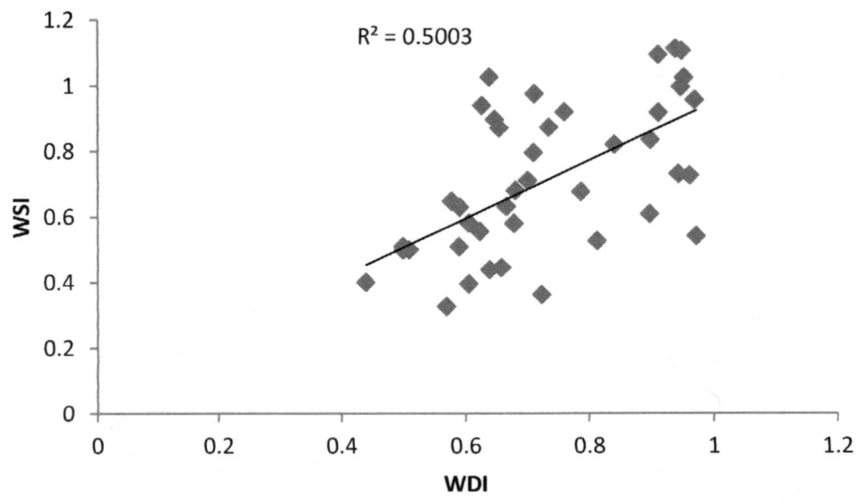

Figure 8. Comparison between WSI and WDI over the study area

4. Conclusions

In this study the water stress for the center of the Nile delta was detected by using WSI and WDI. WSI was ranged from 0.0 to 0.75 for March lower than WDI which was 0.4 to 0.70 due to the difference between ET_o in WDI and E_w in WSI. Otherwise, the WSI was 0.2 to 0.8 in May higher than WSI in March due to the absence of vegetation cover in harvesting season. However, WSI was lower than WDI in May which ranged from 0.7 to 0.9. The correlation between the old approach (WDI) and the new approach (WSI) was 0.5003 (50.03%) which strongly indicate that the WSI could be used for moderate estimation for the water stress location for vegetation at regional scale.

5. References

[1] Khadr, M., B.A. Zeidan, and A. Elnmer, *On-Farm Water Management in the Nile Delta.* 2016.

[2] López López, R., et al., *Water stress index as an indicator of irrigation timing in agricultural crops.* Agricultura técnica en México, 2009. 35(1): p. 97-111.
[3] Kustas, W., et al., *Mapping daily evapotranspiration at field to continental scales using geostationary and polar orbiting satellite imagery.* Hydrology and Earth System Sciences, 2011. 15(1): p. 223.
[4] Olioso, A., et al., *Future directions for advanced evapotranspiration modeling: Assimilation of remote sensing data into crop simulation models and SVAT models.* Irrigation and Drainage Systems, 2005. 19(3-4): p. 377-412.
[5] Liu, J., J. Chen, and J. Cihlar, *Mapping evapotranspiration based on remote sensing: An application to Canada's landmass.* Water resources research, 2003. 39(7).
[6] Pinter Jr, P.J., et al., *Remote sensing for crop management.* Photogrammetric Engineering & Remote Sensing, 2003. 69(6): p. 647-664.
[7] Nouri, H., et al., *A review of ET measurement techniques for estimating the water requirements of urban landscape vegetation.* Urban Water Journal, 2013. 10(4): p. 247-259.
[8] El-Kawy, O.A., et al., *Land use and land cover change detection in the western Nile delta of Egypt using remote sensing data.* Applied Geography, 2011. 31(2): p. 483-494.
[9] Kotb, T.H., et al., *Soil salinization in the Nile Delta and related policy issues in Egypt.* Agricultural water management, 2000. 43(2): p. 239-261.
[10] Serbina, L. and H.M. Miller, *Landsat and water: Case studies of the uses and benefits of Landsat imagery in water resources.* US Geological Survey Open-File Report, 2014. 1108: p. 61.
[11] Du, Z., et al., *Analysis of Landsat-8 OLI imagery for land surface water mapping.* Remote sensing letters, 2014. 5(7): p. 672-681.
[12] Montanaro, M., et al., *Stray light artifacts in imagery from the Landsat 8 Thermal Infrared Sensor.* Remote Sensing, 2014. 6(11): p. 10435-10456.
[13] Irmak, S., et al., *Irrigation efficiency and uniformity, and crop water use efficiency.* 2011.
[14] Sun, Z., et al., *Evapotranspiration estimation based on the SEBAL model in the Nansi Lake Wetland of China.* Mathematical and Computer Modelling, 2011. 54(3): p. 1086-1092.
[15] Singh, R.K., et al., *Application of SEBAL model for mapping evapotranspiration and estimating surface energy fluxes in south-central Nebraska.* Journal of irrigation and drainage engineering, 2008. 134(3): p. 273-285.
[16] Li, H.-j., et al., *SEBAL Model and Its Application in the Study of Regional Evapotranspiration [J].* Remote Sensing Technology and Application, 2005. 3: p. 003.
[17] Su, Z., et al., *Assessing relative soil moisture with remote sensing data: theory, experimental validation, and application to drought monitoring over the North China Plain.* Physics and Chemistry of the Earth, Parts A/B/C, 2003. 28(1): p. 89-101.
[18] Abrishamkar, M. and A. Ahmadi, *Evapotranspiration Estimation Using Remote Sensing Technology Based on SEBAL Algorithm.* Iranian Journal of Science and Technology, Transactions of Civil Engineering, 2017. 41(1): p. 65-76.
[19] Elnmer, A., M. Khadr, and B.A. Zeidan, *Optimal Water Productivity Of Crop Pattern In Central Nile Delta, Egypt.* Conference: Eighteenth International Water Technology Conference, IWTC18- Sharm ElSheikh, Egypt, 12-14 March 2015, At Sharm ElSheikh, Egypt, 2015. 1.
[20] Glenn, E.P., et al., *Relationship between remotely-sensed vegetation indices, canopy attributes and plant physiological processes: what vegetation indices can and cannot tell us about the landscape.* Sensors, 2008. 8(4): p. 2136-2160.
[21] Allen, R.G., et al., *Operational estimates of reference evapotranspiration.* Agronomy journal, 1989. 81(4): p. 650-662.
[22] Sellers, P., et al., *A revised land surface parameterization (SiB2) for atmospheric GCMs. Part I: Model formulation.* Journal of climate, 1996. 9(4): p. 676-705.
[23] Bakry, M.F. and A.A.E.-M. Awad, *Practical estimation of seepage losses along earthen canals in Egypt.* Water resources management, 1997. 11(3): p. 197-206.

Acknowledgements

The author would like to thank Egyptian Ministry of Higher Education (MoHE) for the full–funded of this research and Egypt-Japan University of Science and Technology (E-JUST) for providing the tools that needed for this research.

IOP Conf. Series: Earth and Environmental Science **151** (2018) 012027 doi:10.1088/1755-1315/151/1/012027

Influence of Coastal Measures on Shoreline Kinematics Along Damietta coast Using Geospatial Tools

Mohammed Esmail[1], Wael Mahmod[1,2] and Hassan Fath[1]

1 Environmental Engineering Department, School of EECE, Egypt-Japan University of Science and Technology, E-JUST, P.O.Box 179, New Borg Al-Arab City, Postal Code 21934, Alexandria, Egypt.
2 Civil Engineering Department, Faculty of Engineering, Assiut University, 71515 Assiut, Egypt
E-mail: mohammed.esmail@ejust.edu.eg

Abstract. The shoreline trajectory of Damietta city, located at the Northern coast of Egypt, is dramatically subjected to significant kinematic changes. Several types of coastal measures have been applied substantially along the coastal stretch of Damietta to protect shoreline (detached breakwaters, Jetties, groins, and seawalls). This study focuses in the assessment of shoreline kinematics response due to the existence of these structures during the period from 1990 to 2015 using satellite images and geo-spatial tools. Three semi-automatic extraction techniques are initially tempted for Landsat ETM 2003 imagery namely; Iso cluster technique, threshold method, and onscreen digitizing. The shoreline variation measurement of is described for three sectors based on the allocation of the coastal measures:(1) the western sector encompassing Damietta port with two jetties; (2) the central sector including detached breakwaters; (3) the eastern portion of Damietta estuary passing through a seawall. The results show that, the shoreline has progressed by +10.0 m/year rate west of Damietta port. Behind the detached breakwaters, the shoreline has advanced by +12.0 m/year from 1999 to 2003 and then has decreased gradually to be near the stable state in 2015. The highly eroded by average rate of -78 m/year appeared at end of eastern seawall.

1. Introduction

The Nile Delta is one of the earliest Delta formed in the world. It extends about 240km from west of port said to east of Alexandria at Egyptian northern coast. Rosetta and Damietta promontories and three lagoons, are the most notable topographic features distinction the Nile Delta,. These range from west Idku, Burullus, and Manzala lagoons, taking a large percentage of delta surface area,[1].
Prior to AHD (Aswan High Dam) construction, 100–115 million m^3/y of sediment were discharged into the Mediterranean,[2] and forms headlands into both Damietta and Rosetta promontories and the intervening embayment beaches. The coastline at Damietta branch progressed offshore by 3.0 km between year 1800 and year 1900,[3].
Shoreline is one of the fast varying coastal landforms,[4]. Researcher largely depending on accurate and up-to-date shoreline change information magnitudes in most of coastal studies such as the aspect of accretion-erosion, design of coastal protection structure, shoreline prediction, hazard zoning, environment defense, safe ship navigation, and sustainable resources management at coastal zones,[5].
Breakwaters are defense structures constructed to protect the coastal areas or to safe an anchorage from both longshore drift and weather effects. Breakwaters decrease the wave action strength in

IOP Conf. Series: Earth and Environmental Science **151** (2018) 012027 doi:10.1088/1755-1315/151/1/012027

shallow waters and therefore reduce coastal erosion or provide highly safe harborage. Groins, seawalls, and detached breakwaters were uses along Damietta coast to protect it from erosion.

Shoreline changes were detected and mapped by several approaches. The usual field surveys accomplish high-accuracy measurements but are time consuming and labor intensive, in addition of highly cost, [6]. High-resolution aerial photographs contain many details of topography features, and coastline scan be delineated and spotted with high precision, however, periodic over flights and aerial photograph analyses imply a high cost for updating coastlines, [7]. The Satellite images multispectral nature and repetitive coverage is the most suitable method for digitizing and updating the changes of the shorelines at uniform intervals accurately and cost-effectively , [8]. Over recent decades, Satellite images have become increasingly important in detection and mapping of shorelines, especially because digital images from remote sensing satellites are provided in infrared spectral bands where the water-land interface is well defined. Furthermore, remote sensing information can be integrated with Geographical Information Systems (GIS) as a helpful tool for analyzing and extracting more reliable and consistent information by using satellite imagery as a base data, [4].

In many previous research works, the effect of coastal protection structure have been investigated through both numerical simulation and experimental models, [9]. The performance effect and efficiency of each type has been investigated. In this research, the effect of different types of breakwaters shoreline changes was investigated using Remote Sensing, GIS technique and then compare erosion rates with those collected from previous studies. The considered systems are groins, detached breakwaters, seawalls, submerged breakwaters and shore-connected breakwaters.

Damietta is one of the most important areas for observation beach erosion, since it is used for various activities as transportation, agriculture, fishing and commercial activities represented in Damietta port. The siltation inside the inlet is a huge environmental problem, and the continuous erosion that began after the construction of the Aswan High Dam.

2. Study Area

Damietta coast, 25km length, located at estuary of Damietta branch lies on the north-eastern Nile River delta coast is the study area of the present research. The total length in this study extend from 3.0km western of Damietta port, (13.0km from Damietta promontory) to 12.0 km east of promontory. The shoreline located between E31^0 43' 36" to E31^0 58'(Figure 1).

Ras El Bar resort town, located 145km to the east of Rosetta promontory and immediately to the west of the Damietta entrance, has suffered erosion and shoreline retreat resulting in the loss of buildings and streets. erosion has also adversely impacted recreational resort beaches. The average retreat of this area was about 30 m/year before the construction of the western jetty in 1941, [10]. Erosion continued west of the jetty, so 400 m long seawall was added in 1965. These structures were also experienced to erosion. For this reason, their bases were protected recently by rip-rap and dolos. As erosion still takes place to the west, three groins were built in 1972, [11], directly at the end of the seawall. In 1980, a 350m long rip-rap seawall protection was installed to stop the continued erosion between the three groins. Erosion hits the coastline along the beaches to the west. The recommendations of Delft hydraulics in year 1987, studied the construction of eight offshore detached breakwaters and add the nourishment process to accelerate the process of accretion composing the salient, [9]. A second small jetty was constructed on the eastern side of the Damietta Estuary entrance in 1976 to reduce siltation in the outlet. Two short breakwaters were constructed in 1980 on both sides of the navigation channel at the new Damietta Port Channel which extends in the sea to 15 m contour.

In 1972 a vertical concrete wall with length 1,500 m was installed to protect the coastal road located in the eastern part of the Damietta Promontory. However, erosion continued and the road was damaged and is no longer passable. The erosion continued and threatened the lighthouse which was then protected by an earthen embankment in 1991. This protection was also ineffective against the extreme erosion pressures, and the lighthouse has now been abandoned and replaced by one farther inland, [10]. Fig. (1) shows the study area.

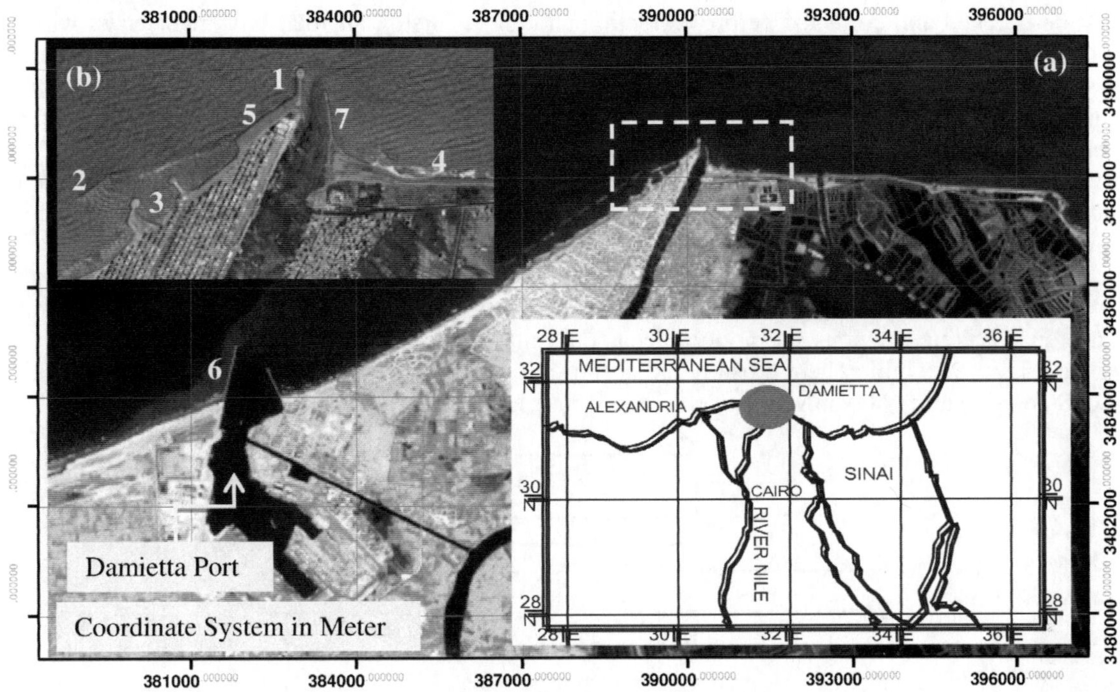

Figure 1. Damietta Promontory Area at the Terminal of Damietta Branch (SPOTimage, 2015);(a)Satellite Image 2015, (b) Google Earth Image, 2017), (1) western jetty of Damietta branch entrance, (2) detached breakwaters, (3) three groins in 1972 and Ras El-Bar now, (4) eastern seawall, (5) rip rap and dolos seawall, (6) Damietta port breakwaters, and (7) eastern jetty of Damietta branch entrance.

3. Study Material and Methods

3.1. Satellite Images

There are four images were acquired for different periods 1990, 1999, 2003 and 2015 for total period of 28 years. All images were selected to be in high quality and the cloud is less than 10% effective and of 30m spatial resolution. They are downloaded in GeoTIFF format at no cost from the US Geological Survey (USGS) Earth Explorer Website (http://www.earthexplorer.usgs.gov). These Landsat data constitute the largest useable database of medium- resolution images that was available, [4]. There are two images 1999, and 1999 from Landsat-1 Multispectral Scanner (MSS), Landsat-5 Thematic Mapper (TM) and other two images 2003, and 2015 were acquired from Enhanced Thematic Mapper Plus (ETM+). The software ENVI v4.8 and ARC GIS v10.1 were used in processing using all mentioned images for extracting shoreline. Table 1 illustrates all images information.

Table 1. Details of the satellite images used.

Acquired Date	SpaceraftID/Sensor	Path/Row	Pixel Size (m)	Coordinate System/Datum	Zone
Aug04, 1990	Landsat_5/TM	176/38	30	UTM/WG84	36
Sep 06, 1999	Landsat_7/TM	176/38	30	UTM/WG84	36
Feb 05, 2003	Landsat_7/ETM	176/38	30	UTM/WG84	36
Nov 29, 2015	Landsat_8/ETM	176/38	30	UTM/WG84	36

3.2. Image Processing

Figure 2 illustrates the pre-processing and post processing operations in this study. Images pre-processing is an operation to prepare images to be more convenient purpose of the study. All images must be checked and corrected before extracting shoreline. Firstly, all images were selected without any stripping defects and were stacked and clipped to the study area, then geometric and radiometric corrections were conducted. Radiometric correction is very important to reduce the atmospheric effects. Landsat satellite sensor type, Sun elevation and data acquisition (Day, month and year for every image) were used in radiance type technique to conduct the radiometric correction.

All images were subjected to Geo-reference process to the World Geodetic System (WGS84) datum and the Universal Transverse Mercator (UTM) projection system. Geometric correction interpretation based on the digital Satellite map as a base map to Landsat image in 2015[12].

For images post-processing, Remote Sensing and GIS technologies have been used for mapping shoreline and for monitoring the coastal processes. In the present study, the accretion and erosion rates and eroded/accreted regions are noticeable in the multi-temporal satellite data (i.e. coastline). GIS thematic layer was built where erosion and accretion rates were calculated. Simply, a transformation of GIS polyline into polygon was carried out to isolate areas of accretion and erosion [13].

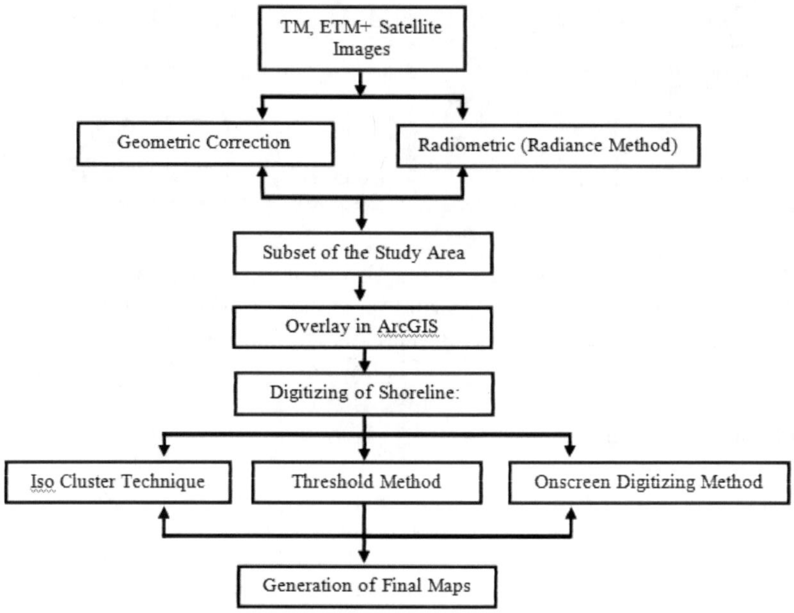

Figure 2. Flow-chart diagram showing the image processing steps applied in the study

3.3. Shoreline Extraction

Three methods were used to extract the studied zone shoreline, Iso cluster technique, threshold method, and onscreen digitizing method. The first and second one are automated shoreline extraction methods that have been developed using ArcGIS 10.1 software from geometrically rectified single band satellite images.

Near-infrared (NIR) wavelengths, water unclouded dark in the image because of its powerful absorbance and exposed soil areas or mainly vegetation that appear brighter because of their strong reflectance, [14]. According to this concept, shoreline can be delineated from a single band image, since the reflectance of water is approximately equal to zero in the reflective infrared bands. This can be completed through measuring histogram threshold for one of the infrared TM or ETM+ bands. According to the histogram threshold technique, water areas can be separated from other land covers and coastline is consequently extracted from Band 5 (mid-infrared) in TM and ETM+ images, [15].

Iso cluster technique is one of unsupervised classification methods which employed a modified iterative optimization clustering processes, also known as the departing means technique. The algorithm disconnect all cells into the specified number of featured unimodal groups in the multidimensional space of the input bands. Two main steps are used to execute this method, firstly is generate clusters, secondly is classify to clusters, [16].

Onscreen digitizing method is considered as one of the most common tools that used specifically for shoreline demarcation, [17], as it provides a highly precision resolution approaches up to 5.0 m. Therewith, this technique depends mainly on the manual tracing of the land-water boundary line in the saturated zone via selecting a path from the Google Earth tool bar. Where, the high-visibility vision of the saturated zone limitation enables the path to be easily traced over the sinuosity of the coastline with the least possible errors.

4. Results and Discussion

4.1. Shoreline Extraction Verification

The shoreline was digitized by the above mentioned three different methods. The results obtained are validated by comparing the data on shoreline change available in situ measurements in year 2003, which indicated a good correlation, (Figure 3).

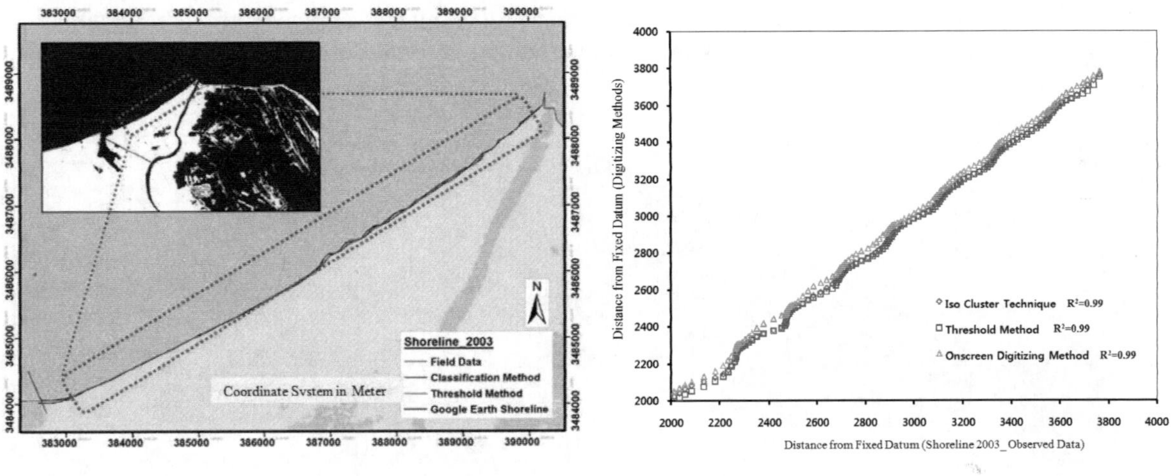

Figure 3. Shoreline in 2003; Field data, Iso Cluster, threshold and Onscreen Digitizing method.

4.2. Comparative Study

For the three suggested classification techniques, it was found that the extracted shorelines have the same trend. From statistical point of view, It is confirmed by the meaning of Coefficient of determination (R^2) which have almost the same value of 0.99 for each technique. Consequently, (RMSE) is precisely selected to assign the optimal one. The obtained (RMSE) values are calculated and presented in Table 2. Accordingly, Iso cluster technique has statistically the superiority of shoreline delineation and attempt to evaluate the transient effect of protection works on the morphology of the shoreline in the study areas.

Table 2. RMSE Values for Every Method.

Digitizing method	Iso Cluster Technique	Threshold Method	Onscreen Digitizing Method
RMSE	22.20	29.50	26.42

Extraction of Damietta shoreline depends on some parameters as; waves, current, grain size, beach slope and beach layout. The shoreline changes are the main indicator for the efficiency of protection works, so digitizing of shoreline at Damietta coast was carried out using spatial tools at four different years 1990, 1999, 2003, and 2015 covered 28 years. Four types of protection works were constructed at Damietta coast; Port breakwaters, seawalls, detached breakwaters and two groins manufactured Ras El Bar bayas shown in Fig. 1, and the results are explained as following:

4.3. Damietta Port Breakwaters

In 1980, two breakwaters were constructed to protect navigation channel for Damietta port. The west one is 1500m length and east one is 800m length as shown in Figure 4. The effect of the two breakwaters are strongly apparent in accretion at west side of port and erosion at other side in all years. The period from 1990to 1999, the effect of Damietta port breakwaters appears in shorelines behavior, accretion occurs western the port by 14.0 m/year, and erosion process occurs at the eastern part of the port by a rate of 10.0 m/year (Figure 4). The accretion still advanced by the same rates to year 2015 but the erosion rate was decreased to be 5.0 m/year. The shoreline changes in west and east of the port represents the dangerous to both port navigation channel and the land east the port.

4.4. Detached Breakwaters front of Ras El Bar Resort

Eight detached breakwaters 200m in length with 200m spaces between each of them were constructed through the period from 1986 to 2000, (Figure 5). The main function of these breakwaters is to protect Ras El-Bar sea resort.The period from 1990 to 1999, the erosion extended from the port east to the area behind the detached breakwater which has been constructed before this period, this part of shoreline had advanced by rate 12.0 m/year.

Thereafter, the pattern of shoreline from year 1999 to year 2003 almost continued in accretion but with lower rates of 6.0 m/year. The changes in shoreline behind detached breakwaters reached to stable state within the period 2003-2015.

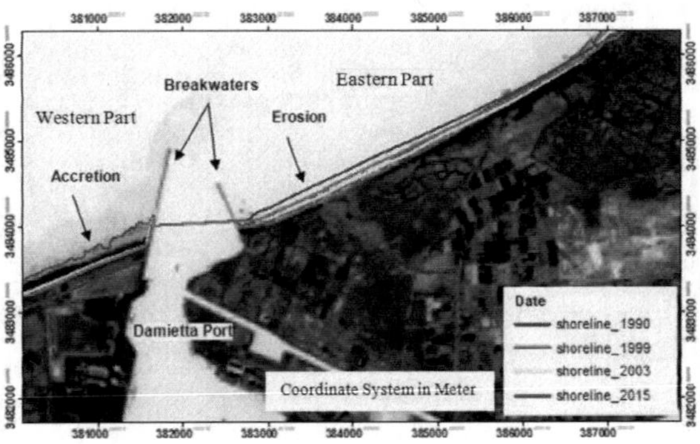

Figure 4. Satallite Image (Landsat_8/ETM); Shoreline at Damietta Port; 1990,1999,2003, and 2015.

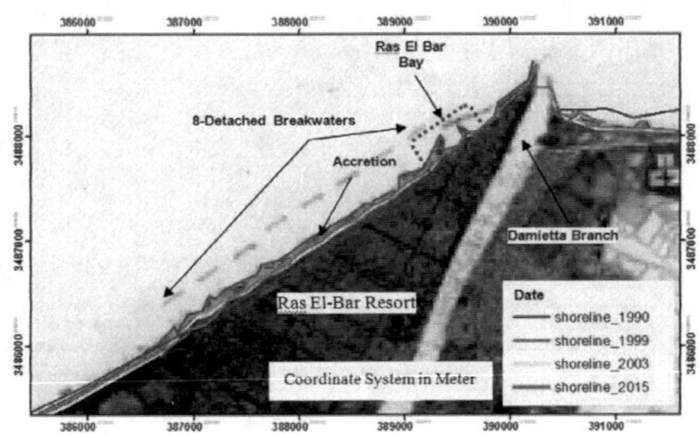

Figure 5. Satellite Image (Landsat_8/ETM); Shoreline at Ras El Bar Resort; 1990,1999,2003, and 2015.

4.5. Seawall Front of Ezbet El Borg

The erosion at east side of Damietta Promontory started after constructed of Aswan High Dam and continued until constructed of 6.50km seawall directly eastern of the Damietta estuary at year 2000, (Figure 6). The erosion rates reached to 60 m/year in period between 1990 to 1999, and then the seawall was successful in stopping this erosion however only in its location. The erosion was shifted to lee side of seawall and shoreline still moved onshore but at higher rates reached to 78m/year. The erosion reached to the old coastal road as shown in Figure 7, that shows the importance of protection works at this area.

4.6. Verification with Previous Studies

In comparison with different remote sensing techniques, the established trend of the one-dimensional shoreline change rate (m/year) along the portions (1), (2), (3) and (4) of Damietta coast that extracted using Iso cluster technique is found to be rather consistent with other earlier studies. This is clear by comparing the present results with the results of El-asmar, [1] and El-zeiny, [13] as consigned in Table (3). However, a quiet disparity between the results is found due to the deference in both the selected elapsed times for each study, and the precision of the utilized coastline demarcation technique.

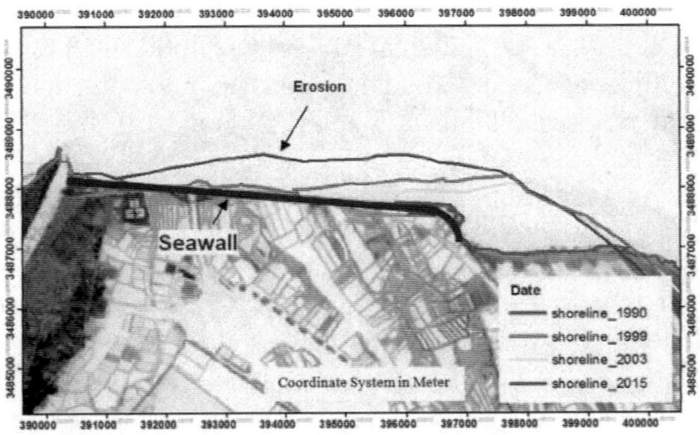

Figure 6. Satellite Image (Landsat_8/ETM); Shoreline at Ezbet El Borg; 1990,1999,2003, and 2015.

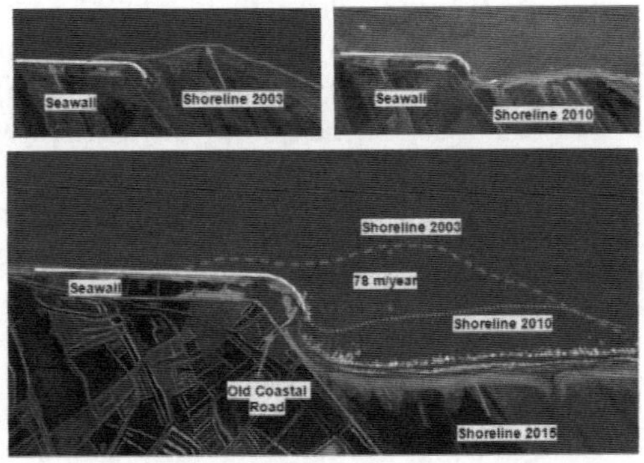

Figure 7. Satellite Image 2003, 2010, and 2015 (Google Earth); Shoreline at Ezbet El Borg; 2003,2010, and 2015.

Table 3. Comparing shoreline changes with another studies

Shoreline Part	Date	Rate(m/year)	Previous Rates	
			Date	Rate(m/year)
Part(1) Damietta Port (west)	1990_1999 1999_2003 2003_2015	+14 +14 +14		
Part(1) Damietta Port (East)	1987_1990 1990_1999 1999_2003 2003_2015	-78 -10 -10 -03	[1]1987_1991 [1]1991_2003 [13]1984_2014	-13.5 -10.3 -29.50 to -16.40
Part(2) Detached Breakwaters	1990_1999 1999_2003 2003_2015	+12 +06	[1]1987_1991 [1]2003_2006	-9.9 +6.7
Part(3) Eastern Seawalls	1990_1999 1999_2003	-60 -78	[1]1987_1991 [1]1991_2003 [13]1984_2014	-27.5 to -67.5 -18.3 to -30 -48

5. Conclusions

Damietta coast suffers from shape changed all time due to effects of waves and currents, so several types of coastal protection structure have been used along this coast (i.e Jetties ,groins, detached breakwaters and seawalls). Four Landsat images, 1990, 1999, 2003 and 2015 were used to investigate the coastal measures effects on temporal shoreline change. Effects of three main protection types were investigates. First one is the port breakwaters, whereas, the predominant character is that the shoreline has progressed west of port by an average beach growth rate of +10.0 m/year. Contrariwise, the shoreline on the east side of port (down drift) has negatively retreated and the corresponding beach is regressed at rates of -5.0 m/year. Second type is the eight detached breakwaters in front of Ras El-Bar resort. The shoreline formation is manifest in the parts directly behind of structure, where the shoreline has advanced by +12.0 m/year from year 1999 to year 2004 and then has decreased gradually to be in the stable state in 2015.The last type is the eastern seawall of Damietta promontory, which succeeded to overcome the shoreline erosion by reflecting the wave energy back into the sea, thus the wave energy available to cause erosion are reduced. However, it accelerate erosion of adjoining, unprotected coastal areas in down drift side, as results a highly eroded area by average rate of-78m/year just eastern of seawall.

6. References

[1] El-asmar HM, El-kafrawy SB, Taha MMN 1993 Monitoring Coastal Changes along Damietta Promontory and the Barrier Beach toward Port Said East of the Nile doi:10.2112/JCOASTRES-

[2] Sestini G 1989 Nile Delta: a review of depositional environments and geological history. *Geol Soc London, Spec Publ.*;41(1):99-127
[3] Syvitski JPM 2008 Deltas at risk. *Sustain Sci.*3(1):23-32. doi:10.1007/s11625-008-0043-3
[4] Louati M, Saïdi H, Zargouni F 2015 Shoreline change assessment using remote sensing and GIS techniques : a case study of the Medjerda delta coast , Tunisia. 4239-4255. doi:10.1007/s12517-014-1472-1
[5] Szmytkiewicz M, Biegowski J, Kaczmarek LM, et al. 2000 Coastline changes nearby harbour structures: comparative analysis of one-line models versus field data. *Coast Eng.* 40(2):119-139
[6] Nadu T, Natesan U, Thulasiraman N 2013 Shoreline change analysis of Vedaranyam coast ,5099-5109. doi:10.1007/s10661-012-2928-y
[7] Zhao B, Guo H, Yan Y, Wang Q, Li B 2008 A simple waterline approach for tidelands using multi-temporal satellite images: a case study in the Yangtze Delta. *Estuar Coast Shelf Sci.* 77(1):134-142
[8] Al-hatrushi SM 2013 Monitoring of the shoreline change using remote sensing and GIS : a case study of Al Hawasnah tidal inlet , Al Batinah coast , Sultanate of Oman. 1479-1484. doi:10.1007/s12517-011-0424-2
[9] Zidan AR, Rageh OS, Sarhan T, Esmail M 2012 Effect of breakwaters on Wave Energy Dissipation (Case Study : Ras El-Bar Beach, Egypt). *Int Water Technol J.* 2(4):268-283
[10] Fanos AM, Khafagy AA, and Dean RG 1995 Protective works on the Nile Delta Coast. *Journal of Coastal Research;* 11(2), f)16 52H. Fort Lauderdale (Florida), ISSN 0749-0208
[11] Mobarek IE 1972 The Nile Delta coastal protection project. In: *Coastal Engineering*. 1409-1426
[12] Esmail M, Masria ALI, Negm A 2016 Monitoring Land Use / Land Cover Changes Around Damietta Promontory , Egypt , Using RS / GIS. *Procedia Eng.* 154:936-942. doi:10.1016/j.proeng.2016.07.515
[13] El-zeiny A 2016 Space-borne technology for monitoring temporal changes along Damietta shoreline , 4(1):459-468
[14] Aedla R, Dwarakish GS, Reddy DV 2015 ScienceDirect ScienceDirect Automatic Shoreline Detection and Change Detection Analysis of Netravati-GurpurRivermouth Using Histogram Equalization and Adaptive Thresholding Techniques. *Aquat Procedia.* 4(4):563-570. doi:10.1016/j.aqpro.2015.02.073
[15] Masria A, Nadaoka K, Negm A, et al. 2015 Detection of Shoreline and Land Cover Changes around Rosetta Promontory, Egypt, Based on Remote Sensing Analysis. 216-230. doi:10.3390/land4010216
[16] Tarabalka Y, Member S, Benediktsson JA, Chanussot J, Member S 2009 Spectral – Spatial Classification of Hyperspectral Imagery Based on Partitional Clustering Techniques. 47(8):2973-2987
[17] Wang C, Johnston J, Vail D, Dickinson J, Putnam D 2015 High-Precision Land-Cover-Land-Use GIS Mapping and Land Availability and Suitability Analysis for Grass Biomass Production in the Aroostook River Valley, Maine, USA. *Land.* 4(1):231-254

Acknowledgements
The first author would like to gratitude Egyptian Ministry of Higher Education (MoHE) for awarding him the Ph.D. scholarship. Also, thanks and appreciation is due to Egypt-Japan University of Science and Technology (E-JUST) and JICA for their subsidize and for offering the tools needed for this research.

IOP Conf. Series: Earth and Environmental Science **151** (2018) 012028 doi:10.1088/1755-1315/151/1/012028

Methodological Aspects of Building Information System for Monitoring of Landslide Hazards

V Nikolova[1] and P Zlateva[2]

1 Department of Geology and Geoinformatics, University of Mining and Geology, Sofia, Bulgaria
2 IR, Bulgarian Academy of Sciences, Sofia, Bulgaria
E-mail: plamzlateva@abv.bg

Abstract. The aim of the current paper is to present some methodological aspects of building monitoring information system for landslides. Having regard the importance of data in functioning of the monitoring system a special attention is given to the landslide factors, determining the weights of different factors and methods for making a landslide susceptibility map which is a base for selecting monitoring points and assessment of landslide hazard. The paper is not about the monitoring equipment and technical issues of building monitoring system. It concern landslide factor analysis and characteristics of spatial data in geographic information system (GIS) environment used for assessment of landslide susceptibility as a part of monitoring system. Based on ArcGIS Spatial analyst tools (ESRI Inc.) two overlay techniques – fuzzy logic and weighted sum are suggested for making a landslide susceptibility map. Application of GIS allows processing of a large amount of data and easily updating of information. It helps to better communicating the information about landslide hazards to the public by the relevant visualisation of the spatial data and also facilitate decision makers in case of natural hazard.

1. Introduction

Landslides are a movement of earth masses or unconsolidated deposits on a slope surface. These are one of the most often natural hazards. In many cases they have high intensity and sudden occurrence, in others they develop slowly. However in the both cases these natural phenomena can be assessed as hazard and risk regarding the probability of affecting human life. The landslide hazard is usually defined as the probability of occurrence of a potentially damaging phenomenon within a certain area and in a given period of time [1]. The risk is considered as a combination of the probability of a hazard event and of the potential adverse consequences for human health, the environment,, economy and social sphere.

Landslides triggered by earthquakes, intensive rainfall or intensive snowmelt has rapid development and can destroy transport infrastructure and buildings, and also they can decrease the quality of life and even to take lives. The slow movement of surface land masses, activated by changes in groundwater level or rainfall, causes damage gradually and if the appropriate measures to stop the process are not taken into account it can cause irreversibly negative environmental and social impact.

Landslides are wide spread natural hazard and limiting factor for anthropogenic activity and life, and they are studied for a long time. As a result of these researches we can say that landslides are well known and people are informed about the behaviour of landslides. The knowledge of landslides is important in the construction of roads, hydro-technical facilities and the construction of buildings. In this relation the geomorphological researches are related to the geotechnical and hydrogeological ones.

Factors influencing on landslide occurrence and development as lithology, tectonic features, slope angle, road network, drainage network, land use and rainfall are considered in many publications [2]-[5]. A special attention is given to the degree of slope as a main factor of slope surface stability [6]-[8]. Though the wide investigation of landslide factors, the complex character of the landslides and interaction between these factors, and variation of their occurrence and influence, depending of the particular situation, future researches in this field, collecting and analysing new data, as well as applying new approaches of researches are needed. The publications in the last several years, concerning landslides show rapid development of GIS and remote sensing analyses in determining the landslide areas, analysing landslide activating factors, elaborating landslide inventory, susceptibility and hazard maps and landslide monitoring [1], [9]-[14]. Literature review shows application of different approaches in landslide susceptibility map as a first step in assessing landslide hazard and a base for monitoring Using wide scope of spatial information in digital format and computer technology allow processing of big volume of data and applying variety of methods (fuzzy logic, logistic regressions, analytical hierarchy process, weighted overlay) for analyzing landslide factors and making a susceptibility map [5], [10], [15]-[17]. Determining of landslide susceptibility is essential for organising landslide monitoring and particularly for selecting the locations of monitoring points.

In order to mitigate the negative impact of landslide hazards monitoring of the hazardous events should be organized and regularly held. This is essential for predicting the behaviour of landslides and forecasting which storms can trigger large numbers of landslides [18]. A good landslide monitoring system should have the following features: quick communication in case of emergency, reduced burden of observation and monitoring, affordably a low-cost system, and real time acquisition and transfer of the data [19].

The aim of the current paper is to present some methodological aspects of building monitoring information system for landslides. The paper is not about the monitoring equipment and technical issues of building monitoring system. Having regard the importance of data in functioning of the monitoring system a special attention is given to the landslide factors, determining the weights of different factors and methods for making a landslide susceptibility map which is a base for selecting monitoring points and assessment of landslide hazard.

2. Determining Landslide Triggering Factors

The first step in building monitoring system for geomorphological hazards is clarifying the identity of the natural phenomena, subject of the monitoring, and identifying the factors determining the occurrence of natural hazards. When we talk about geomorphological hazards the most important factors are geological – geomorphological and hydroclimatological conditions. The role of other natural components as soils and vegetation is also important. Anthropogenic activity could influence the mentioned factors in different rate and contribute to the development of the natural hazard. The analysis and assessment of factors should be done taking into account the identity and peculiarities of the particular geomorphological hazard which is subject of the monitoring. For the aim of building monitoring system for landslide hazard we consider the following factors: lithology (rock composition), fault tectonics, slope of the topographic surface, rainfall, distance from streams, soils and vegetation.

The type of rocks and the rate of consolidation of rocks fragments determine the water permeability of the rocks and susceptibility of sliding. The area is more susceptible to sliding when it is built by unconsolidated rock materials or earth masses, located on solid or non-permeable rock and slope surface. In the event of heavy rainfall or rapid snow melting, surface water penetrates into the unconsolidated deposits, saturates the surface horizon, and makes the underlying solid rock substrate slippery. In such a situation, even on small slopes, the earth masses slides down the slope of the topographic surface. The sliding speed depends on the slope of the surface and the resistance on the sliding surface. In this relation physical-mechanical properties of soils and particularly soil permeability have strong impact on the landslide occurrence and development. Intensive stream erosion and high groundwater level also contribute to the landslide activation. Land cover / land use influence on the infiltration of surface water in soil, possibility for development of erosion and could cause changes in the groundwater level. Because of this land cover / land use types are considered as landslide triggering factors.

In order to assess landslide susceptibility landslide factors have to be evaluated according to their importance for landslide occurrence. Having regards the characteristics of each one of the landslide triggering factors we suggest the following rates: 1 - very low; 2- low; 3 – moderate; 4 – high; 5 – very high. The rates of susceptibility of each one of the landslide factors should be considered in relation to the other factors. For example unconsolidated deposits have very high sliding susceptibility but if they are located on flat areas the total landslide susceptibility will be lower. Slopes greater than 45° are more prone to slide than 10° slopes, but if the steeper slopes are built by solid rocks and less steep areas are covered by unconsolidated rocks materials, the less steep slopes will have higher landslide susceptibility. If we have two areas with a slope of 15° and they are covered by one and the same moderately solid rock or unconsolidated rock but the first one have high humidity, intensive rainfall or rapid snowmelt and the second is dry, the first area will be more prone to landsides.
Having regard the landslides factor analysis the landslide susceptibility should be assessed taking into account the interaction between the factors and their cumulative impacts. For this purpose Analytic Hierarchy Process [20] can be applied (Table 1). The values in rows of the matrix show the importance of the given factor to that written in the relevant column. For example the type of rocks is evaluated as 5 times more important in comparison to the distance from faults, and we consider the distance from fault (row 2) against the rock type as the reciprocal value (1/5). The rates of different factors are determined by expert's view and assessment but summarizing the rates and using the relatively share for the weights minimise the subjectivity of the evaluation. The analysis of the landslide triggering factors and the weights given in the matrix show that slopes, rainfall, rocks type and soils permeability are the most important factors for landslide occurrence which confirm the above considered example.

Table 1. Matrix of landslides triggering factors and factor weights

Factors	Rocks type	Distance from faults	Slope	Rain-fall	Distance from streams	Stream network density	Ground-water level	Soils permeability	Land cover	Total	Weights (%)
Rocks type	1	5	1	2	4	4	2	1/2	3	22.50	16.15
Distance from faults	1/5	1	1/4	1/4	1/2	1	1/2	1/3	1/2	4.53	3.25
Slope	1	4	1	3	4	3	4	4	4	28.00	20.09
Rainfall	1/2	4	1/3	1	5	5	4	4	4	27.83	19.97
Distance from streams	1/4	2	1/4	1/5	1	1	1/3	1/4	1/3	5.62	4.03
Stream network density	1/4	1	1/3	1/5	1	1	1/3	1/4	1/3	4.70	3.37
Ground-water level	1/2	2	1/4	1/4	3	3	1	1/4	3	13.25	9.51
Soils perme-ability	2	3	1/4	1/4	4	4	4	1	4	22.50	16.15
Land cover	1/3	2	1/4	1/4	3	3	1/3	1/4	1	10.42	7.48
										Total:	100

1 - equal importance; 2 – moderate importance; 3 – strong importance; 4 – very strong importance; 5 – extreme importance

3. Building a GIS Database

3.1. Organising the Data

The GIS database consists graphical and attributive information. All available data for each one of the landslide triggering factors have to be entered as a separate layer in the database. An important condition to do correct spatial analyses is all layers to be in one and the same coordinate system, and particularly the analyses in ArcGIS environment require projected coordinate system. The necessary graphical information with the relevant attribute table is as follow:

- Relief – digital elevation model (DEM) – georeferenced raster file or vector file, presenting isolines of topographic map. It is preferable the map to be enough detailed (for example in a scale 1: 5000; 1:10000; 1:25000).
- Rivers – polyline. Polygon buffer layer is created on the base of rivers lines.
- Meteorological stations – point layers. Discontinuous point data for precipitation should be presented as continuous surface by an appropriate interpolation method. The output file is in raster format. The literature review and authors' tests show that the inverse distance weighted interpolation is more appropriate than the other ones for climate and hydrology data. Regression analysis of precipitation and altitude is also applicable and give good results. However the interpolation method should be chosen taking into account the identity of modelled indicator, number and allocation of the measurement points.
- Groundwater level – polygon layer or interpolated raster surface.
- Soils – polygon layer containing soil types polygons and soil indicators.
- Land cover – polygon layer presenting the land cover (including vegetation types) and land use types.
- Roads, railways – polyline layers.
- Settlements – polygon layer.
- Observed cases of occurrence of landslides – polygon or point layer depending on the map scale. This layer is to be used for the validation of the landslide susceptibility map.

3.2. Assessment of Landslide Susceptibility – Susceptibility Map

The assessment of landslide susceptibility of the observed area can be done on the base of landslide susceptibility map. The susceptibility is considered as a sensitivity degree to sliding. The first step in elaborating a susceptibility map is evaluation of landslide triggered factors (considered above). New layers of evaluated factors are generated in GIS environment. The susceptibility map can be elaborated using an overlay analysis. The vector files should be converted in raster format in order to proceed spatial overlay. We suggest to make the susceptibility map by two overlay techniques – fuzzy logic and weighted sum. The fuzzy logic used as an overlay technique gives good results in modelling natural hazards, particularly when there are short data records or few measuring points [21]. The sequence of making landslide susceptibility map is given on Figure 1.

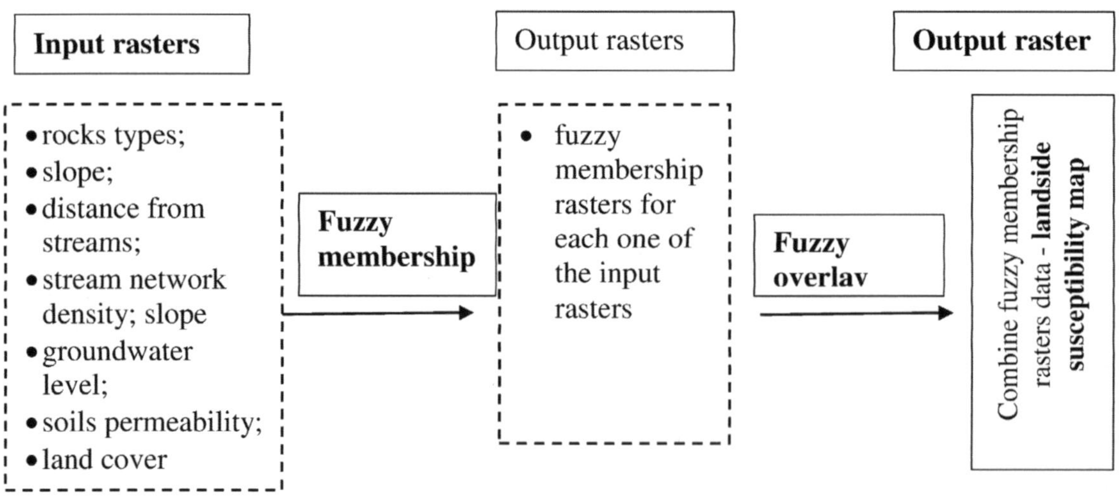

Figure 1. Chart flow of applying fuzzy overlay for making a landslide susceptibility map (based on the ArcGIS Spatial analyst tools)

Weighted sum can be applied using the weights determined in factors analysis. Rasters layers presenting the landslide factors are overlaid, multiplying each ones by the given weight and summing the layers together (Figure 2)

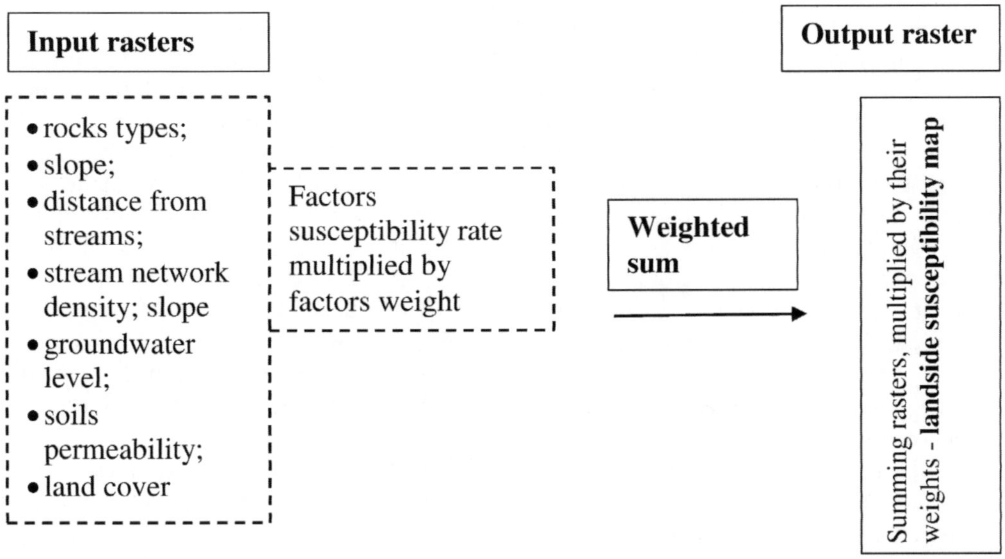

Figure 2. Chart flow of applying weighted sum overlay for making a landslide susceptibility map (based on the ArcGIS Spatial analyst tools)

The next step after making landslide susceptibility maps is to validate the models and to choose the most appropriate method. The validation of the model have to be done on the base of the map of observed cases of landslide in the investigated area.

4. Characteristics of the Monitoring

Landslide monitoring aims to collect information about the landslide occurrence and development, and to warn for possible landslide activation. An important point in organising the natural hazard monitoring is selection of monitoring locations and indicators to be monitored. Landslide monitoring includes observation and measurement of earth mass and deluvial - proluvial materials movement. Rainfall and soil moisture data also have to be monitored. In this regard the monitoring equipment should include landslide speed sensors (landmass moving sensors), rain-gage sensors or stations (automatic weather stations), soil moisture sensors and piezometers in shallow boreholes to observe the changes in water table depth. Ground-based remote monitoring systems consist field sensors but they also employ data acquisition units to record sensor measurements, remote telemetry (such as radio, satellite, or cell phone links), automated data processing, and displays of current conditions (often via the Internet). There is always some delay between sampling remote conditions and displaying these conditions to users, and the term "near-real-time monitoring" is commonly used [22].
About the time and periodicity of monitoring there are continuous monitoring which gives an information about the monitored phenomenon in real time and periodic monitoring which gives information at the particular moment. Continuous monitoring of landslide triggering factors and particularly rainfall is essential for areas estimated as high and very high susceptible to slide, while movement of slope materials could be monitored seasonally.
Selection of the monitoring points location should be done on the base of landslide inventory map (presenting the locations of observed landslides) and landslide susceptibility map (presenting areas prone to slide).

5. Updating of the Susceptibility Map

The landslide susceptibility map should be updated on the base of the monitoring data and other geological-geomorphological researches of the area. The most changeable components of the monitoring system are precipitation, humidity conditions and vegetation. This requires regular updating of the components maps which to be reflected to the susceptibility map. If new areas of high landslide susceptibility are determined after the updating they should be checked and of case of need a new monitoring point to be established.

6. Possibilities of GIS in Building and Operation of the Monitoring System

Building and operation of landslide monitoring system require collecting and organising big volume of various information about landslide activating factors and infrastructure. Geographic information systems give big opportunities in entering and analysing spatial data. Application of GIS allows processing of a large amount of data and easily updating of information. GIS are also an important part of real-time (near-real-time) monitoring. It helps to better communicating the information about landslide hazards to the public by the relevant visualisation of the spatial data. Landslide inventory and landslide susceptibility maps elaborated in GIS environment as well as the all data collected in the monitoring process can be used in territorial planning and could help the decision making.

Though the advantages of GIS technology there are some uncertainties. They are mostly related to the quality of the initial data about the landslide factors. For example the data from different sources and in different scales can present the modelled phenomena in varying degrees of detail or there would be some gaps in the data. Often remote sensing data is used in the spatial analyses and the quality of the images are not always good because of the clouds conditions.

The capacity of the GIS software and hardware facilitate the statistical and spatial analyses but on the other side they could be considered as a limiting factor concerning the selected techniques of the analyses, having regard the comprehensive character of landslides as a natural phenomenon which depends of a complex of interacting factors.

7. Conclusion

Monitoring of landslides is an important part of preventative activity and mitigating the negative effects of sliding. It is a powerful tool for territorial planning authorities in decision making and disaster reduction. For the proper functioning of the monitoring, the information system should contain reliable information about the factors that determine the occurrence of landslides. These factors can be grouped in the following groups: geology-geomorphological; hydro-climatological and land cover and anthropogenic factors. The analysis of the factors considered in the current paper shows the highest importance of slope, rainfall, lithology and soils permeability. The first stage in mitigating and tackling the landslide problem is collecting data, making landslide inventory and landslide susceptibility maps. Application of information technology provide opportunity for using different approaches and techniques for elaborating of landslide susceptibility map. Two overlay techniques (fuzzy overlay and weighted sum) are suggested in this research. However the spatial analyses should be chosen taking into account the availability of the data, number and location of the monitoring points as well as the complex of the landslide factors. In this relation more investigations are needed in removing data imperfections and adding more landslide factors in consideration, in order to have more reliable monitoring system.

8. References

[1] Varnes D J 1984 *Landslide Hazard Zonation: A Review of Principles and Practice* (UNESCO, Paris) pp 1-63.

[2] Bathrellos G D, Kalivas, D P and Skilodimou H D 2009 GIS-based landslide susceptibility mapping models applied to natural and urban planning in Trikala, Central Greece *Estudios Geológicos,* 65(1) pp 49-65, ISSN: 0367-0449, doi:10.3989/egeol.08642.036

[3] Guimarães R F, Machado V P, Abílio de Carvalho J O, Montgomery D R, Trancoso Gomes R A, Greenberg H M, Cataldi M and Mendonça P C 2017 Determination of Areas Susceptible to Landsliding Using Spatial Patterns of Rainfall from Tropical Rainfall Measuring Mission Data, Rio de Janeiro, Brazil *International Journal of Geo-Information*, 6, p 289

[4] Magliulo P, Di-Lisio A, Russo F, et al. 2008 Geomorphology and landslide susceptibility assessment using GIS and bivariate statistics: a case study in southern Italy. *Natural Hazards* 47 pp 411-435. https://doi.org/10.1007/s11069-008-9230-x

[5] Sahana M, Sajjad H 2017 Evaluating effectiveness of frequency ratio, fuzzy logic and logistic regression models in assessing landslide susceptibility: a case from Rudraprayag district, India. *Journal of Mountain Science* 14: 2150 https://doi.org/10.1007/s11629-017-4404-1
[6] Lee S, Min K 2001 Statistical analysis of landslide susceptibility at Yongin, Korea. *Environment Geology* 40 pp 1095-1113 https://doi.org/10.1007/s002540100310
[7] Saha A K, Gupta R P, Sarkar I, et al. 2005 An approach for GISbased statistical landslide susceptibility zonation—with a case study in the Himalayas. *Landslides* 2 pp 61-69.
[8] Sitanyiova D, Vondrackova T, Stopka O, Mysliveckova, M, Muzik, J 2015. GIS Based Methodology for the Geotechnical Evaluation of Landslide Areas. *Procedia Earth and Planetary Science*, 15 pp 389-394.
[9] Gupta R P, Joshi B C 1990 Landslide Hazard Zonation using the GIS Approach – A case Study from the Ramganga Catchment Himalayas *Engineering Geology* 28 pp 119–131
[10] Mancini F, Ceppi C, Ritrovato G 2010 GIS and statistical analysis for landslide susceptibility mapping in the Daunia area, Italy *Natural Hazards Earth System Sciences* 10 pp 1851–1864.
[11] Mahler1 C F, Varanda E, De Oliveira L C D 2012. Analytical Model of Landslide Risk Using GIS. *Open Journal of Geology* 2 pp 182-188, http://dx.doi.org/10.4236/ojg.2012.23018
[12] Ciampalini A, Raspini F, Bianchini S, Frodella W, Bardi F, Lagomarsino D, Di Traglia F, Moretti S, Proietti C, Pagliara P et al. 2015. Remote sensing as tool for development of landslide databases: The case of the Messina Province (Italy) geodatabase *Geomorphology* Vol 249 pp 103–118, http://dx.doi.org/10.1016/j.geomorph.2015.01.029
[13] De la Cerna M A and Maravillas E A 2015 Landslide Hazard GIS-based Mapping Using Mamdani Fuzzy Logic in Small Scale Mining Areas of Surigao del Norte, Philippines, *Proceedings of the World Congress on Engineering and Computer Science* Vol II. pp 901 – 906 (WCECS 2015, October 21-23, 2015, San Francisco, USA)
[14] Casagli N, Cigna F, Bianchini S, Hölbling D, Füreder P, Righini G, DelConte S, Friedl B, Schneiderbauer S, Iasio C et al. 2016 Landslide mapping and monitoring by using radar and optical remote sensing: Examples from the EC-FP7 project SAFER, *Remote Sensing Applications: Society and Environment* 4 pp 92–108
[15] Ayalew L, Yamagishi H 2005 The application of GIS based logistic regression for landslide susceptibility mapping in the Kakuda–Yahiko Mountains Central Japan *Geomorphology* 65 (1/2) pp 15–31
[16] Champatiray P K, Dimri S, Lakhera RC, Sati S 2007 Fuzzy-based method for landslide hazard assessment in active seismic zone of Himalaya *Landslides* 4 pp 101–111
[17] Kayastha P, Dhital M, De Smedt F 2013 Application of the analytical hierarchy process (AHP) for landslide susceptibility mapping: a case study from the Tinau watershed, West Nepal. Computers & Geosciences vol 52 pp 398–408
[18] USGS Landslide monitoring https://landslides.usgs.gov/monitoring/
[19] Tanaka T 2013 *Int. Journal of Landslide and Environment* 1(1) pp 101-102
[20] Saaty R W 1987 The analytic hierarchy process – what it is and it is used Math Modelling vol 9 number 3-5 pp 161 - 176
[21] Nikolova V, Zlateva P 2017 Assessment of Flood Vulnerability Using Fuzzy Logic and Geographical Information Systems. In: Information Technology in Disaster Risk Reduction. 2016. IFIP Advances in Information and Communication Technology, vol 501 (Springer, Cham)
[22] Reid M, LaHusen R G, Baum R L, Kean J W, Schulz W H and Highland L M 2012. Real-time monitoring of landslides. USGS. Factsheet. pubs.usgs.gov/fs/2012/3008/contents/FS12-3008.pdf

Acknowledgments
The authors express their gratitude to the Bulgarian National Science Fund for the partial financial support under Grant № DNTS/ China 01/6 from 2014, Bulgarian–Chinese scientific cooperation

IOP Conf. Series: Earth and Environmental Science **151** (2018) 012029 doi:10.1088/1755-1315/151/1/012029

A Fuzzy Logic Model for Assessment of Socio–Economical Consequences for Household in Case of Power Outage due to Natural Disasters

P Zlateva[1], S Tzvetkova[2] and D Velev[2]

1 IR, Bulgarian Academy of Sciences, Sofia, Bulgaria
2 University of National and World Economy, Sofia, Bulgaria
E-mail: plamzlateva@abv.bg

Abstract. The paper propose a fuzzy logic model for level assessment of the negative socio–economical consequences to the household and each of its members in case of power cuts due to natural disasters. The appropriate preventive activities are proposed to help to reduce the vulnerability and to increase the preparedness against the negative impact of power cuts on households. The fuzzy logic model is designed as a three-level hierarchical system with four inputs and one output. The system is designed in Matlab computer environment using Fuzzy Logic Toolbox. The designed fuzzy logic model is part of the Cloud Integrated Information System for risk management of natural disasters, which are developed.

1. Introduction

In recent years, there has been an increase in the number of extreme weather related disasters throughout the world. Natural disasters have become the primary threats for the continuity of electricity service in Europe and in the rest of the world [1]. The electricity has now become a vital part of modern societies. The electrical equipment is involved in every aspect of people's lives. Therefore, the power outage severely disturbs the normal course of daily activities [2].

It is useful for every household to have more information about level of the potential negative socio-economical consequences in case of power cuts due to natural disasters [3]-[5]. This is especially important in case of natural disasters characterized by extreme temperatures [6], [7].

Therefore, each household should plan and implement appropriate vulnerability reduction activities according to the type of expected extreme temperatures, whether they are extremely high or extremely low [8]-[11]. It should be noted that the level of negative socio-economical consequences for the household depends strongly on the duration of the power cut in extreme temperature conditions.

There are various qualitative and quantitative methods for assessment of negative socio-economical consequences in case of power cuts. However, there are not enough methods to take into account the subjective perception of the household and each of its members about the level of negative consequences in case in case of power cuts due to natural disasters [12].

The intelligent methods are an appropriate tool for assessment of these negative consequences. These methods using the fuzzy logic theory provide adequate processing of the expert knowledge and uncertain quantitative data [13]. The concept of the fuzzy logic is proposed by Zadeh in 1965 [14]. It is developed around the basic idea of so-called fuzzy sets or membership function. Fuzzy set theory is regarded as an extension of classical set theory. It enables the processing of imprecise information by means of membership functions, in contrast to Boolean transformations. Usually the membership functions is assigned 0 to false values and 1 to true ones, but fuzzy logic also allows in-between values [15].

The aim of this paper is to propose a fuzzy logic model for level assessment of the negative socio-economical consequences to the household and each of its members in case of power cuts due to natural disasters. The appropriate preventive activities are proposed to help to reduce the vulnerability and to increase the preparedness against the negative impact of power cuts on households.

2. A Three-level Hierarchical Fuzzy System for Assessment of Negative Socio–Economical Consequences for Household in Case of Power Outage

The general idea of this research is to propose on fuzzy logic model for assessment of socio – economical consequences for the whole household or individual household member, which may occur during power outage due to natural disasters.

Here, the fuzzy logic model is designed as a three-level hierarchical system with four inputs and one output. The system inputs correspond to the linguistic variables (indicators), which described four types negative consequences (negative effects) by reason of household power outage. The system output represents an assessment of negative socio–economical consequences for household or its members in case of power outage due to natural disasters.

The each system level of the designed system includes one fuzzy logic subsystem with two inputs and one output. The output of the third subsystem is output of the designed fuzzy logic system. A scheme of this three-level hierarchical fuzzy system is presented on Figure 1.

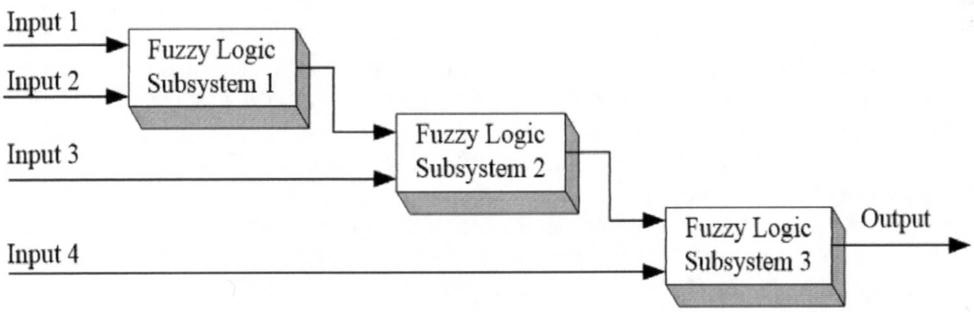

Figure 1. Three-level hierarchical fuzzy system.

Those four input variables of the designed fuzzy logic system are defined using the expert knowledge and research studies related to various types of household members, as follow:

Input 1 "*Negative effects on domestic activities*": Inability to use all electrical appliances to carry out household daily activities related to the normal operation and maintenance of the house - cleaning, ironing, washing, cooking, etc.; Difficulties caused by lack or reduced lighting; Blocking access to private cars that are located in garages with electrical power supply and remote control.

Input 2 "*Negative effects on ICT activity*": Interruption of all information sources - internet connection, radio and television channels; Difficulty in using home phone; Inability to use mobile phones after exhausting of batteries if there is no extra charger, etc.

Input 3 "*Negative physical and psychological consequences*": Difficulties in daily work obligations, stress and moral damages of household members, etc.

Input 4 "*Indirect economic costs*": Spoilage of food products after defrosting refrigerators and freezers for prolonged power outage; Wasting the batteries of the alarm system, Unforeseen future payments related to the incurred (example, damages in various serving systems), etc.

Two intermediate linguistic variables are defined in the design of the model as follows:

Intermediate variable 1 "*Negative effects on daily activity*";

Intermediate variable 2 "*Negative social consequences*".

These two intermediate variables are the outputs of the first and second subsystems respectively.

IOP Conf. Series: Earth and Environmental Science **151** (2018) 012029 doi:10.1088/1755-1315/151/1/012029

In proposed model, the inputs of the first fuzzy logic subsystem are Input 1 "*Negative effects on domestic activities*" and Input 2 "*Negative effects on ICT activity*", and the output variable is the Intermediate variable 1 "*Negative effects on daily activity*".

The inputs of the second fuzzy logic subsystem are Intermediate variable 1 "*Negative effects on daily activity*" and Input 3 "*Negative physical and psychological consequences*", and the output variable is the Intermediate variable 2 "*Negative social consequences*".

The inputs of the third fuzzy logic subsystem are Intermediate variable 2 "*Negative social consequences*" and Input 4 "*Indirect economic costs*". The output of this fuzzy subsystem is output of the whole fuzzy logic system. The linguistic output variable is defined as "*Negative socio-economical consequences*".

Here, the value of this output variable gives an assessment for the level of the "*Negative social-economical consequences*" for household or its members in case of power outage due to natural disasters. The higher value of the level corresponds to greater severity of the negative socio-economical consequences.

It has to be emphasized that each estimated level of consequences refers to a certain time interval of power outage. It is logical to have different levels for the negative socio-economical consequences for periods of 3, 9, 12, 24 hours. Nevertheless, there is a direct positive correlation between the level of the consequences and the duration of the time-free interval. This dependence is not necessarily linear.

Direct economic costs in cases of a sudden power outage of the household normally associated destruction or damage to the various electrical devices (some of which are very expensive). These direct costs can be very high compared to other costs of the household or some individual members. Generally, the direct costs can be objectively calculated without taking into account the subjectivity and personal attitude of the individual members of the household to the electrical devices compared to other costs as *Indirect economic costs and Social cost*. For this reason, the proposed fuzzy logic model does not include the *Direct economic costs*.

3. Design of Fuzzy Logic Model for Assessment of Socio–Economical Consequences for Household in Case of Power Outage

In proposed fuzzy logic model the input linguistic variables (four indicators and two intermediate) are represented by three fuzzy membership functions: "Low", "Middle", and "High". The all input variables are assessed in the interval [0, 10] using trapezoid membership functions.

The fuzzy logic system output (Negative socio-economical consequences) is described by five fuzzy membership functions: "Very low", "Low", "Middle", "High", and "Very high". The level of negative socio-economical consequences is assessed in the interval [0, 100] using triangular membership functions. The input and output membership functions are shown on Figures 2-3.

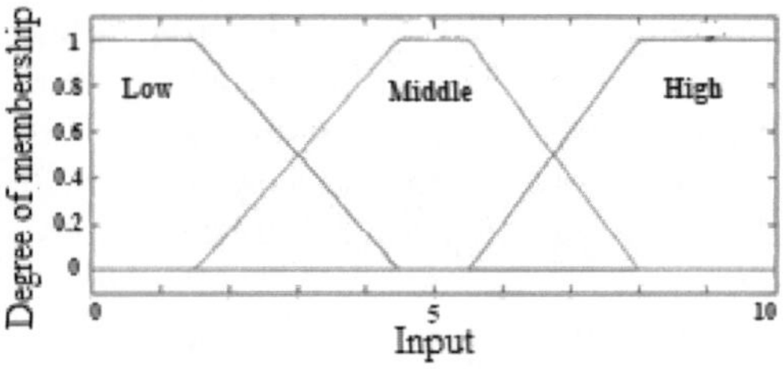

Figure 2. Membership functions of the input variables.

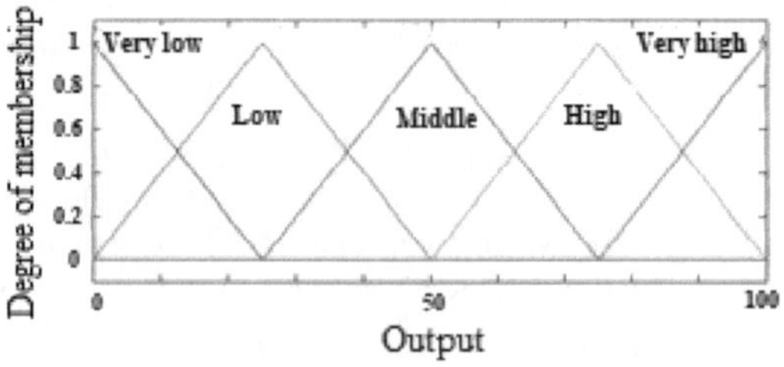

Figure 3. Membership functions of the fuzzy system output.

The inference rules in the fuzzy logic system and subsystem are defined through "IF - THEN"-clause. Rule numbers of the knowledge base per each of the fuzzy logic subsystems are 9. Some of the inference rules are defined as follow:

IF "*Negative effects on domestic activities*" are "*Low*" and "*Negative effects on ICT activity*" are "*High*" THEN "*Negative effects on daily activity*" are "*Middle*";

IF "*Negative effects on domestic activities*" are "High" and "*Negative effects on ICT activity*" are "Middle" THEN "*Negative effects on daily activity*" are "*Middle*";
IF "*Negative effects on daily activity*" are "*Low*" and "*Radius of turn*" are "*Middle*" THEN "*Intermediate variable 2*" are "*Low*";
IF "*Negative effects on daily activity*" are "*Middle*" and "*Negative physical and psychological consequences*" are "*High*" THEN "*Negative social consequences*" are "*Middle*";
IF "*Negative effects on daily activity*" are "*High*" and "*Negative physical and psychological consequences*" are "*Middle*" THEN "*Negative social consequences*" are "*High*";

……

IF "*Negative social consequences*" are "Low" and "*Indirect economic costs*" are "Low" THEN "*Negative socio-economical consequences*" are "Very low";
IF "*Negative social consequences*" are "Low" and "*Indirect economic costs*" are "High" THEN "*Negative socio-economical consequences*" are "Middle";
IF "*Negative social consequences*" are "High" and "*Indirect economic costs*" are "High" THEN "*Negative socio-economical consequences*" are "Very high".

The hierarchical fuzzy logic system is designed in Matlab computer environment using Fuzzy Logic Toolbox [16]. The output of the fuzzy logic model is calculated as a weighted average of all the inference rules, which are defined in the fuzzy logic matrix. The fuzzy logic system is based on Mamdani's inference machines, max/min operations and center of gravity defuzzification. The inference surfaces in 3D for the three fuzzy logic subsystem with two input variable and one output fuzzy model are shown on Figures 4-6. The logic inference matrix of each fuzzy logic subsystem are given in Tables 1-3, respectively.

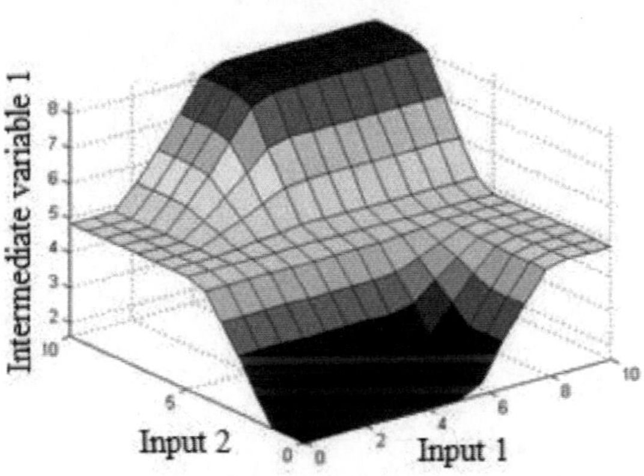

Figure 4. Surfaces of the Fuzzy logic subsystem 1.

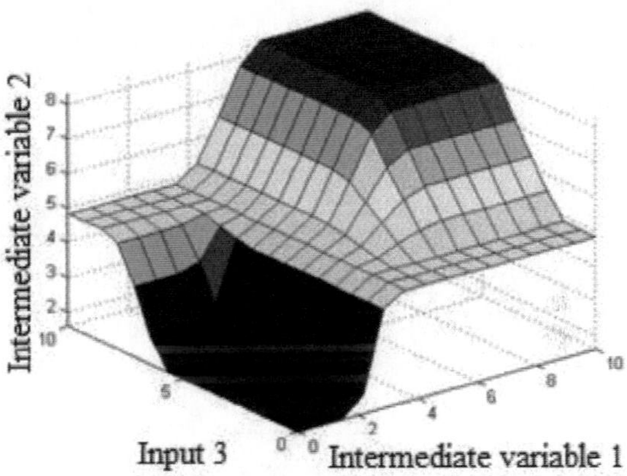

Figure 5. Surfaces of the Fuzzy logic subsystem 2.

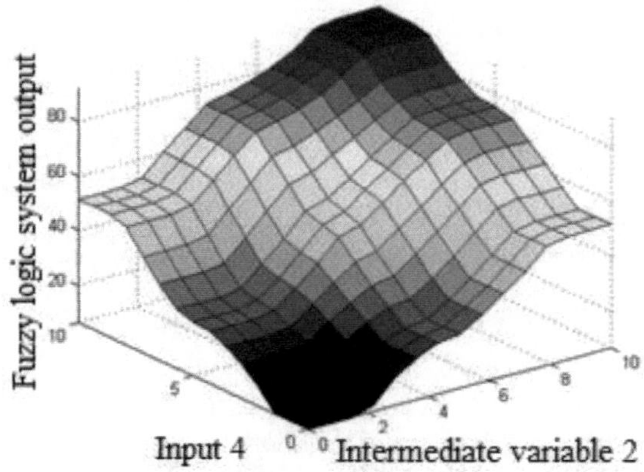

Figure 6. Surfaces of the Fuzzy logic subsystem 3.

Table 1. The logic inference matrix for "*Negative effects on daily activity*".

	Input 2		
Input 1	Low	Middle	High
Low	*Low*	*Low*	*Middle*
Middle	*Middle*	*Middle*	*Middle*
High	*Middle*	*High*	*High*

Table 2. The logic inference matrix for "*Negative social consequences*".

	Input 3		
Intermediate variable 1	Low	Middle	High
Low	*Low*	*Middle*	*Middle*
Middle	*Low*	*Middle*	*High*
High	*Middle*	*Middle*	*High*

Table 3. The logic inference matrix for "*Negative socio-economical consequences*".

	Input 4		
Intermediate variable 2	Low	Middle	High
Low	*Very low*	*Low*	*Middle*
Middle	*Low*	*Middle*	*High*
High	*Middle*	*High*	*Very high*

The interval [0, 100] of the system output variable "*Negative socio-economical consequences*" is divided into five parts to assessment the level of the negative social-economic consequences for household or its members because of power outage due to natural disasters:
IF the estimated output value ∈ [0, 20) THEN *Negative socio-economical consequences* are *Very low.*
IF the estimated output value ∈ [20, 40) THEN *Negative socio-economical consequences* are *Low.*
IF the estimated output value ∈ [40, 60) THEN *Negative socio-economical consequences* are *Middle.*
IF the estimated output value ∈ [60, 80) THEN *Negative socio-economical consequences* are *High.*
IF the estimated output value ∈ [80, 100] THEN *Negative socio-economical consequences* are *Very high.*

4. Activities to Reduce the Vulnerability of Households to Power Outage in Case of Power Outage due to Natural Disasters

4.1. Availability of a Generator

Owning a generator (a portable generator) is a key element in reducing the vulnerability of the household to power outage due to natural disasters (for example, extreme high and low temperatures). The generator enables basic electrical appliances to remain switched on. It can keep works of the lights, heating system, refrigerators, TV, modem, mobile devices, and other electrical appliances. It keeps also the sump pump running to reduce basement flooding during strong precipitation and storms [17].

The generator is necessary to testing and preparing for emergencies time to time. Safety should always come first and for this reason, the household needs to follow a few safety measures when operating a portable generator. Always it is necessary a carbon monoxide (CO) alarm to been turned on. The generator has to use only outside, never uses it indoors or in any enclosed area. The generator's exhaust has to keep away from windows, doors, or any air leaks i order to avoid the risk of carbon monoxide gas entering in house.

4.2. Availability of Emergency Heating Equipment and Fuel

In seasons with low temperatures under conditions of prolonged power failure, the household is advised to have additional heating equipment and fuel (a gas fireplace, a cooker or a fireplace) in order to maintain at least one heated room. In such cases, it is important this room will be well ventilated. During natural disaster as storms or windy weather, the household needs to check for hot air leaks from windows and doors. Then, if necessary, it does a special isolation, which gives extra heat protection.

It is very useful for the household to have alternative sources of energy as solar, wind or water energy

4.3. Availability of Flashlights and Extra Batteries

Every household should have flashlights on batteries and extra batteries in case of power outage. It is also advisable to have a portable radio in the house. The radio is an important source of weather and emergency information during natural disasters.

4.4. Additional Useful Activities, if for Long Period the Power Outage is Expected

The cell phones and any battery powered devices must been charged. It is useful the household to own a cell phone with an emergency charging option (car, solar, hand crank, etc.) in case of a power failure. If a power outage is 3 hours or less, you need not be concerned about losing your perishable foods. For prolonged power outages, the household can take useful activities to minimize food loss. In advance, the household should turn on refrigerator and freezer to their coldest settings. During power outage, the refrigerator or freezer door do not open. Most food requiring refrigeration can be kept safely in a closed refrigerator for several hours. An unopened refrigerator will keep food cold for about 4 hours. A full freezer will keep the temperature for about 48 hours. Also it can be filled a gallon container with water and place them in the freezer to help keep food cold. The household can buy special cooling bags.

If a person has medications that require refrigeration, he must to check with pharmacist for guidance on proper storage during an extended power outage.

It is necessary to review the process for manually operating an electric garage door.

The households should have an alternative to the house's electric security alarm.

The households should purchase ice or freeze water-filled plastic containers to help keep food cold during a temporary power outage.

Each household should insure your house against natural disasters, including the subsequent power outage.

5. Conclusion

A fuzzy logic model for level assessment of the negative socio-economical consequences to the household and each of its members in case of power cuts due to natural disasters is designed. The fuzzy logic model is designed as a hierarchical system with four inputs and one output in Matlab software environment using Fuzzy Logic Toolbox and Simulink. The four inputs are "Negative effects on domestic activities", "Negative effects on ICT activity", "Negative physical and psychological consequences" and "Indirect economic costs". The two intermediate variables are defined "Negative effects on daily activity" and "Negative social consequences". The fuzzy logic system output give information for level of Negative socio-economical consequences consequences to the household and each of its members in case of power cuts due to natural disasters. The appropriate preventive activities are proposed to help to reduce the vulnerability and to increase the preparedness against the negative impact of power cuts on households.

6. References

[1] Schmidthaler M, J Reichl, J Cohen 2014 Assessing the Socio-Economic Effects of Power outages in the European Union Ad Hoc Using www.blackoutsimulator.com, CIRED Workshop - Rome, 11-12 June 2014, Paper 0486.
[2] Anderson S., 2016 School Power Outages Mean More than Canceled Classes, Last updated on September 23, 2016, http://www.powerprotectionresource.com/articles/425439-school-power-outages-mean-more-than-canceled-classes.htm
[3] What You Need to Know When the Power Goes out Unexpectedly, https://www.cdc.gov/disasters/poweroutage/needtoknow.html
[4] Power Outage, Disaster Survival Resources, 2017, http://www.disaster-survival-resources.com/
[5] Massachusetts Emergency Management Agency (MEMA) 2006 Winter Power Outage Tips, www.mass.gov/mema. http://emergency.tufts.edu/weather/winter-power-outage-tips/
[6] Snowstorms & Extreme Cold, America's PrepareAthon!, https://www.ready.gov/winter-weather
[7] Klinger C, Landeg O, Murray V. 2014 Power Outages, Extreme Events and Health: a Systematic Review of the Literature from 2011-2012. PLOS Currents Disasters, http://currents.plos.org/disasters/article /power-outages-extreme-events-and-health-a-systematic-review-of-the-literature-from-2011-2012/
[8] Prepare Your Home for Winter Weather, Page last updated: December 20, 2016, https://www.cdc.gov/disasters/winter/beforestorm/preparehome.html
[9] Heatilator, September 2015 Prepare Your Home with this Power Outage Checklist, http://www.heatilator.com
[10] Generator Mag Blog 2017 Prepare for a Power Outage Checklist, Last updated on October 12th, 2017, https://www.generatormag.com/prepare-for-a-power-outage-checklist/
[11] Power Outage Checklist 2009, BeRedCrossReady, American National Red Cross, RedCross.org
[12] Martin Cerny 2013 Economic and Social Costs of Power Outages, Charles University Prague, Faculty of Social Study, Thesis.
[13] Zlateva P, V Velev 2013 Complex risk analysis of natural hazards through fuzzy logic. *Journal of Advanced Management Science*, 1 (4), pp 395-400.
[14] Zadeh L 1965Fuzzy sets. Information and Control 8, pp 338–353.
[15] Zimmerman H 1996 Fuzzy set theory and it applications. Kluwer Academic Publishers, Norwell MA, USA.
[16] Sivanandam S, S Sumathi, S Deepa 2007. *Introduction to Fuzzy Logic using MATLAB*. Berlin, Springer pp 277–320.
[17] Generator Mag Blog, Prepare for a Power Outage Checklist 2017 https://www.generatormag.com/prepare-for-a-power-outage-checklist/, Last updated on October 12th, 2017.

Acknowledgments
The authors express their gratitude to the Bulgarian National Science Fund for the partial financial support under Grant № DNTS/ China 01/6 from 2014, Bulgarian–Chinese scientific cooperation

Comparator Design in Sensors for Environmental Monitoring

Jingtao Li[1], Hua Fan[1,*], Yuhan Wu[2], Quanyuan Feng[3], Dagang Li[4], Daqian Hu[4], Yuanjun Cen[4], and Hadi Heidari[5]

1 State Key Laboratory of Electronic Thin Films and Integrated Devices, School of Electronic Science and Engineering, University of Electronic Science and Technology of China, Chengdu, China
2 The seventh high school campus of high and new area, Chengdu, China
3 The school of information science and technology, Southwest Jiaotong University, Chengdu, China
4 Electronics and Nanoscale Engineering Division, School of Engineering, University of Glasgow, G12 8QQ, Glasgow, UK
5 Chengdu Sino Microelectronics Technology Co.,Ltd Chengdu, China

Abstract. This paper presents circuit design considerations of comparator in analog-to-digital converters (ADC) applied for a portable, low-cost and high performance nano-sensor chip which can be applied to detect the airborne magnetite pollution nano particulate matter (PM) for environmental monitoring. High-resolution ADC plays a vital important role in high performance nano-sensor, while high-resolution comparator is a key component in ADC. In this work, some important design issues related to comparators in analog-to-digital converters (ADCs) are discussed, simulation results show that the resolution of the comparator proposed can achieve 5 μV, and it is appropriate for high-resolution application.

1. Introduction

In the real world, all signals in the nature environment are essentially analog signal. Once, almost all of the signal processing was carried out in the analog domain, which lasted until very large scale integrated circuits have appeared, because noise suppression capacity of digital circuit is much better than that of the analog circuit, various signal processing functions can be performed with higher resolution, higher stability, and much lower price in the digital domain than analog domain.

In modern electronic system, including environmental monitoring, telecommunications, voice, video, computer and many other electronic systems, in order to make full use of these advantages of digital signal processing, analog signals are often converted into digital signals by analog to digital converters (Analog-to-digital, converter, ADC), which are essential components for analog and digital domains.

ADC is mainly divided into Over-sampling ADC and Nyquist-Rate ADC. The basic principle of Over-sampling ADC is over-sampling and noise shaping. Band quantization noise is eliminated by digital decimation filter, as a result, the quantization noise can be reduced dramatically. By altering the number of order sampling rate and transfer function, its precision can reach 30 bits or more [1], but analog components in Over-sampling ADC called modulators and digital components in Over-sampling ADC used to implement decimation and digital filtering lead to a lot of hardware consumption. Typical Nyquist ADC can be divided into many types according to their operating principles, for example, Flash, Pipelined, Sub-arrangement, successive approximation, etc. They are able to work at a fast rate with very large bandwidth. The standard CMOS process allows the ADC to achieve GHz levels of bandwidth. However, with the rapid development of digital signal process technology, the requirement of the accuracy of an analog to digital data converter becomes more and more critical especially on environmental monitoring applications, the disadvantage of Nyquist ADC lies in the fact that their accuracy cannot be very high. This is due to the limited process matching of analog components. If no correction or trimming technology is applied, it is difficult for the Nyquist ADC to achieve an accuracy more than 12~14 bits. Another drawback of Nyquist ADC is its complex hardware structure. For example, flash ADC, mainly due to the large number of comparators, additional bit will lead to the hardware consumption increases exponentially. Although the use of interpolation structure can reduce the number of comparators, it is still very large. High performance Pipelined ADC exploits multi stages of operational amplifiers to realize high sampling rate and high resolution, it is not appropriate for the low power sensor design. SAR ADC, which makes a trade-off between speed and accuracy, can achieve an ideal power efficiency by its simple architecture.

Table 1. Comparison of ADC with different architectures.

Type	Resolution	Speed	Power	Area
SAR	Medium to high	Medium	Ultra-low	Small
Flash	Low	Ultra-high	High	Big
Pipeline	Medium to high	High	High	Big
ΣΔ	Ultra-high	Ultra-low	medium	medium

Table 1 concludes the performance of Nyquist-Rate ADC with different architectures. It is worth to mention that the ADC with the lowest FOM is SAR ADC [2]-[11]. Figure 1 shows a simplified representation of a SAR ADC, whose main components are a Sample/Hold stage, a SAR logic stage, a N-bit DAC, and a comparator. It converts an analog value into its digital counterpart via a binary search algorithm. Nowadays, SAR ADC has become a practical solution for high efficient ADC design. In the process of designing a SAR ADC, comparator is the main functional part in it, in this paper, one novel time-domain comparator is designed. The remainder of this paper is organized as follows. Section II concludes some important progress of comparator in recent years, and section III describes design considerations, circuit architecture and simulation results. The conclusions are finally drawn in section IV.

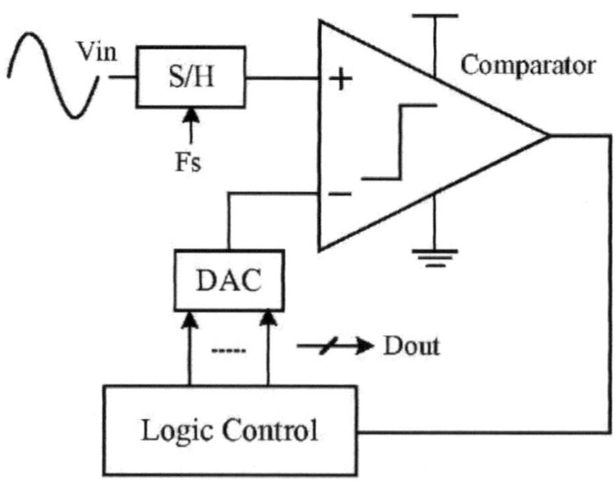

Figure 1. Typical architecture of SAR ADC.

2. Overview of Progress in Comparators in Analog-to-Digital Converters

Generally speaking, comparator can be classified into two classes: voltage comparator and current comparator; voltage comparator is more commonly used, basic function of voltage comparator is to

IOP Conf. Series: Earth and Environmental Science **151** (2018) 012030 doi:10.1088/1755-1315/151/1/012030

compare two voltage signals, output 1 or 0 according to the comparison result. For conventional voltage comparator, the most commonly used structures can be roughly divided into two types: the open-loop comparator and latch, as shown in figure 2. In figure 2(a), essentially, open-loop comparator is an operational amplifier which works in saturation region. Because the high gain of open-loop operational amplifier, the output reaches saturation even if there is only small difference exists between the two inputs, the advantage of open-loop comparator is high resolution, but limited by response time and power consumption; Latch in figure 2(b) does not consume static power and the speed of latch reaches GHz, but limited by the resolution and noise, the latch is usually cascaded by two or three pre-amplifiers to decrease the kickback noise influence of positive feedback latch, as shown in figure 3, but preamplifiers still consume non-ignorable power consumption.

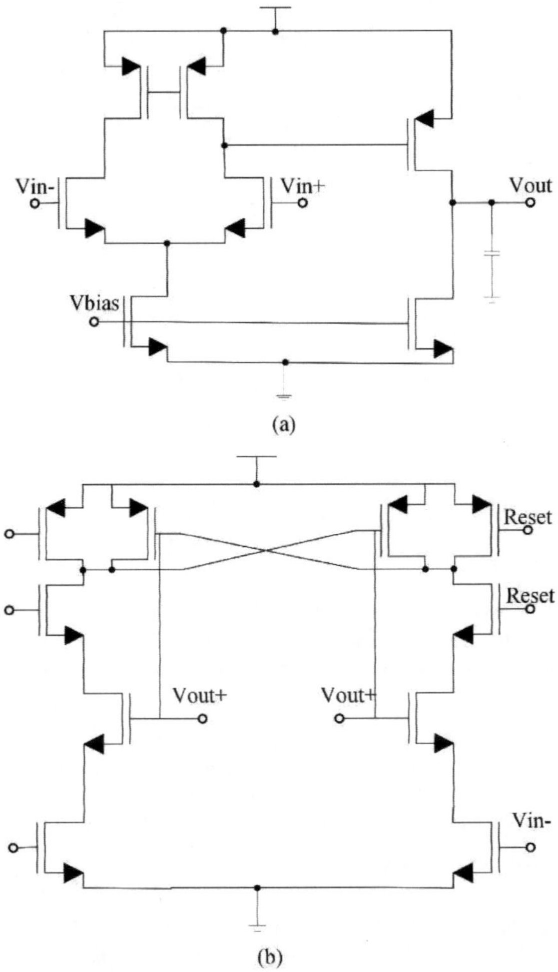

Figure 2. Typical voltage comparators: (a) open-loop comparator; (b) latch .

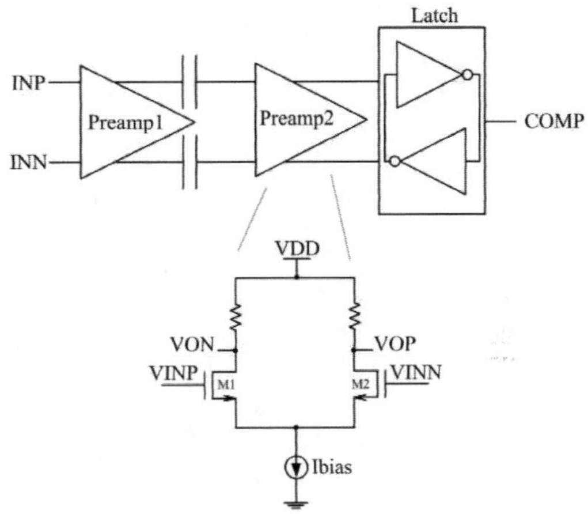

Figure 3. Two stages of preamplifiers with latch.

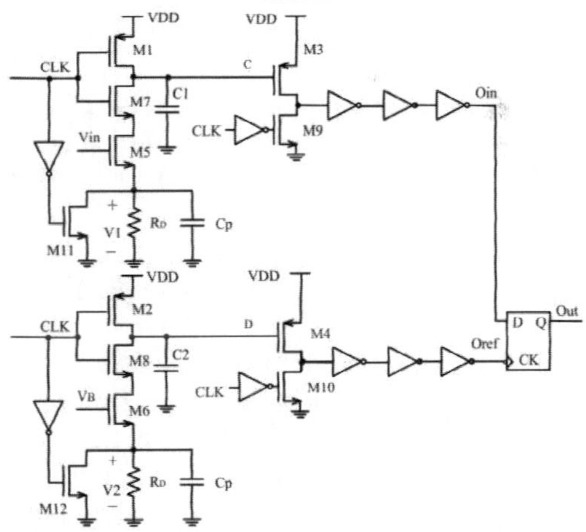

Figure 4. Time-domain comparator proposed in [12].

The time-domain comparator proposed in [12] obviates the need of preamplifier that dissipates static power, but it can only be applied to single-ended SAR ADC, because one of the inputs of the time-domain comparator in figure 4 is connected to a fixed voltage, VB, which must be large enough to provide the reference pulse, as a result, differential signal cannot be applied to VB. The operation of the time-domain comparator is as follows: when the signal CLK is low, transistors M1 and M2 charge the nominally equal capacitors C1 and C2 to VDD, when CLK rises, transistors M5 and M6 begin to discharge capacitors C1 and C2, producing two pulses with different durations T1 and T2. The flip-flop reflects the pulse that ends first and provides the comparator output.

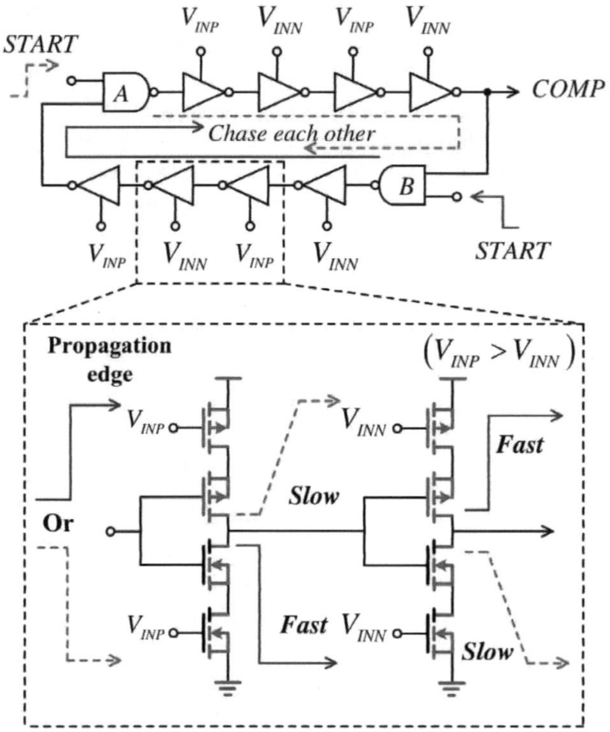

Figure 5. Structure of the Edge-Pursuit Comparator in [13].

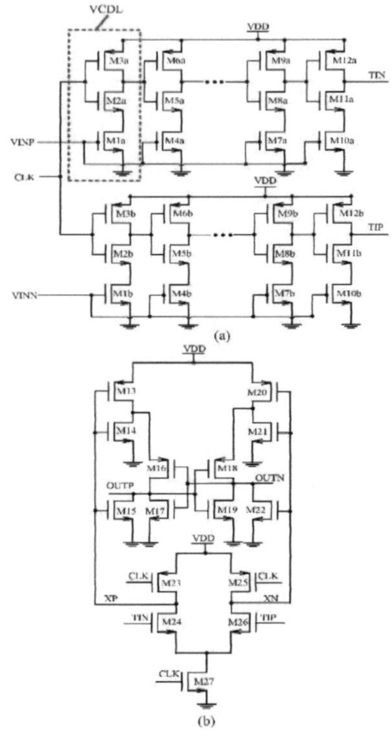

Figure 6. Time-domain comparator proposed.

Inverter based comparator was invented in [13] to further decrease power consumption. As shown in figure 5, it is composed of two NAND gates and inverter delay cells, automatically scales comparison energy according to its input voltage difference without external control, eliminating unnecessary energy. Initially, as the signal START is low, the comparator is in the reset state, the comparator begins to compare when the signal START goes high, which injects two propagating edges into the oscillator. The drawback of the comparator proposed in [13] lies in the fact that the propagation path of rising edge is different from that of falling edge, and positive terminal voltage V_{INP} and negative terminal V_{INN} appear alternately, leads to cross coupling of signals and layout between V_{INP} and V_{INN}.

3. Design and Realization of Time-Domain Comparators in SAR ADC

Figure 6 shows the structure of the high-resolution time domain comparator proposed, which improves the conventional time domain comparator proposed in [12] and [14], and it can be applied to both single-ended SAR ADC and full differential SAR ADC. As shown in figure 6, it is composed of a voltage-time converter and a time-digital converter, and the voltage-time converter is composed of a series of inverters which are current limited [14]. In reset mode, CLK is high, the output voltage TIP and TIN of the voltage-time converter reset to low level. In comparison mode, CLK becomes high, then TIP and TIN transfer to high, too. The speed of TIP and TIN go from low to high is related to the magnitude of input voltage VINP and VINN. The greater the input voltage, the shorter the time for TIP and TIN transfer from low to high. The outputs (OUTP and OUTN) of time-digital converter [15] are determined by the time difference between TIP and TIN change from low to high. If VINP is greater than VINN, OUTP is high and OUTN is low. On the contrary, if VINP is lower than VINN, OUTN goes high and OUTP goes low.

Time-domain Comparator proposed in this work is simulated in $0.18\mu m$ technology, simulation results show that the resolution is $5\mu V$, the highest speed of comparator reaches 200 MHz, while the time-domain proposed in [14] is simulated with the same technology and same simulation conditions. As shown in table 2, simulation results show that the comparator proposed achieves higher speed, higher resolution and wider input range, and the time-domain comparator proposed can be applied to both single-ended and differential SAR ADC while the comparator proposed in [14] can only be applied in single-ended SAR ADC.

Table 2. Simulation summary of comparator.

	[14]	Proposed
Technology(μm)	0.18	0.18
Single-ended/Differential	Single-ended	Both Single-ended and Differential
Highest Speed(MHz)	166	200
Resolution(μV)	122	5
Input voltage Range	0~ 1	-1~ 1
Simulation/Measure	Simulation	Simulation

4. Conclusion

This paper presents circuit design considerations of time-domain comparators in data converters applied for SAR ADC, some important progress related to comparators in analog-to-digital converters (ADCs) are introduced. One practical design of high resolution time-domain comparator design is given in this work and the design considerations mentioned in this paper are useful for the practice and high resolution data converters applications in environmental monitoring fields.

5. References

[1] ADI, "32-bit, 10 ksps, sigma-delta adc with 100 μs settling and true rail-to-rail buffers," *ANALOG DEVICES DATASHEET*

[2] Tai H-Y, Hu Y-S, Chen H-W and Chen H-S 2014 A 0.85 fj/conversion-step 10b 200ks/s subranging sar adc in 40nm cmos *Digest of Technical Papers of IEEE International Solid-State Circuits Conference(ISSCC)* pp 196–7

[3] Liu M, van Roermund A and Harpe P 2016 A 7.1 fj/conv.-step 88db sfdr 12b sar adc with energy-efficient swap-to-reset *European Solid-State Circuits Conference 42nd* pp 409–12

[4] Kim M, Ha U, Lee Y, Lee K and Yoo H-J 2016 A 82nw chaotic-map true random number generator based on sub-ranging sar adc *European Solid-State Circuits Conference* 42nd pp 157–60.

[5] Lin J-Y and Hsieh C-C 2015 A 0.3 v 10-bit 1.17 f sar adc with merge and split switching in 90 nm cmos *IEEE Transactions on Circuits and Systems I: Regular Papers* vol 62 no 1 pp 70–9

[6] Chung Y-H, Wu M-H and Li H-S 2015 A 12-bit 8.47-fj/conversion-step capacitor-swapping sar adc in 110-nm cmos *IEEE Transactions on Circuits and Systems I: Regular Papers* vol 62 no 1 pp 10–8

[7] Sanyal A and Sun N 2014 An energy-efficient low frequency-dependence switching technique for sar adcs *IEEE Transactions on Circuits and Systems II: Express Briefs* vol 61 no 5 pp 294–8

[8] Harpe P, Dolmans G, Philips K and de Groot H 2012 A 0.7v 7-to-10bit 0-to-2ms/s flexible sar adc for ultra low-power wireless sensor nodes *European Solid-State Circuits Conference(ESSCIRC)* pp 373–6

[9] H. Fan, H. Hadi, F. Maloberti, D. Li, D. Hu, and Y. Cen, "High Resolution and Linearity Enhanced SAR ADC for Wearable Sensing Systems," in *IEEE International Symposium on Circuits and Systems(ISCAS)*, 2017, pp. 180–3.

[10] van Elzakker M, van Tuijl E, Geraedts P, Schinkel D, Klumperink E and Nauta B 2010 A 10-bit charge-redistribution adc consuming 1.9μw at 1ms/s *IEEE Journal of Solid-State Circuits* vol 45 no 5 pp 1007–15

[11] H. Fan and F. Maloberti, "High-Resolution SAR ADC with Enhanced Linearity," IEEE Transactions on Circuits and Systems II: Express Briefs, vol 64, no 10 pp 1142–6, 2017.

[12] Agnes A, Bonizzoni E, Malcovati P and Maloberti F 2008 A 9.4-enob 1v 3.8μw 100ks/s sar adc with time-domain comparator *Digest of Technical Papers of IEEE International Solid-State Circuits Conference(ISSCC)* pp 246–610

[13] Shim M, Jeong S, Myers P D, Bang S, Shen J, Kim C, Sylvester D, Blaauw D and Jung W 2017 Edge-pursuit comparator: an energy-scalable oscillator collapse-based comparator with application in a 74.1 db sndr and 20 ks/s 15 b sar adc *IEEE Journal of Solid-State Circuits*

[14] Kobenge S B and Yang H 2008 Ultra-low-power high-speed vcdl based time domain comparator *Chinese Patent* ZL200810114514.4 pp 1–10

[15] Schinkel D, Mensink E, Kiumperink E, Van Tuijl E and Nauta B 2007 A double-tail latch-type voltage sense amplifier with 18ps setup+ hold time *Digest of Technical Papers of IEEE International Solid-State Circuits Conference(ISSCC)* pp 314–605

Acknowledgments

The work of Hua Fan was supported by the National Natural Science Foundation of China (NSFC) under Grant 61771111 and 61401066, as well as supported by China Postdoctoral Science Foundation under grant 2017M612940 and Special Foundation of Sichuan Provincial Postdoctoral Science Foundation.

The work of Quanyuan Feng was supported by the National Natural Science Foundation of China (NSFC) under Grant 61531016, supported by the project of Science and Technology Support Program of Sichuan Province under Grant 2015GZ0103, and in part by the Sichuan Provincial Science and Technology Important Projects under Grant 2017GZ0110.

IOP Conf. Series: Earth and Environmental Science **151** (2018) 012031 doi:10.1088/1755-1315/151/1/012031

Improving Traffic Safety Towards Sustainable Built Environment in Dammam City, Saudi Arabia

Umar Lawal Dano

Assistant Professor, Department of Urban and Regional Planning, College of Architecture and Planning, Imam Abdulrahman Bin Faisal University, P.O. Box 2397, Dammam 31451, Saudi Arabia
E-mail: uldano@uod.edu.sa

Abstract. Traffic safety is vital to the socioeconomic wellbeing and sustainable development of any society. This study assesses factors undermining traffic safety in Dammam city of Saudi Arabia. The study utilized expert-based questionnaire survey, and employed Analytic Hierarchy Process (AHP) in developing an experts'-based traffic safety model. Six factors believed to be triggering road traffic accident were analyzed, and weights (W) were assigned to each factor using Saaty's 9-point scale of reference. The weights were further normalized through the AHP. The study found traffic lights violations (W=0.37), over speeding (W=0.18), and wrong turns (W=0.17) to be the most critical factors responsible for road traffic accident in the study area. Speed cameras, speed bumps, and speed limits signage were evaluated by experts as means of minimizing road traffic accidents. Speed cameras (W=0.54) and speed bumps (W=0.30) are perceived to enhance road traffic safety over speed limits signage. The study stresses that benefits attached to speed cameras and speed bumps are yet to be fully implemented within the city and outskirts. The study concludes by highlighting how the model can minimize road traffic accidents in the city.

1. Introduction

Improving traffic safety is increasingly important, yet challenging for governments, urban planners and traffic engineers, especially in rapidly urbanizing cities. Traffic safety is here described as a system that prevents road users from being killed or injured and it is a vital component of sustainable transportation system [1]. An improved transportation system is that which succeeded in minimizing traffic accidents, traffic congestion, pollution, and improving public transport infrastructure [1]. However, in low and middle-income countries traffic accidents claim about 1.25 million lives per annum and have a tremendous effect on health and development [1]. In these countries where urbanization has been accompanied by soared motorization (with roughly 54% of the world's vehicles) and road traffic accidents, youngsters between 15 and 29 of age were killed, and cost governments roughly 3% of GDP [1]. Youngsters are more vulnerable to road traffic accidents due to: inexperience, risk-taking, over speeding, distractions, and drugs and alcohols among others [1]. Similarly, rapid transformations of cities and living standards of people is partly responsible for the dramatic increase in road traffic accidents which accounts for 90% of the world's mortality rates occurring [1]. Due to human, technological, and environmental factors, road traffic accident has become epidemic worldwide, causing tremendous increase in death and casualties.

In Saudi Arabia, road traffic accident is a serious socioeconomic and public safety challenge. Since the oil boom of the 1980s, there is massive increase in urban growth and associated transportation networks, land uses and infrastructural facilities. High living standard in Saudi Arabia has led to an increase in car ownership rate to 219 cars for every 1000 people [2]. Thus, this has contributed to road traffic accidents which is the second major health and economic challenges that annually cost Saudi government about 21 billion Saudi riyals [3]. According to Saudi Traffic Authority's statistics, on average over 4100 people died with 28,000 being injured yearly from road traffic accidents in the country [4]. A study conducted in 2008/2009 revealed that road traffic accidents have killed 6,485 people and injured more than 36,000 in just more than 485,000 traffic accidents in Saudi Arabia [5]. Therefore, road traffic accident will continue to be one of the contending challenges causing many injuries and deaths. The impact of road traffic accident on societies and the built environments has been widely acknowledged. Despite its importance, traffic safety was not included in the Millennium Development Goals (MDGs). However, upon recognizing its importance in improving the wellbeing of the citizens and promoting sustainable built environment, traffic safety is recently included in the Sustainable Development Goals (SDGs) under Goal 3 [1]. The target is to reduce road traffic fatalities by 50% by 2020. However, achieving this noble Goal is impossible without a much clear focus on assessing the factors influencing road traffic accidents. Identifying and understanding these factors offers targeted mediations towards realizing the Goal.

Many studies on road traffic accidents have been carried out in some developed and developing countries. These include nationwide studies in Thailand [6], England [7], Singapore [8], Iran [9], and Saudi Arabia among others [3], [10], [11]. A review of the literature shows a relationship between road traffic accidents and seatbelt utilization [11], gender, age [9], [12], education attainment, status [9], age [8], season [13], tire failure, high speed [3], [13], sleeping, loss of control, pedestrian [10], wrong turns, wrong overtaking, traffic lights violations [3], [6], wrong parking, alcohol [3], weather condition [7], road geometries, traffic signs, traffic signals, and rainwater drainage [6]. However, the few national and sub-national studies conducted in Saudi Arabia assesses different road traffic causative factors [10]-[13], without exploring the major and common road traffic causative factors as studied by Ansari et al. [3]. However, Ansari et al. [3] conducted their studies at a national level. Thus, there is a need to study these road traffic accidents causative factors at a local level as each geographical region within the country has its own socioeconomic and demographic background peculiar to it. Similarly, it could help in identifying and understanding areas within a city or region that needs urgent government interventions and assistance in terms of traffic safety enhancement. Therefore, this study considered to build upon the previous studies by employing AHP to evaluate the road traffic accident causative factors using the Saaty's 9-point scale of preference.

The use of AHP in addressing transport related issues has been conducted by various authors. It has emerged as a multipurpose decision support tool and evaluation technique with extensive application areas. Its applications in transportation planning are quite effective and diverse including projects for the enhancement of the transportation system [14]. Likewise, it was used in road accidents studies (see [6]). Another modern application of AHP is its integration with a GIS in freight terminal location and in transit-oriented development (TOD) [15], [16]. It helps planners in structuring a multifaceted decision making process. Podvezko & Sivilevičius [17] employed AHP in investigating the relationship between transport system factors and traffic accident rate as well as the effect of these factors on road traffic accident generation. The authors found the relationship between vehicles and motorists to be the key component in ensuring road traffic accidents are minimized.

2. Study Area

This study is carried out for an area comprising three major conurbations in the Eastern Province of the Kingdom of Saudi Arabia: Dammam, Khobar, Dhahran and the evolving areas of Aziziyah and the Half Moon located in the south of Khobar. Bigger Dammam encompasses Dammam Metropolitan Area (DMA) and the adjoining cities of Qatif, Safwa and Ras-Tanura. The total area of Dammam is approximately 380,000 ha, accommodating roughly 1.8 million people with substantial proportion of expatriates that increases the number of vehicles in the area [18]. The Kingdom's economy largely relies on the exploration, extraction and processing of oil and the exportation of petroleum products. The Kingdom's Eastern Province is mainly the heart of Saudi Arabian oil production. Thus, the economy as well as the physical development of Dammam is largely reliant on the petroleum industry. Figure 1 shows the components of Dammam and its external connections. Therefore, the scope of the study involved Municipals of DMA, Khobar, and Dhahran.

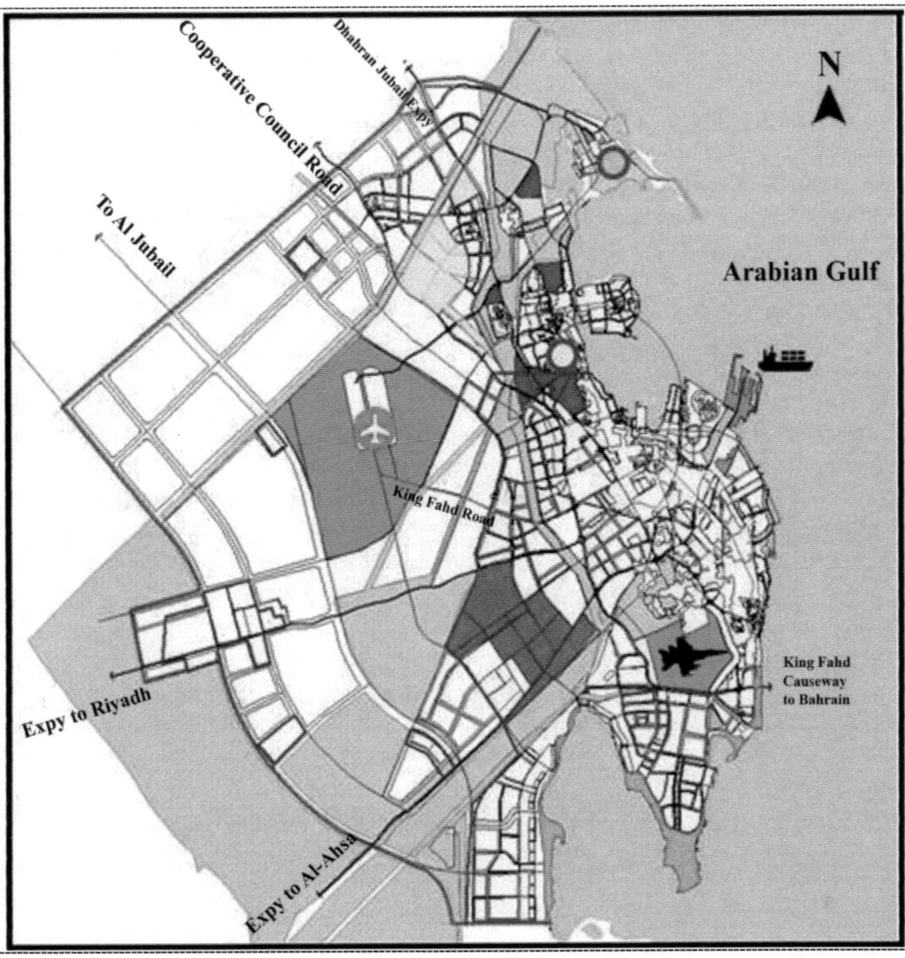

Figure 1. *Dammam Metropolitan Area: Components and External Connections* [19].

There is incessant increase in the number of cars in the study area. The average car ownership per household in DMA alone is very high, which is 6.1, compared to other metropolitan areas across the world [20]. It is high because in Dammam cars are affordable, available and inexpensive. As shown in Table 1, approximately only 5% of Saudi families do not have private cars, while about 95% of them have at least one car. Compared with non-Saudi families, only 29% do not have private cars, while 71% have at least a car per family. These facts revealed that most residents depend on private mode of transportation for their daily commuting.

Table 1. Car Ownership in Dammam City (Compiled from [20].

	Saudi families (%)	Non-Saudi families (%)
No car	4.97	28.72
One car	48.21	58.86
Two cars	25.28	7.78
Three cars and more	21.54	4.63
Total	100	100

3. Study Methodology

Analytic Hierarchy Process (AHP)

This study used AHP mathematical model in ranking the importance of road traffic accidents influencing factors in DMA. The AHP technique supports Multi-criteria Decision Analysis (MCDA) through calculation of ratio scales based on pairwise comparison of the factors under examination. The study's inputs and outputs are based on the subjective judgments derived from 7 experts through the AHP questionnaire survey. These experts were selected according to their expertise in transportation planning and relevant fields. Priority weights were calculated and further validated by calculating the Consistency Ratio (CR). The rule of thumb in the CR is that, a CR greater than 0.1 is implies unacceptable level of the priority weights calculated, and therefore the pairwise comparisons of the factors should be revised until an acceptable level of the CR is reached. Similarly, a CR less than or equals to 0.1 implies an acceptable level of the calculated priority weights.

AHP Questionnaire

AHP-based questionnaire was designed and distributed among 7 experts. Considering more than one expert is meant to avoid bias in calculating the experts' priorities [21], otherwise a judgment from a single expert is enough [22]. The design of the questionnaire was carried out in a way that each individual question was clearly worded to avoid vagueness and checked for expression, objectivity and applicability to the goal being investigated. The respondents were requested based on their expertise to kindly tick the degree of preference of each criterion/factor with respect to road traffic accident generation on a 1 to 9 Saaty's scale of preference as shown in Table 2. These criteria/factors were derived from literature survey.

Table 2. Interpretation of the Saaty's scale of measurement (Compiled from [21].

Degree of importance	Definition	Interpretation
Scale 1	Equal importance	Two element making equal contribution to the goal
3	Somewhat more important	Moderate importance of element over the other element
5	Much more important	Essential or strong importance
7	Very much important	Very strong importance
9	Extremely important	Extreme importance
Scale 2,4,6, and 8	Intermediate values	These are required when comparison between two adjacent judgment is needed
Reciprocals	If v is the judgment value when i is compared to j, then 1/v is the judgment value when j is compared to i.	

Figure 2 presents the schematic diagram of the experts-based road traffic safety model for sustainable Dammam city. The concept was developed by Thomas L. Saaty in 1980 using the MCDA principles, which help in decomposing complex and difficulty in decision making into units and subunits. First of all, the problem is structured into a hierarchical structure (Figure 2): goal, criteria, sub-criteria, and alternatives. The second phase comprises pair-wise comparisons of the road traffic accident causative factors and the alternatives on a 9-point scale of preference in a matrix structure. In each group of comparisons, there are $n (n - 1)/2$ judgments [21], where n represents the matrix's size. Through the comparison, most important factors and alternatives are determined. Comparison is carried out in such a way that 2 criteria are compared at a time within the hierarchy or sub-hierarchy levels. The third phase comprises the use of the weights of the matrix to determine the strength of the AHP methodology, which is the calculation of the consistency ratio (CR).

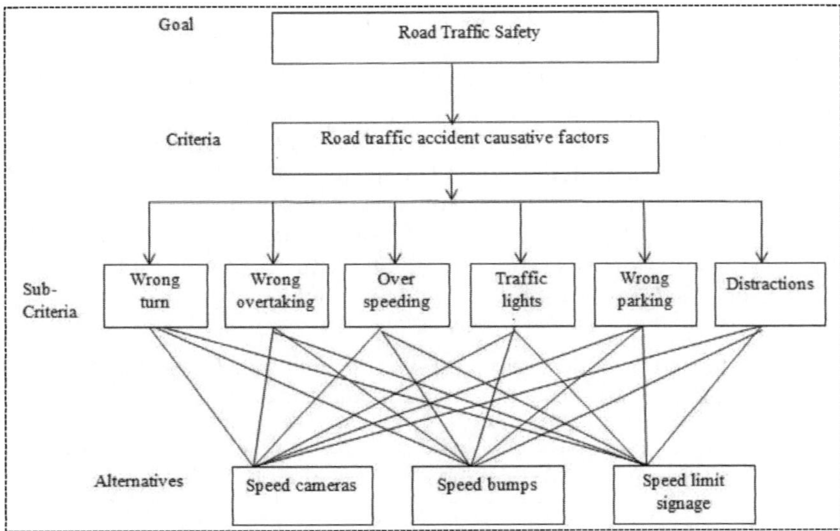

Figure 2. *Experts-based road traffic safety model for sustainable Dammam city (Source: Author, 2017)*

The experts' priorities were aggregated and calculated using a Geometric Mean method as shown below:
Geometric Means = $((X_1)\,(X_2)\,(X_3)....\,(X_N))^{1/N}$:

where,
X = Individual score; and N = Sample size (Number of scores). Experts' preferences helped in performing the pairwise comparison matrix.

Pairwise Comparisons
The design of the questionnaire used in this study was based on pairwise comparisons of the road traffic accidents criteria used herein. Pairwise comparison assists in examining the way in which the respondents assigned their priorities to each factor with reciprocal matrices. The experts were requested to give judgment while carrying out each pairwise comparison based on the scales explained in Table 2. However, the pairwise comparisons of all the criteria under consideration are considered as the inputs whereas the priority weights calculated are the outputs; which were validated using a CR. The weights were further normalized and the normalized values of the eigenvector were used as the final outputs. The weights of the factors were generated based on the steps described by [22]. Table 3 presents the average random index values representing the number of the comparison elements.

Table 3. Average random consistency index values (Compiled from [23])

Size of matrix	1	2	3	4	5	6	7	8	9	10
Random Consistency	0	0	0.58	0.90	1.12	1.24	1.32	1.41	1.45	1.49

The Consistency Index (CI) was calculated using the following equation:

CI= $\frac{\lambda - n}{n-1}$: Where, n = number of criteria (i.e. 6); λ = average value of the consistency vector.

Moreover, the following equation was used in calculating the CR:

$CR = \frac{CI}{RI}$: Where, RI = random inconsistency index whose value rest on the number (n) of criterions being compared; for n = 6, RI = 1.24 ([22]).

4. Results and Discussion

This part presents the study findings in line with the objectives of the study. It analyzes the road traffic accident causative factors in DMA. Each factor was evaluated with respect to its influence in triggering road traffic accident using pairwise comparison matrix. Numbers from 1 to 9 were assigned to each factor/criterion as mentioned earlier herein by the experts to reflect their relative importance. The findings of the study are as Table 4.

Table 4. Impact of the alternatives in reducing road traffic accidents

		STEP 1			STEP 2			
		SC	SB	SL	SC	SB	SL	Priority
SC	Speed cameras	1.00	2.00	3.00	0.55	0.57	0.50	0.54
SB	Speed bumps	0.50	1.00	2.00	0.27	0.29	0.33	0.30
SL	Speed limits	0.33	0.50	1.00	0.18	0.14	0.17	0.16
Sum		1.83	3.50	6.00	1.00	1.00	1.00	1.00
		Consistency ratio is 0.01						

The influence of speed cameras in reducing road traffic accidents is perceived higher than speed bumps. So, 2 is entered in the first row cell of the second column, and its reciprocal is entered in the transpose position. Therefore, the weighted impact is rated highest as illustrated in the priority column. The same was applied to the rest of the comparisons. Similarly, the pairwise comparisons of the sub-criteria are presented in Table 5.

Table 5. Judgments matrix on the influences of road traffic accident causative factors for sustainable road transportation system in Dammam city, Saudi Arabia

		STEP I						STEP II						
		C1	**C2**	**C3**	**C4**	**C5**	**C6**	**C1**	**C2**	**C3**	**C4**	**C5**	**C6**	**Priority**
C1	Wrong turns	1.00	5.50	0.83	0.14	6.00	4.00	0.09	0.30	0.11	0.06	0.25	0.23	0.17
C2	Wrong overtaking	0.39	1.00	0.24	0.36	4.50	2.83	0.04	0.27	0.03	0.16	0.19	0.16	0.14
C3	Over speeding	1.50	5.00	1.00	0.40	5.00	2.33	0.13	0.30	0.13	0.17	0.21	0.13	0.18
C4	Traffic lights violations	7.50	5.50	4.00	1.00	5.50	6.00	0.64	0.02	0.53	0.44	0.23	0.35	0.37
C5	Wrong parking	0.35	0.41	0.37	0.20	1.00	1.17	0.03	0.05	0.05	0.09	0.04	0.07	0.05
C6	Distractions	0.40	0.97	1.13	0.19	1.83	1.00	0.04	0.05	0.15	0.08	0.08	0.06	0.08
Sum		11.14	18.38	7.57	2.29	23.83	17.33	1.00	1.00	1.00	1.00	1.00	1.00	1.00
		Consistency ratio is 0.10												

Where, C1 = wrong turns, C2 = wrong overtaking, C3 = over speeding, C4 = traffic lights violations,

C5 = wrong parking, and C6 = distractions.

The priories assigned by the experts using pairwise comparisons of the sub-criteria along with the sub-criteria weightings (Table 5) were used in calculating the CR. Thus as CR = 0.10, it expresses that, there is an accurate amount of consistency in the pairwise comparisons and therefore, the weights 0.17, 0.14, 0.18, 0.37, 0.05, and 0.08 can further be assigned to wrong turns, wrong overtaking, over speeding, traffic lights violations, wrong parking, and distractions respectively. The findings of the present study are in line with those reported in similar studies. In Riyadh city for example, a study reported that 52.5% of fatal collisions and roughly 45.9% of injury collisions are caused by traffic lights violations and over speeding [10]. Similarly, Ansari, et al. [3] revealed that more than 50 percent of road accidents in Saudi Arabia are because of over speeding (which is about 3.5 more times than in the US), sudden brakes/stops, and distractions: speaking on phones, text messaging, et cetera while driving. Twenty years ago, Saudi Arabia has documented 4 million road traffic accidents that led to 86,000 deaths and 611,000 casualties, 7 percent of which led to permanent disabilities [24]. A lot of advancement with regards to technology services, infrastructure, trade and industry to transform Dammam to an international standing city is witnessed. However, the magnitudes of road traffic accident predicaments as well as the extent of human and economic damages caused are gloomy. It is therefore necessary for the decision makers to understand the basic road traffic accidents causative factors within and in the outskirts of the city to enable them take sustainable strategies in curtailing the frequencies and fatalities of road traffic accidents in the study area

5. Conclusion

This paper highlights the impacts of road traffic accident causative factors in Dammam city. Although many factors are responsible to the overall severity of road traffic accidents, this study concentrated mainly on the human causative factors: wrong turning, wrong overtaking, over speeding, traffic lights violations, wrong parking, and distractions as the major factors contributing to road traffic accidents in Dammam city. Significant findings include that traffic lights violations (37%), over speeding (18%), and wrong turns (17%) are the most critical factors responsible for road traffic accidents in the area, according to experts. The experts also rated speed cameras (54%) and speed bumps (30%) as sustainable strategies for enhancing road traffic safety, while the use of speed limits signage is less preferred. Key lessons from these findings include the need to install more speed cameras and speed bumps at major and minor junctions, roundabouts, and pedestrian crossing among others within and in the outskirts of the city including rural areas. This will help in promoting healthy lives and well-being of all residents and to reduce road traffic fatalities by 50% by 2020. Therefore, there is an urgent need for government interventions to implement traffic safety strategies in the city, more especially speed cameras and bumps. Though there is a dilemma of balancing contending challenges with regards to urban transportation initiative [25]. This is because the citizens are used to traveling in their private cars and therefore may not instantly accept public transportation that might assist in minimizing road traffic accidents, energy consumption and pollution.

6. References

[1] World Health Organization 2015. Global status report on road safety 2015. *World Health Organization.*

[2] Gately, D., Al-Yousef, N., & Al-Sheikh, H. M. H. 2013. The rapid growth of OPEC's domestic oil consumption. *Energy Policy, 62, 844–859. doi:10.1016/j.enpol.2013.07.044*

[3] Ansari, S., Akhdar, F., Mandoorah, M., & Moutaery, K. 2000. Causes and effects of road traffic accidents in Saudi Arabia. *Public health, 114(1), 37-39.*

[4] General Directorate of Traffic, (2003). Annual Traffic Report. *Ministry of Interior, Riyadh.*

[5] Green Prophet, 2010. Saudi Arabia Has the Highest Road Accident Death Toll in the World. <http://www.greenprophet.com/2010/03/saudi-arabia-death-toll-driving/> *Accessed 02.11.2017*

[6] Temrungsie, W., Raksuntron, W., Namee, N., Chayanan, S., & Witchayangkoon, B. 2015. AHP-based prioritization on road accidents factors: A case study of Thailand. *International Transaction Journal of Engineering, Management, Applied Sciences & Technologies, 6, 135-144*

[7] Edwards, J. B. 1999. The relationship between road accident severity and recorded weather. *Journal of Safety Research, 29(4), 249-262.*

[8] Wong, E., Leong, M. K., Anantharaman, V., Raman, L., Wee, K. P., & Chao, T. C. 2002. Road traffic accident mortality in Singapore. *The Journal of emergency medicine, 22(2), 139-146.*

[9] Montazeri, A. 2004. Road-traffic-related mortality in Iran: a descriptive study. *Public health, 118(2), 110-113.*

[10] Al-Ghamdi, A. S. 2003. Analysis of traffic accidents at urban intersections in Riyadh. *Accident Analysis & Prevention, 35(5), 717-724.*

[11] Shanks, N. J., Ansari, M., & Ai-Kalai, D. 1994. Road traffic accidents in Saudi Arabia. *Public health, 108(1), 27-34.*

[12] Bendak, S. 2005. Seat belt utilization in Saudi Arabia and its impact on road accident injuries. *Accident Analysis & Prevention, 37(2), 367-371.*

[13] Al-Zahrani, A. H., Jamjoom, M. M. O., & Al-Bar, H. O. 1994. Traffic Accident Characteristics in Jeddah, Saudi Arabia. *Engineering Sciences, 6(1).*

[14] Banai, R. 2000. Public Transportation Decision-Making: A Case Analysis of the Memphis Light Rail Corridor and Route Selection with Analytic Hierarchy Process. *Journal of Public Transportation, Vol. 9, No. 2, 2006.*

[15] Banai, R. 2000. Transit station area land use/site assessment with multiple criteria: An integrated GIS-expert system prototype. *Journal of Public Transportation 3 (1): 95–110.*

[16] Dantas, Sergio, and Yaeko. 2001. Integrating GIS and AHP for bus route definition. *Proceedings Symposium on ASIA GIS, Tokyo, Japan*

[17] Podvezko, V.; Sivilevičius, H. 2013. The use of AHP and rank correlation methods for determining the significance of the interaction between the elements of a transport system having a strong influence on traffic safety. *Transport 28(4): 389–403.*

[18] Abou-Korin, A.A. and Al-Shihri, F.S., 2015. Rapid Urbanization and Sustainability in Saudi Arabia: The Case of Dammam Metropolitan Area. *Journal of Sustainable Development, 8(9), p.52.*

[19] General Administration of Urban Planning 2017. Construction and Project Agency, Amana, Eastern Province, Saudi Arabia.

[20] Al-Sheriff, M. 2010. The problem of public transport within the cities of Saudi Arabia. *Alegt Journal, No. 5999, online, retrieved 15 March 2010, from* http://www.aleqt.com/2010/03/15/article_363931.html. *Accessed 03.03.2017.*

[21] Ishizaka, A. and Labib, A. 2011. Review of the Main Developments in the Analytic Hierarchy Process. *Expert Systems with Applications, 38(11), 14336-14345.*

[22] Saaty, T. L. and Sagir, M. 2009. Extending the Measurement of Tangibles to Intangibles. *International Journal of Information Technology and Decision Making, 8(01), 7-27.*

[23] Lawal, D. U., Matori, A. N., Yusof, K. W., Hashim, A. M., Aminu, M., Balogun, A. L., & Mokhtar, M. R. M. 2014. Group-based Decision Support for Flood Hazard Forecasting: A Geospatial Technology-based Group Analytic Hierarchy Process Approach. *Research Journal of Applied Sciences, Engineering and Technology, 7(23), 4838-4850.*

[24] Arab News, 2013. Road accidents cost KSA SR 87 billion annually. http://www.arabnews.com/saudi-arabia/road-accidents-cost-ksa-sr-87-bn-annually *Accessed 03.11.2017.*

[25] Abubakar, I. R. & Aina, Y. A. 2016. Achieving sustainable cities in Saudi Arabia: Juggling the competing urbanization challenges. *In Population Growth and Rapid Urbanization in the Developing World. Hershey, PA: IGI Global.*

Guidelines for Improving the Quality of Urban Passenger Transport in the City of Sofia within the Context of Stable Urban Mobility

S D Tzvetkova

University of National and World Economy, Sofia, Bulgaria
E-mail: svetlatzvetkova@abv.bg

Abstract. The problems regarding the necessity for improving the quality of urban passenger transport trips have become especially relevant in recent years due to the fact that an increasingly large portion of Europe's population now lives in major cities. They comprise about 85% of the EU's GDP. They are the engine of European economy, attracting investments and providing new jobs. At the same time, their dynamic development has led to multiple negative consequences regarding urban traffic. Despite the progress of automobile technology, the increased number of personal vehicles, the street traffic, the harmful emissions and noise pollution have become one of the biggest challenges for most cities, including Sofia. This has led to negative economic, social and ecological consequences and the destruction of the natural urban environment. Therefore, the European Commission has composed the Green Paper of Transport which outlines the primary guidelines for ensuring stable urban mobility by facilitating traffic in cities, improving the accessibility, security and safety of public transport and reducing the harmful impact of urban traffic on the environment.

1. Introduction

The policy for stable urban mobility is a priority for the EU's member states, but the goals outlined in it cannot be achieved without the functioning of quality urban passenger transport.

Every major city, including Sofia, needs an effective transportation system that will secure access to workplaces and services for all its citizens, guarantee safety and security during travel, minimize pollution, greenhouse effects and energy consumption [1, 2]. It has to guarantee quality transportation by public transport and it should be aimed at increasing the appeal of the urban environment. The evaluation of transport expenses and benefits has to be based primarily on social value in a wide range. The measures that need to be taken have to be aimed primarily at increasing the social effectiveness of urban passenger transport. It is crucial to carefully estimate the costs and the benefits, including the ones that are difficult to measure – for example, the ones regarding the quality of the air [3].

Increasing the social effectiveness of public transport plays a defining role in the social and economic development of major cities. Quality transport helps improve the cultural and economic relations between separate regions, smooth passenger transportation for industrial and personal needs and it becomes a basis for improving the living conditions and standards of citizens [4].

2. Analyzing the Mobile Situation in the City of Sofia

As the capital city of the Republic of Bulgaria, Sofia holds a key place in the country's economy. By 2016, it is inhabited by 1 304 772 people and it produces about 1/3 of the country's GDP. The city is defined by a higher employment rate and a lower unemployment rate, compared to the average factors for the country. It is the largest cultural, educational and health centers and about half of all the

institutes of higher education are concentrated in it. In recent years the city's population has drastically begun to increase due primarily to the city's appeal regarding education, professional development and employment, in addition to demographic processes, natural migration and the still untapped potential of the territories and zones around it.
In 2016 over 475 000 000 travels in the capital have been done using public transport. The types of transport used are as follows: 54% of the passengers have traveled by bus, 22% – by tram, 13% – by trolley and only 11% of Sofia's citizens use the subway (Fig. 1.).

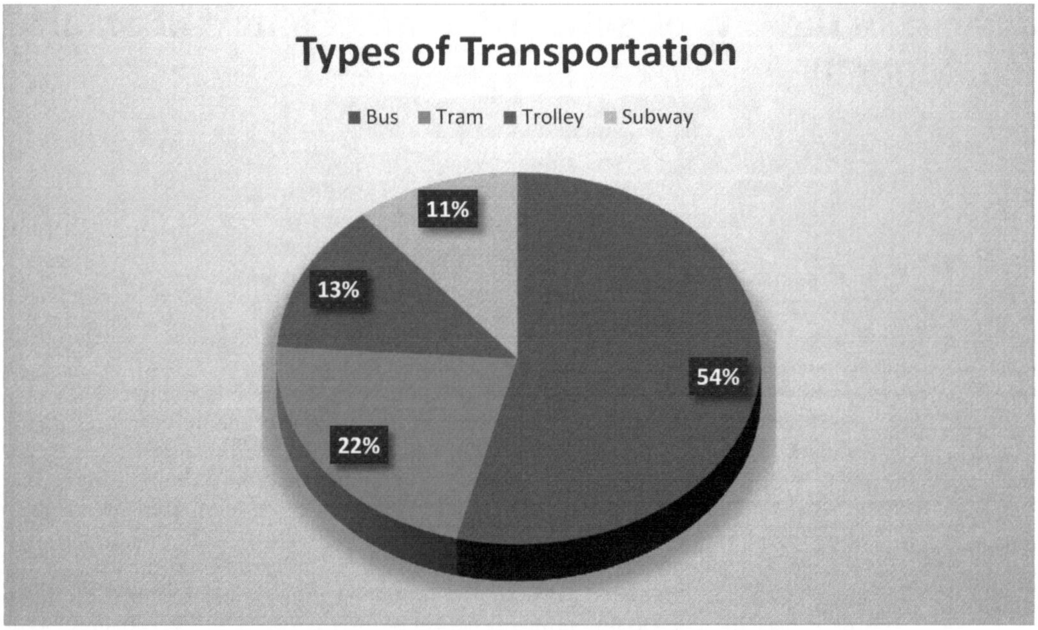

Figure 1. Correlation between the types of transport and the number of travels in percentage [5]
Source: Sofia Municipality's Program for the Development of Public Transport 2012-2015

The capital's transport services are provided by all main types of transport (bus, trolley, trams, subway, taxis and personal vehicles), which, in their entirety and interaction, form the transport system within the city's boundaries.
Buses are the primary and most preferred means of transportations for the capital's citizens. About 260 million passenger travels are done by bus every year. Urban trolleys are secondary and an alternative to buses for urban public transport users. The number of travels by trolley are about 65 million every year. The trolley route network comprises only 13% of Sofia's total urban transport route network; a large portion of it competes with bus transport. The available rolling stock of trolley transport has significantly decreased in recent years, with only 128 trolleys moving on the streets of Sofia in 2016. Tram transport in Bulgaria is less developed because of the poor condition of railways. In the past 20 years the number of trams in the capital has decreased doubly. With some tram lines, waiting intervals reach 30 minutes, which is utterly unsatisfactory when it comes to the quality of the transport service. The subway in the capital has the highest speeds, which is of the utmost significance for the conditions of the big city. It surpasses all other means of public transport with its high transportation capacity, which varies between 21 000 to 50 000 passengers per hour, in one direction. However, the construction of the subway requires the largest capital investments and therefore it is only effective in the directions with the strongest passenger flows. This is the only type of transport in the capital which has shown a significant growth in every aspect since its commission; 500 000 people use the subway in the capital on a daily basis. Taxi travels in the capital are a serious competitive alternative of public transport. They do not adhere to the regulations of stopping at bus stops; they copy public transport routes, offer higher transportation quality and attract a high-payment portion of urban public transport users. According to expert evaluations of operators, an unregulated taxi transportation leads to a 20%

loss for mass public transport [6]. Primarily, it satisfies the episodically arising needs of the urban population for higher-quality and more comfortable public transportation and it complements mass urban passenger transport during rush hour and during the night when the need for transportation decreases and mass urban transport ceases activity. It becomes an inseparable part of mass urban transport due to the increasing demands of the city's population. In recent years the number of transportations via personal automobiles in Sofia has increased rapidly. In 2016 passenger cars have the largest share of the total number of registered vehicles in the country – 82.5%, followed by trucks and towing vehicles – 11.6%. Buses comprise 0.6% and other vehicles, which include special automobiles, mopeds and motorcycles – 5.3% of the park [7]. Every day over 1 million automobiles pass through the city's streets, which causes more and more problems regarding traffic, parking and environmental pollution. The increased traffic also leads to another equally significant problem – noise pollution, which is the result of the poor quality of public transport and the lack of bicycle infrastructure. These issues are deepened by heavy-freight transit traffic in the central part of the city, which causes further noise pollution, harmful emissions in the atmosphere and risks of an increased number of severe traffic accidents within the capital.

3. Guidelines for Improving the Quality of Urban Passenger Transport

In order to improve the quality of urban passenger transport in practice, it is necessary to devise a uniform plan for its future development. It has to be based on good European practices and include measures of technical, operative and organizational nature. It is the only way to fully satisfy citizens' travel needs and guarantee a higher quality of life in the city.

3.1. Further Development of High-Speed Transport in the Capital

Further development of high-speed transport in the capital, by constructing the third subway section in the city of Sofia. Global practice shows that a functioning and developed underground mass urban transport which offers comfort and convenience during travel, high velocity and freight capacity is key to the quickly and effectively servicing large passenger flows and relieving traffic in major cities. After the completion of the 1st and 2nd metro diameter in Sofia, public benefits were indicated in the city of Sofia, such as a 25% reduction of ground traffic, a 22% reduction of traffic accidents, a reduced amount of harmful gases in the atmosphere – by 90,5 thousand tons per year, a 20% reduction of noise pollution in the capital and an overall increase of the subway's share in the total number of public transport travels. Because of this, a significant number of bus lines that copy subway routes were eliminated. This is a very good tendency, considering the fact that the functioning of bus transport is linked to the largest amount of harmful emissions in the environment. In recent years the subway's development in the city of Sofia has become apparent. A quick glance at numbers and facts provides concrete evidence for it. The capital's subway fully fits in with the European vision for developing intelligent, environmentally friendly and integrated transport. The European examples and the EU's Common Transport Policy puts special emphasis on the development of fast, comfortable and environmentally friendly urban transport. Therefore, underground railways are currently being constructed or expanded with European funds in a number of cities on the Old Continent. The future of Sofia's subway involves the construction of a third metro diameter, which is expected to be finished by the end of 2018. It will have 20 subway stations and it will link the northeastern and the southwestern neighborhoods of the capital. The rail-cars will be completely automated.

3.2. Incorporating Environmentally Friendly, Electrically Driven Means of Transport with Modern Equipment in the city's Transport Network, as well as Gradually Substituting Trams and Modernizing Tram Railways

The technical condition of vehicles and the comfort they offer is crucial to increasing the quality of transportation. Modern environmentally friendly vehicles are guaranteed to reduce noise and dust pollution as well as ensure the tranquility of citizens whose place of abode is located right next to elements of public transport infrastructure. Incorporating new technology in electric transport guarantees that the kinetic energy it gives off will be converted and used repeatedly. For its part, this leads to savings and reduction in losses. Regarding the process of renewing vehicles that service public transport lines in Sofia, the implementation of European legislation, especially the parameters

and criteria foreseen in Directive 2009/33/EO of the European Parliament and the Council from April 23, 2009 for advocating clean and energy-efficient vehicles should be taken into account. Stability must be present when planning and securing the overhaul, modernization and development of the infrastructure that provides for electric transport (see Table 1).
Incorporating electric buses into the city's transport network is undoubtedly the most environmentally friendly solution for public transportation in an urban transport, with no contact network, the lowest noise reduction, no harmful impact on people's health and reduced CO_2 emissions.

Table 1. Reducing CO_2 emissions from the operation of electric buses with an annual run of 50 000 km.

Type of bus	CO_2 (tones per bus)	Reducing Chariot e-bus CO_2 compared to other vehicles
Chariot e-bus with an ultra-condenser	32,0	-
Electric bus with batteries	40,0	26%
Trolley	42,0	31%
Diesel bus EVRO VI	74,6	57%
Methane bus (EVV standard)	62,6	48%

Note: For the calculations, the CO2/MWh emission factor of Bulgaria for the production of electric energy in tones has been used. The emissions are lower compared to the emissions that would be given off from the direct burning of fuels by diesel and methane buses. Source: DENKSTATT

3.3. Intelligent Transportation Systems

Intelligent Transportation Systems (ITS) are a modern approach for observing, managing and helping the functioning of the transport system and, partially, automobile transport.
The incorporation of intelligent transportation systems covers various types of technology, whose use will become widespread in years to come. They are created by realizing interconnected telematics decisions, which include a set of instruments based on information technology, wireless communication and electronics. They allow more effective management of transport infrastructure and vehicles, which helps increase safety, reduce traffic jams, shorten travel time and reduce fuel consumption [8].
They can be used to identify problematic areas in the transportation system, but for this purpose, it is necessary to define its parameters in function of time, place and other alarming events, such as bad atmospheric conditions, emergency and repair activities. Measuring the pressure, determining the places where traffic jams and accidents are frequent and identifying the main automobile flows during rush hour will significantly help discover the reasons and solve the existing transport problems [9].
With the help of the amassed operative experience, some of the benefits from ITS are measured quantitatively with regard to the following key aspects of exploitation that concern the city of Sofia: Environment and Road Safety.
The key activity of the network management is optimizing the effective use of the available resources and capacity, shortening the travel time, delays and stoppings of vehicles.
By using ITS, network management could reduce travel time by about 10% on the average and by 29% at most. Delays on the road could be reduced by about 25% on the average, traffic capacity could be increased by about 20% and compulsory stoppings could be reduced by about 40% at most (Fig.2).
There is also a direct dependency between the environment and the improved network management, which leads to the realization of positive effects such as reduced fuel consumption, greenhouse gases and emissions by about 30-60% (Fig. 3).

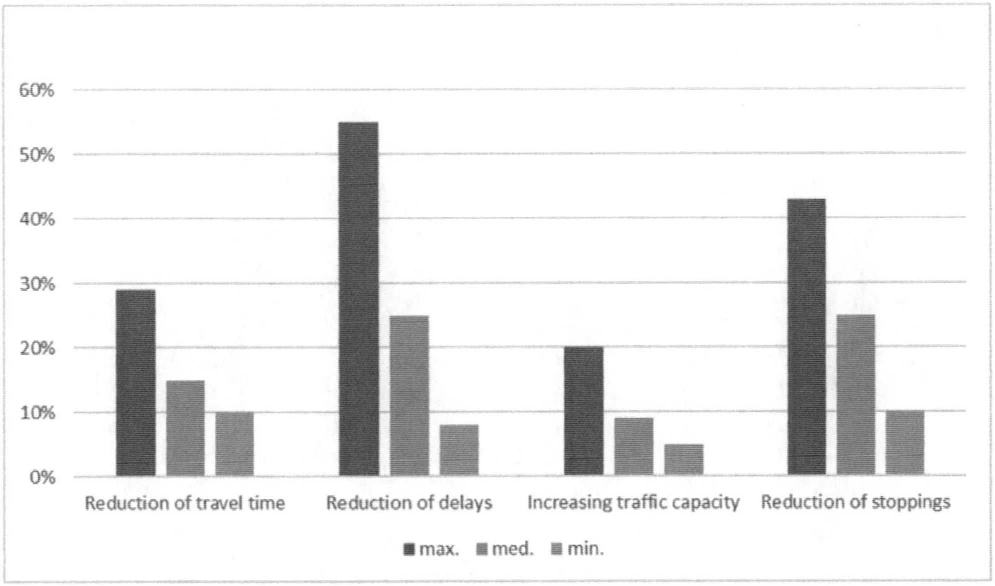

Figure 2. Benefits of incorporating ITS in network management

Source: Strategy for ITS Development as part of Sofia Municipality's Integrated Information System, November 2010

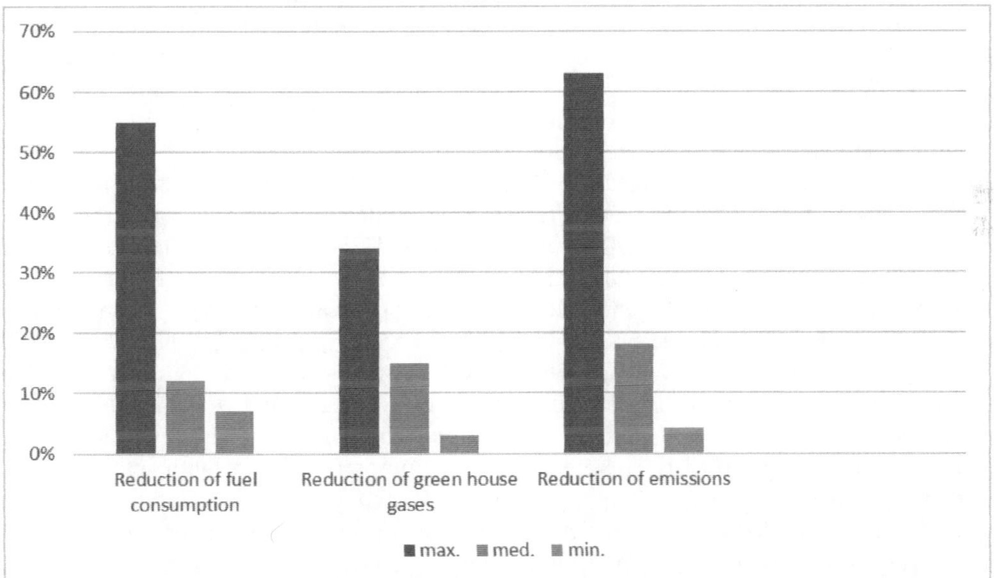

Figure 3. Reducing the harmful impact on the environment

Source: Strategy for ITS Development as a part of Sofia Municipality's Integrated Information System, November, 2010.

The incorporation of ITS in the city's transport system also has undeniable benefits for the reduction of traffic accidents and crashes with a lethal outcome by about 50%, as well as for improving the situation on the road by improving the control for observing the necessary limits and better reaction on the part of drivers (Fig.4).

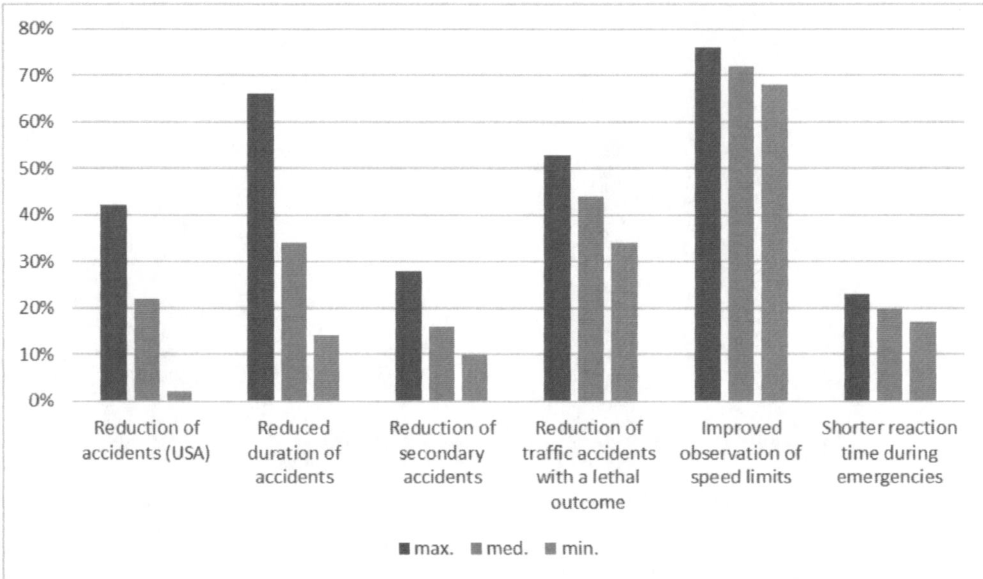

Figure 4. Benefits for road safety

Source: Strategy for ITS Development as part of Sofia Municipality's Integrated Information System, November 2010

The main goals of ITS incorporation in the capital's public transport system have to be based on the issues that impact the current transport situation, especially the significant increase of personal cars, as well as the seriously overladen exploitation of the provided opportunities for parking in the city, the deteriorated indicators for travel time of private cars and public transport, and safety-related considerations [10].

3.4. Increasing the Quality of Taxi Services

Increasing the Quality of Taxi Services. In order to fully satisfy Sofia citizens' needs for public transportation, it is necessary to improve the service and raise the requirements regarding the technical and operating qualities of the used cars. The unorganized and random offers of taxi transport with all of their negative consequences can be avoided by using taxi services with "smart mobility". Ordering a taxi service in the future has to be done solely by smartphones and GPS, thus locating the taxi and sending the fastest car to the address. Such "smart apps" already operate in some European cities like Zurich, Berlin, Vienna, Barcelona, Warsaw, etc. The effect of it is two-fold: the order time is reduced and it creates conditions for more effective use of taxi cars by time and run. In order to reduce the harmful influence on the environment, conventional taxi cars should be gradually substituted with electric ones, which use electrically-powered engines. Compared to cars with internal combustion engines, they possess the following advantages: higher efficiency, higher acceleration, quietness, low prices per kilometre, significantly simplified car service, fewer and more durable parts and, most importantly, they do not pollute the environment.

3.5. Developing Bicycle Transport in Sofia

The European Commission advocates policies for reduction of motorized traffic by stimulating cycling and walking. In practice, increased bicycle traffic will improve conditions for the other traffic participants. One bicycle equals one less car on the road. Cycling in the city is often regarded as a dangerous activity. However, statistics indicate that an increased number of cyclists on the road reduces the number of dead cyclists per kilometre of travel distance. The reason is that drivers become more cautious in the presence of multiple cyclists. Thus, the stimulation of bicycle transport increases road safety in the whole city and for every traffic participant. According to the World Health Organization, cycling 30 minutes a day makes people healthier and happier. So far, a bicycle infrastructure has not been established in the city of Sofia. The development of bicycle traffic within the city's structure is imperative and is a good reason to improve the urban environment. In turn, a high-quality urban environment will attract more investments and tourists [11]. When designing the bicycle infrastructure, road safety factors should be taken into account. The safe behaviour of all traffic participants has to be guaranteed through road tolerance and safe cycling campaigns. Successful cycling policies and the construction of a bicycle infrastructure in Sofia should be inseparable parts of the strategy for stable urban mobility.

3.6. Automated Traffic Control System

The integration of this system into the existing system for territorial positioning and control of mass urban transport vehicles will ensure a more equal distribution of all vehicles in the travel network, which in turn will create opportunities for unimpeded mass urban transport traffic. These two systems will allow flexible and effective development of mass urban transport by providing reliable, accurate, timely and comprehensive information for the processes related to the traffic conditions and the activities of transport operators.

3.7. Effective Organization and Management of the Transportation Process

The effective management should consist of building optimal links and high coordination between the different types of urban transport, guaranteeing security, speed, comfort and stability. As the biggest environmental pollutant, bus transport should service regions and provide links in zones which are not covered by tram, trolley or subway transport. The crucial gradual reorganization of tram and trolley transport as the most environmentally friendly types of ground transport is imperative to the effective management and organization of urban transport in the capital. The purpose of this restructuring is to adapt and integrate these transport types into the available and future design subway lines. Sofia's subway should become primary and structurally defining for transporting large groups of people to relatively long distances due to its indisputable technical and economic features such as environmental friendliness, high velocity, high passenger capacity and effectiveness.

4. Conclusion

Designing a uniform program for the future development and improvement of the quality of urban transport passenger transport within the context for stable urban mobility in the city of Sofia will guarantee the functioning of environmentally friendly, high-quality urban passenger transport, which provides opportunities for attracting more people and fully satisfying their needs for urban travels, thus reducing citizens' dependence on their cars, with subsequent environmental benefits. Accessibility to all urban zones and services will be improved, better environmental parameters such as quality of the air and noise will be reached and the population's health status will be improved. The quality of life for people with impeded mobility and people living in regions which have not been connected with the urban transport network in the past will be improved significantly. The overall image of the city of Sofia will be improved – it will have a new vision of an innovative and progressive modern European city.

5. Reference

[1] Dimitrov D, N Georgiev, E Hadjiev, B Galev 2008 *Safety and health at work in transport*, Todor Kableshkov University of Transport, Sofia.

[2] Dimitrov D, E Hadjiev 2008 Regulation Strategies of Work in Transport Providing Development and Labour Motivation. *Mechanics, Transport, Communications - Academic journal*, Issue 3, pp IV-35 IV-42

[3] Tzvetkova S 2017 Improving the Quality of Urban Passenger Transport in the City of Sofia by Incorporating Intelligent Transport Systems, *Economic and Social Alternatives*, Issue 4, p 32

[4] Tzvetkova S 2017 Increasing the social effectiveness of public transport, CBU, *International conference on Innovations in Science and Education*, March 22-24, 2017, Prague, Czech Republic, p 486

[5] Tzvetkova S 2016 Analysis of Urban Passenger Transport in the City of Sofia and Directions for Its Development, UNWE, p 96

[6] Gatovski I 2016 Analysis of the state and primary trends in the development of public transport in Sofia, *Mobility for a connected world-Research-and-Practice Conference*, UNWE, p 332

[7] www.nsi.bg
[8] Minkov T 2016 General Measures and Potential for Limitation the Harmful Impact of the Transportation on the Urban Environment, *Proceedings of the Scientific Practical Conference on the Mobility for A Connected World*, October 28, 2016, Sofia, Bulgaria, pp 135-139
[9] Tzvetkova S 2017 Improving the Quality of Urban Passenger Transport in the City of Sofia by Incorporating Intelligent Transport Systems, *Economic and Social Alternatives*, Issue 4, p 37
[10] Strategy for ITS Development as part of Sofia Municipality's Integrated Information System, November, 2010, Sofia;
[11] Velikova, E., The Influence of the Events in Tourism, *Journal of Science and Research*, Volume 1, 9/ 2016, pp. 83-91.

Ruralization vs. Urbanization Sprawl as Guiding Regional Planning: Development Scenario for Rivers Watershed in the Southern Syrian Coastal Region

T Rahmoun[1, 3], W Zhao[1], M Hammad[2], M Hassan[1, 3]

1 Faculty of Architecture & Urban Planning, Chongqing University, Chongqing, P.R. China.
2 Department of Physical Geography and Geoinformatics, Szeged University, Szeged, Hungary.
3 Faculty of Architectural Engineering, Tishreen University, Latakia, Syria.
E-mail: tarek.rahmoun@qq.com

Abstract. This study aimed at opening discussions concerning new ideas of suggesting sustainable scenario based on the principle of an integrated spatial development ring between urban-rural areas along and between AL-Abrash and AL- Hseen rivers watershed, as an application of the bottom-up planning model, seeking to achieve ruralization in parallel with urbanization. This paper adopted data collections and analysis through using a step-by-step approach. Firstly, investigated the land-cover change (LCC) during 30-years. Therefore, using multi-temporal satellite data from different dates for the same study area to create thematic land cover maps which can be used for land cover change detection. Three Landsat satellite images from1987, 2002 and 2017 were classified separately using the supervised classification method in ArcGIS, to provide an economical way to quantify, map and analyse changes over time in land cover. Then, SWOT analysis for the possibilities and determinants within the two-way flow of the current and futuristic economic activity, besides discussion the opportunities of the land-use (LU) taking into account slop map to achieve conservation priority for natural resources. Finally, evaluate results and establishing a sustainable spatial scenario approachable to upgrade into scaling up/out to covering the coastal region watersheds, can support regional planning and decision-making in the future.

1. Introduction

Urbanization is one of the leading causes of damage and loss of natural vegetation, arable land, and habitat [1]. Prodded by the desire for development, converting rural areas into urban areas at an unparalleled rate in recent history. This unrelenting effort to urbanize is showing its effects on the natural functioning [2]. Monitoring changes in land cover and understanding the spatial properties of it will lead to an insightful appreciation of how human activity affects the general ecological state of the urban environment [3]. Developing countries are showing a significant human activity-attributed land use and land cover changes than the more developed countries [4]. Historically, people in many countries marked preference for settling in mountain areas due to favourable climate and the easy access to water. These days, mountain areas are affected by tourism, transportation, and industrialization activities. Hence, more people now living in the mountain regions led to being depleted of their natural resources. Consequently, people need to understand the relationship between development and land dynamics [5].

On the other side, rivers watershed offers an essential unit for sustainable development; considering, human settlements built within a watershed area record stronger physical, social, and economic ties with each other compared to settlements found outside the watershed area. In this unit, rural development programs should take into account, especially to establish the linkage between land resources, populations and interests [6]. Nevertheless, the deterioration of watershed sources seems to be going through quickly, particularly in developing countries [7]. Watersheds vary in size. It can measure a few hectares to be as large as a thousand square kilometres. Regardless of size, a watershed is a complex system of water, wildlife, forests, cultivation, and human practices [8], Watershed management previously focused on the biophysical aspects. Nowadays planners recognize that watershed management should take into consideration the fact that a watershed is a combination of both human and biophysical processes [9]. The intricate interactions can have harmful far-reaching effects on the environment, as well as on economic activities [10]. Hence, Watershed management today requires cooperation from among varied sectors including tourism, hydraulic engineering, transport, spatial planning, and agriculture – all of which have a stake when it comes to the sustainable development in the rural areas, particularly in the mountain regions [11].

Geographic information systems ArcGIS tools extensively used to map and analyses land cover changes in a lot of watersheds [12]. GIS and remote sensing techniques are also employed to prioritize watersheds that need conservation as tools that have proven very useful for planning and development. Furthermore, multi-temporal remote sensing data help to accurately map the extent of the different land cover types of a particular area and to monitor the changes over a particular period of time. GIS has spatial tools that help to monitor land use and land cover LULC changes [13]. In the same time, there is increased interest in using sustainable activities in the rural area to be a gateway toward achieved ruralization as a planning tool aimed at re-polarization with a stronger interest in using agropolitan strategies for regional development by spur agro-economic growth. These solutions should include infrastructure, protection of settlements, security of landscape, and respect for the values of biodiversity and fair allocation of economic benefits [14].

In Syria, the coastal region considered the main water tower at the national level. Since the 1980s government adopted engineering solutions by building large and small dams as a tool to water resources management to storage of water for power generation, agriculture, irrigation and domestic use. Coastal geography was divided into 21 river basins in Lattakia and Tartous governorates, based on river sources, feeder springs, and slope direction. However, the idea of river-watershed management as an integrated socio-economic planning unit between urban-rural areas and between mountain-coastal areas is a new down-up planning approach, as long as most of the researchers adopted administrative unit's basis to achieve the required development. For that, this paper discusses new down-up model using a step-by-step approach, begins in the mountainous rural areas, then applies the scaling-up progression for upland and lowland interactions, Taking into considering three priority fields "Tourism - Agriculture - Transport" as the basic assets of the Mediterranean coastal regions. The scaling-out processes are used to include the adjacent watersheds for an integrated regional planning.

This empirical study attempts to identify the spatiotemporal pattern of LCC for rivers watershed in the study area using geospatial data so that both the scientific community and decision-makers can assess the various dynamics affecting in this environment. The objectives of this study were thus to explore the characteristics of current land use and LCC and characterize the underlying driving forces in the study area by making use of remotely sensed data and socioeconomic information. Specifically, the objectives are: (a) to elucidate and evaluate the LCC between 1987 and 2017; (b) to analyze the driving forces of land use change and urban expansion, and (c) to propose a futuristic scenario to achieve ruralization in the face of urbanization.

2. Research Overview and the Study Area

There is very little data available regarding the temporal and spatial aspects of LULC changes that have influenced the urban development of the coastal region of Syria.

Tartous province area is 190.000 HA, 1% of the total area of Syria, and the population of 780.000 in 2010 and 3.8% of the total population of the Syria. Tartous population increased from 383,000 in the early 1970's to 663,000 in 1990 and 938,000 in 2010, an average annual increase of 14,000 between

1970-1990 and 13.800 between 1990 and 2010.Moreover, expected to reach more than one million by 2030, according to the average natural growth rate of 1.79% between 2000-2010, which recede from 3.16% during 1970-1981 to 2.19% during the period 1981-1994 due to socio-economic and cultural developments, in additional to Internal and external migration force [15]. All previous statistics recorded until 2010, while there are no official statistics yet after 2011 because of the war, where a huge displacement recorded to the coastal region as a safe area reached to more than 1.5 million populations.

2.1. The Study Area: River Basin (watershed).
The study area is a part of the Syrian coastal region, it is located between 34°40′ to 35°3′north latitude and 35°51′to 36°19′east longitude, covering an area of 1040.50 sq. km. Four rivers watershed in Tartous province 1- watershed Abrash river WS/A, 2- watershed Hseen river WS/H as major river basins and 3- watershed Ghamqa river WS/G, 4- watershed Mntar river WS/M as secondary river basins, with about 77% of province population. Characterized by a Mediterranean climate with limited impact and geographic variability. Mean elevation is 473 m above sea level, with a maximum of 1100 m a.s.l. In the north-west areas, a strong influence from the Mediterranean climate brings high precipitation. For the period 1987–2016, the mean annual air temperature across the target basins was 15 °C, ranging from -4 to 38 °C. The mean annual precipitation was 1100 mm/y, ranging from 900 mm/y near the river mouth to 1200 mm/y in the western mountains. The area under study has the following primary sections; Lower, Middle, and Upper Sections 'Figure 1'. These sections are further subdivided on the basis of land-cover and topography into basin socio economic management regions.

2.1.1. The Upper Area. The Upper Area runs between 600 and 1100 m a.s.l. It includes the water sources found within the Western Forest Mountains – the Hseen and Abrash rivers in Masyaf and Al-Kafrun. In this section, the basins maintain a source character, where small mountain settlements with a simple rural agricultural lifestyle are engaged in planting vegetables and fruit trees and raising livestock. While a kind of ruralization based on winter tourism and eco-tourism parallel with agricultural in the south.

2.1.2. The Middle Section. The Middle section runs between 200-600 m a.s.l. The plateaus region that forms the wider home of the coastal rural-area, while bed slope decreases. River banks become sandier and bounded by forests of large-leaved and coniferous trees. Olive trees are abundant. At the end of 1997, Al- Abrash dam built to the south of "Safita district" as the most important project in the coastal region and put under investment in 2000. The dam measures 733 meters in width and 50 meters in height. The lake at the rear of the dam runs 32 km and has a storage capacity of 689/ha. Moreover, irrigation network and pumping stations contribute to irrigation of 10160 HAs of agricultural land [16].

2.1.3. The Lower Section. The Lower Section extends between 0 and 200 m a.s.l. You will see a huge contrast in land use in this section. In the north lies the biggest commercial port in the region. You will also find Tartous City, recognized as the principal urban area of the south of the coastal region. In the south lies a fertile plain filled with field crops, fruit trees, and green-house produce. Along the coast line, founded a kind of tourist industry. There is a proposal to declare the area a free economic zone, in parallel to upgrade the agricultural airport to a civil airport requirements. Regional decision makers now see this area as a gateway to maritime economic development. On the other hand, the main highway and railway which penetrate coastal plain considered the vital commercial artery connecting the coastal region and the interior.

3. Database

Landsat Surface Reflectance data of the study area from the Landsat 5 TM (Thematic Mapper), Landsat 7 ETM+ (Enhanced Thematic Mapper Plus) and Landsat 8 OLI/TIRS (Operational Land Imager) satellites were ordered as higher-level products through the United States Geological Survey (USGS) http://earthexplorer.usgs.gov/. These three processed Landsat images were used to create land cover changes (LCC) maps for 1987, 2002 and 2017 (Table 1).

IOP Conf. Series: Earth and Environmental Science **151** (2018) 012033 doi:10.1088/1755-1315/151/1/012033

Table 1. Details about images used in the study.

Image acquisition date	Sensor (Image ID)	Resolution (m)	Bands	Product
28 May 1987	Landsat5 TM (p174r36_5t19870526)	28.5	2,4,3	GeoTiff
11 May 2002	Landsat7ETM+ (LE71740362002131SGS00)	30	2,4,3	GeoTiff
26 May 2017	Landsat 8 OLI/TIRS (LC81740362017148LGN00)	30	3,4,5	GeoTiff

Survey of Syria (SOS) toposheet numbers NI37-M-3-d, NI-37-M-3-c, NI-36-R-4-d, NI-37-S-1-a and NI-36-R-4-b covers the study area on 1:50,000 scale. Base map of the study area was prepared from the SOS toposheets. Moreover, the height information to generate the Triangulated Irregular Network (TIN) and slop maps 'Figure 1', from the free DEM data of the SRTM 1 arc-second (30 meters resolution).

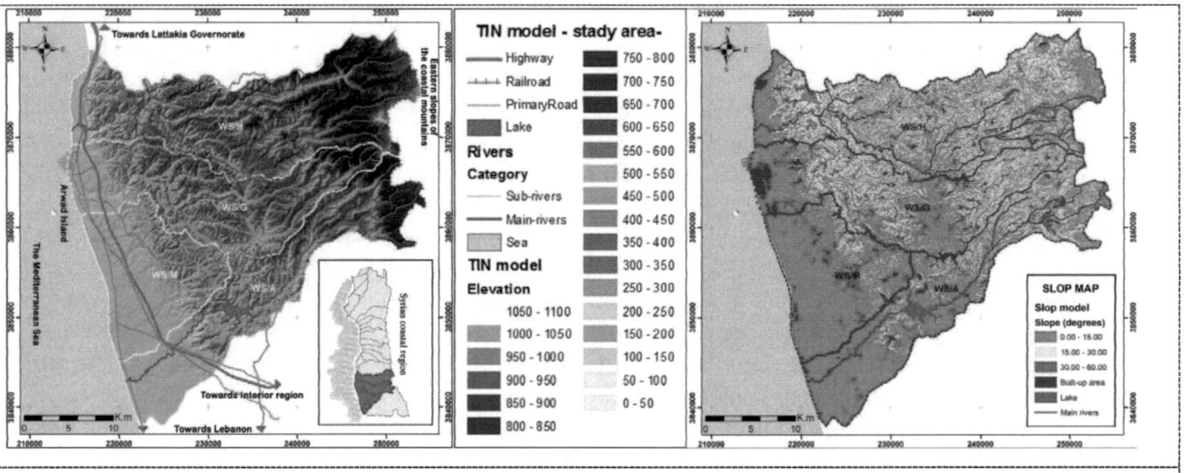

Figure 1. TIN model and slop map for the study area

4. Methodology

This paper uses a mixed methodology. Both quantitative and qualitative techniques are used to analyses the available data and information within the study area. Additionally, an abundance of published literature related to functionally and spatially integrated river watershed planning has been reviewed.

In this study, the supervised classification method was chosen and applied for images of three different years using the ArcGIS software. There are different algorithms for supervised classification. One of them most common algorithm of the supervised classification is Maximum Likelihood Classifier algorithm which was used in this study. This algorithm examines the probability function of a pixel for each of the classes and assigns the pixel to the class with the highest probability. Determination of classes is a very important step in supervised classification. Four classes – Built-up Area & Quarries, Forest & Dense Vegetation, Agriculture Land & Open Vegetation and Water Body – were created for each image 'Figure 2'. We attempted to select the best samples representing the classes; however, some mistakes may have occurred because the spatial resolution of the images was 30 meters.

A total of 180, 125 and 100 well distributed training sites representing all LC classes were selected for classifying the 1987, 2002 and 2017 images respectively. As there was no ancillary data available for 1987 and 2002, we resorted to visual interpretation of the Landsat image based on the authors' knowledge of the study area. A total of 160 sampling points were created using random points tool and used for validating the 2017 land cover map. The overall accuracies were 83.7%. An accuracy matrix was created using the observed and the classified land cover map showing each class′s and overall accuracy as confusion matrix. (Table 2). Frequency tabulations of classification results alongside referenced ground cover classes were developed. These were converted to pivot tables from where

IOP Conf. Series: Earth and Environmental Science **151** (2018) 012033 doi:10.1088/1755-1315/151/1/012033

error/confusion matrixes detailing the overall accuracy and user accuracy for each class were calculated.

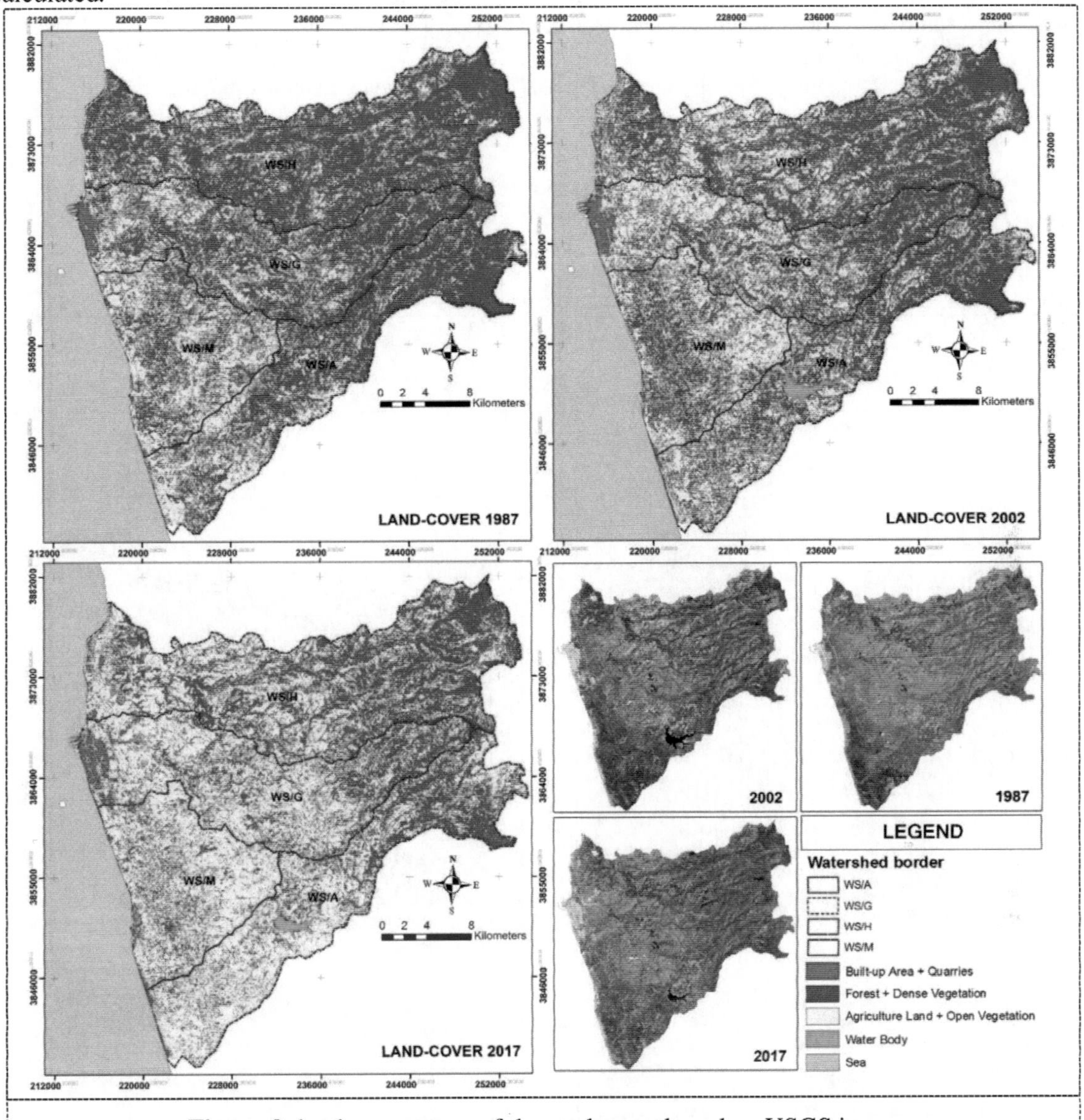

Figure 2. land cover maps of the study area based on USGS image

Table 2. The accuracies of classification for the 2017 land cover map.

Class Name	class	1	2	3	4	User Accuracy	Number of validation points
Built-up Area + Quarries	1	3	0	1	0	75%	4
Forest + Dense Vegetation	2	0	60	2	0	96.80%	62
Agriculture + Open Vegetation	3	0	23	70	0	75.30%	93
Water Body	4	0	0	0	1	100%	1
Overall Accuracy						83.75%	160 Total

The user accuracy represents the number of correctly classified pixels in one class divided by the total number of sets recognized in that class. While the overall accuracy represents the sum of all of the correctly classified pixels divided by the total number of all sets recognized in all classes.

Spatial analysis on land cover classes of 1987, 2002 and 2017 respectively was performed in the ArcGIS environment. The area under each category calculated (in sq. km), areas and percentages were compared to stand at changes during the monitoring period. (Table 3).

Table 3. The land cover change analysis of the study area between 1987 and 2017.

	On 1987		On 2002		On 2017		LCC (1987-2017)	
LC classes	Area (sq.km)	Area (%)	Area (sq.km)	Area (%)	Area (sq.km)	Area (%)	Change (sq.km)	Change (%)
Built-up Area + Quarries	23.914	2.34	30.058	2.94	46.689	4.57	22.775	2.23
Forest + Dense Vegetation	651.844	63.75	579.794	56.70	389.436	38.08	-262.41	-25.67
Agriculture + Open Vegetation	346.803	33.92	407.649	39.87	582.705	56.98	235.902	23.06
Water Body	0	0	5.059	0.49	3.73	0.36	-1.329 [a]	-0.13 [a]

[a] Between 2002-2017. Dams were built after 1994.

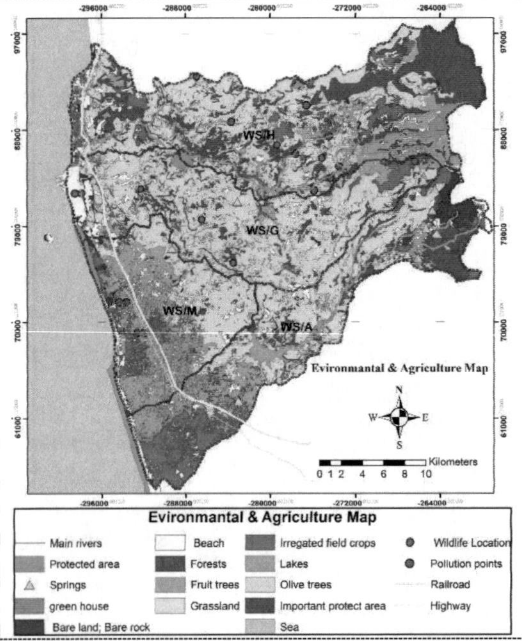

Figure 3. Environmental and agricultural land-use map, where classes describe the type of use of agricultural and open vegetation.

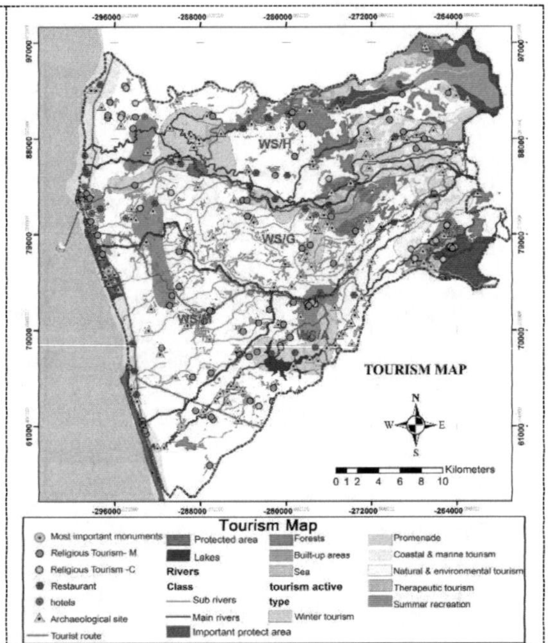

Figure 4. Tourism land-use map shows the classifications of tourism types. In additional to main tourist sites and services.

The internal affecting factors were classification into three classes due to current land use, the tourism map, the demographic and industrial spread, and the environmental and agriculture map 'Figure 3,4,5'. SWOT "Strengths, Weaknesses, Opportunities, Threats" analysis to approved maps, taking into account the connection routes between them, That's to understand the impact/risk factors that relate to the current situation and the future of the study area. Based on data and information from the relevant government departments and interviews with regional decision-makers during the summer of 2016. Furthermore, to suggest a ruralization scenario more suitable to the research goal, in parallel with the results of monitoring LCC due to climate and human factors. As a new step for the balanced permanent development of the coastal region starting from rivers watershed units. Given that each river watershed has a distinctive character regarding the proportion of the urban-rural population and places of settlement spread between the three sections of each watershed through projection data on the slope map. In addition to the difference in the human activity affecting. Hence, SWOT analysis used for each basin taking into account the most prominent features (Table 4).

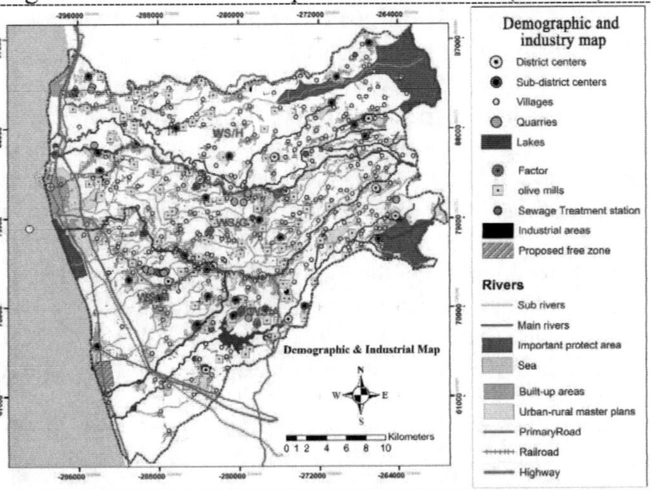

Figure 5. Demographic and industry map to identify the current and futuristic spread urban-rural settlements. Also, determination of the industrial activity locations and main transport network

Table 4. SWOT analyse for Land-use of the sub-watersheds.

	STRENGTHS – S	WEAKNESSES – W	OPPORTUNITIES – O	THREATS– T
WS/H	▪ Forests & nature reserve areas.	▪ Weak tourism infrastructure.	▪ Integrated environmental tourism.	▪ Urban expansion on agricultural land.
	▪ Diversified agricultural production.	▪ Cement factory at the marine basin window.	▪ The possibility of integrated agro-tourism.	▪ River pollution due to waste olive mills.
	▪ Low population density.	▪ A narrow coastal plain.	▪ Rural Development "Agropolitan centre".	▪ Loss of traditional rural lifestyle.
	▪ Important points for wildlife.	▪ Weak services in the upper section.	▪ Vertical intensification of agro-production.	▪ Decreased vegetation due to climate change.
WS/G	▪ Marine tourist gateway, Tartous port.	▪ Weak of the direct link with other basins.	▪ Agro-industry strategy for marketing olive oil.	▪ Transformation towards urbanization.
	▪ Includes the biggest. urban centre "Tartous"	▪ High population density.	▪ A traditional maritime tourism industry.	▪ Increase fragmentation of agricultural property.
	▪ The strategic olive product at the regional.	▪ Loss of runoff water despite heavy rainfall.	▪ Build a set of storage dams.	▪ The phenomenon of real estate development.
	▪ Launch of the University of Tartus.	▪ Decline of olive quantity production.	▪ Expand the scope of therapeutic tourism.	▪ Polarization towards the city of Tartous.
WS/M	▪ Wide seafront + great archaeological sites.	▪ Expansion of housing within agro-land.	▪ Establishment of a free economic zone.	▪ Industry condensation at the future.
	▪ Vast plain + intensive agro-production.	▪ Weak tourist services.	▪ Establish agropolitan centre, Yahmur area.	▪ Beach environmental degradation.
	▪ Low population density.	▪ Citrus-producing non-competitive quality.	▪ Agro-tourism strategy.	▪ Random polarization of real estate investment.
	▪ Transport links at the national level.	▪ Reduced marine fish wealth.	▪ Establishment of marine fish farms.	▪ Expansion of licensing of quarries
WS/A	▪ Important lake and natural site.	▪ Weak tourism marketing.	▪ Establishment of fish farms beside the lake.	▪ Pollution of lake water due to sewage.
	▪ Archaeological water mills along the river.	▪ Shortage of sewage treatment plants.	▪ Dam Investment in Electric Generation.	▪ Decline of the lake's level.
	▪ Integrated tourist destination.	▪ Weak infrastructure in the lower section.	▪ Invest In wind energy in the upper areas.	▪ Targeting natural areas by extractive industries
	▪ Integrated field crop production	▪ Subsistence farming non-competitive.	▪ Starting for a future national tourist route.	▪ Pressure on mountain natural resources.

5. Results and Discussion

5.1. Land Cover Change Analysis

As far as land cover changes from 1987 to 2017 is concern, the study area as a whole presents a general degradation of land resources. It has been observed that during 30 years the Forest and Dense Vegetation decreased by 25.66%. At the same time, the Agriculture Land and Open Vegetation increased by 23.07%. This expansion in the depletion of arable land at the expense of dense vegetation is due to an attempt to increase agricultural production, whether as agriculture of subsistence or as a major source of income for rural areas. Furthermore, the amount of arable and fallow land in Tartus Governorate fell to zero in 2005 [17]. But in parallel with the fragmentation of agricultural holdings as a result of inheritance factors, the average tenure is currently only 1.3 HA and is subject to fragmentation [18], the loss was double. On the one hand, large areas of dense green cover have been lost. On the other hand, agriculture in many areas has not achieved the expected material return. Whereas, the Built-up Area and Quarries area increased by 2.23% at the expense of forest and agricultural lands. Sub-watershed wise details of the land cover and the changes in area under each class from 1987 to 2017 (Table 5).

Table 5. The land cover analysis in the years 1987 and 2017 of the sub-watersheds.

	On 1987		On 2017		LCC (1987-2017)	
LC classes	Area (sq.km)	Area (%)	Area (sq.km)	Area (%)	Change (sq.km)	Change (%)
WS/A Basin (274.9 sq.km)						
Built-up Area + Quarries	2.98339	0.012138	8.2971	0.033754	5.313706	0.0216154
Forest + Dense Vegetation	159.935	0.650745	83.7981	0.340907	-76.137174	- 0.3098384
Agriculture + Open Vegetation	82.8535	0.337115	150.508	0.612296	67.654339	0.2751811
Water Body	0	0	3.2058	0.013042	3.2058	0.01304183
WS/H Basin (354.5 sq.km)						
Built-up Area + Quarries	6.98616	0.019524	8.5194	0.023812	1.533238	0.0042878
Forest + Dense Vegetation	270.8521	0.756951	191.3616	0.534861	-79.490473	-0.2220898
Agriculture + Open Vegetation	79.98145	0.223524	157.5072	0.440237	77.525755	0.2167129
Water Body	0	0	0.3897	0.001089	0.3897	0.0010892
WS/G Basin (246.5 sq.km)						
Built-up Area + Quarries	10.293644	0.041343	18.9225	0.076021	8.628855	0.0346784
Forest + Dense Vegetation	149.00482	0.598459	79.1244	0.317884	-69.880425	-0.2805752
Agriculture + Open Vegetation	89.68215	0.360197	150.8625	0.606094	61.180353	0.2458967
Water Body	0	0	0	0	0	0
WS/M Basin (164.6 sq.km)						
Built-up Area + Quarries	3.6567495	0.021505	10.9566	0.064473	7.2998505	0.0429675
Forest + Dense Vegetation	72.075003	0.423883	35.1522	0.206850	-36.922803	-0.2170329
Agriculture + Open Vegetation	94.303037	0.554610	123.831	0.728675	29.527962	0.1740654
Water Body	0	0	0	0	0	0

Of course, climate change cannot be ignored during the period of monitoring changes, especially during the last decade. Despite the mean annual precipitation within the overall average, but heavy rainfall during short periods wasted as a surface scrub without interest because of the lack of small storage dams. Compensated by long and concentrated periods of heat during the summer and increase the rate of frosts during the winter. Again, increase in the agricultural land at the expense of forest and dense vegetation cover can be taken as negative change as it indicates the growth of built area later in transfer phenomenon toward real estate development especially in the hills area. Similarly, an increase in the settlement/ built-up area is a negative change on land resources can be doubling from two to three times over the next three decades according to the rate of natural growth of population.

5.2. Integrated Analysis of SWOT, LU/LC and Slop Map

Tripartite analysis forms an integrated analysis and a database of crossed information's and data. This will give a chance to the decision-makers to draw more accurate and harmonized futuristic and development scenarios. Hence, the readings of the changing curve in the classifications of the main land-cover and its projection on the slop map will ensure stricter conservation conditions in the areas facing an environmental degradation of the natural sources, like the forest areas and dense vegetation. Moreover, to steer the expanding of built-areas towards the sloping areas which are less than 30%. Where the slope is always a controlling factor in the development of land cover on the surface. Moreover, development of LU and LC over the area with unfavourable slope condition is always prone to potential environmental hazards [7]. In the other hand, SWOT analysis gives a logic reading of the probable risks facing the direction towards investing available chances, especially in the rural mountain areas. Because the rural development should consider all these analysis intersections which give clear indicators that SW/H+ SW/M can ensure the possibility to build the core of the ruralization scenario as a proposed agropolitan centres. Besides the qualitative data and the other agricultural and industrial activity distribution, also of the available building areas and its density which can found through the Land-use maps. At the same time, we can see that the resulting database of information can give an additional indicator towards suggesting a horizontal transportation network connects SW/H+SW/M with its surroundings. Due to these areas with an expanding residential curve through the last 30 years, that holds linked features on social, touristic and agricultural levels, with a growing phenomenon of transfer to random fast development which must administrate rationally.

5.3. Built Ruralization Scenario Based on Current Planning Data: Learning from the Past.

Taking into consideration spatial land-cover changes, areas of urban expansion and its reasons. The proposed scenario mainly based on transfer polarization towards rural areas through secure jobs and increase income and economic asset development consistent. As a roadmap for urban-rural balanced development. Geographical regional planning units "river watershed" to create the enabling environment for expanded economic networks. Cluster tourism networks in parallel with the networks required for promising industries. Since, spatial rural-urban linkages, which are flows of people, goods, money and information between urban centres and rural regions, are important drivers of economic activities [19]. Therefore, propose vital centres within rural areas to enhance agricultural growth to find a new method to increase production, export and agro-industrial products alike. Additionally, increase the employment because poor rural residents can combine peri-urban agriculture with other urban-based, non-farm occupations as agro-tourism services and transportation services.

5.4. Integrated Cluster-Tourism with Agriculture for Ruralization.

The geographical spread of the river watershed between upstream to downstream provides the suitable environment for establishing an integrated cluster tourism where the leisure tourism, nature tourism, cultural, archaeological tourism and rural-agro-tourism form together tourism network. The problem of small agricultural holdings considered the first barrier to establish a suitable agro-tourism, as well as the lack of experience in agrotourism, marketing, and management among the rural residents. Moreover, before investing in agrotourism, you must carefully analyse your potentials as a host and available resources for investment in an agro-tourism venture [20]. Therefore, propose a new style of agricultural cooperatives are the best solution, where a group of farmers contribute to form a modern agricultural project supported by government investment in enhancing infrastructure requirements. The management will combine representatives of farmers and local government to achieve an integrated agro-industries. It begins with high quality intensive agricultural production using modern agricultural techniques to improved varieties. Moreover, rural families form local units for small industrial lines based on advanced technology which will be managed by scientific expertise from the new generation of young rural. Further, the responsibility of the government sector is securing transport to the neighbourhood and abroad through the agropolitan centre. In the other side, agricultural units will contribute as a new tourist destination to experience the rural agricultural lifestyle and enjoy the original product. This style will lead to a two-way flow relationship between agriculture as a profession, agro-industries, transportation, and agrotourism as an additional source of income, seek to attract young rural away from migration. That can be applied to the vertical upgrading of the strategic agricultural production of olive, tobacco, citrus, and greenhouses according to spatial distribution and agricultural density. Its bottom-up strategy Pro-poor operates at a local level, that generates net benefits for the poor, not only economic benefits but also positive socio-cultural and environmental benefits.

5.5. Scaling-up Processes and Scaling-out by Comparing Watersheds and Learning from Each Other.

The importance of the agropolitan approach to achieve a new kind of ruralization within the river watershed planning units must be completed by scaling-up progress cover the rural-rural and urban-rural linkages with all the economic assets. In another meaning cover other relationships with non-agriculture activities. In this step, we need to determine the possibilities of various areas targeted by scaling-up processes depend on SWOT analysis. For example, in mountainous rural areas where livestock is raised, it's possible to imitate the Chinese experience of rural-urban construction plans and environment-friendly agriculture in Chongqing province "modern agriculture zones in Changshou District: Modern Stock Raising Zone." The partnership between the public sector and a group of livestock breeders and farmer families in the establishment of livestock breeding projects is the core idea, seek to increase the efficiency of animal products and increase their numbers in private sheds created by the government sector with modern technology. This example repeatable to increase fisheries and beekeeping. Such projects targeting different rural sectors will strengthen rural-rural relations through the participatory process from the source of the product, and also rural-urban relations through the daily mobility towards the agropolitan centres, product marketing, and tourism marketing.

As shown above, the process of scaling-up scenario to include the entire geographic river watershed area is based on an interconnected clustered structure non-linear process; the challenge is how to address cumulative effects over different temporal and spatial scales. Therefore, we can classification the 'Success' of the scenario by monitored the causes of land-cover changes to develop environmental alternatives instead of the rapid consumption of natural resources. Subsequently, the results of the scenario are assessed from socio-economic and cultural perspectives on rural areas. The scenario can be converted into a measurable and repeatable plan in scaling-out process for sharing analytical methods. Thus, turning the studied area into a database and a platform for sustainable regional environmental planning, and marketing the strategy of sustainable rural communities as long-term investments.

6. Conclusion

The study demonstrated a new model of regional planning approach based on scaling up/out for geographic planning units "rivers watershed" upper off the administrative levels. Based on land-cover analyses during 30 years and integrated analysis of SWOT with slop vs. land-use. Therefore, the core role was of remote sensing and GIS in classification and combination for prioritizing the rivers watersheds in management and conservation of the natural resources to achieve the main idea of ruralization. Based on this study, it has been observed that SW/H and SW/A are the two main-watersheds fall under the category of high priority to establish the development scenario in the rural areas of the watershed. Moreover, SW/G consider as a dynamic basin for accommodate future economic activities with high population concentration. Whereas SW/M with SW/H sharing identified the success of the ruralization scenario through the availability of the requirements for the establishment of the agropolitan centres in the upper and middle sections alike. Therefore, this scenario likely that will provide a sustainable solution to the problem of the rapid transition to urbanization and the loss of rural identity of large sections of the coastal region. Again, this study needs to supplement the results by specialization in the economic income and the long-term implications of the phenomenon of climate change, with constant updating of the database, especially post-war and the reconstruction phase.

Acknowledgments

The authors acknowledge the financial support of Chongqing University. We are also immensely grateful to Tartous governor Mr. Safwan Abu Saada for interviews and administrative support. In addition, to Eng. Hassan AL-Abbas, Regional Planning Directorate, Tartous.

7. Reference

[1] Y. Liu, Y. Wang, J. Peng, Y. Du, X. Liu, S. Li and D. Zhang, "Correlations between urbanization and vegetation degradation across the world's metropolises using DMSP/OLS nighttime light data," *Remote Sensing,* vol. 7, no. 2, pp. 2067-2088, 2015.

[2] L. Jiang, X. Deng and K. C. Seto, "The impact of urban expansion on agricultural land use intensity in China," *Land Use Policy,* vol. 23, pp. 33-39, 2013.

[3] B. Pandey, P. Joshi and K. C. Seto, "Monitoring urbanization dynamics in India using DMSP/OLS night time lights and SPOT-VGT data," *International Journal of Applied Earth Observation and Geoinformation,* vol. 23, pp. 49-61, 2013.

[4] J. L. Baker, "Dhaka: Improving living conditions for the urban poor.," World Bank, Dhaka, Bangladesh, 2007.

[5] M. U. Shafiq, A. A. Mir, P. Ahmed and P. A. Bhat, "Landuse/ Land cover Analysis in Hamal Watershed of North western Himalaya's using Remote Sensing & GIS," *International Research Journal of Engineering and Technology,* vol. 3, no. 4, pp. 2799-2805, 2016.

[6] T. B. H. Nyerges, C. Steinitz, T. Canfield, M. Roderick, J. Ritzman and W. Thanatemaneerat, "Geodesign dynamics for sustainable urban watershed development," *Sustainable Cities and Society,* vol. 25, pp. 13-24, 2016.

[7] S. Sarma and T. Saikia, "Prioritization of sub-watersheds in Khanapara–Bornihat area of Assam–Meghalaya (India) based on land use and slope analysis using remote sensing and GIS," *Journal of the Indian Society of Remote Sensing,* vol. 40, no. 3, pp. 435-446, 2012.

[8] P. E. Black, Watershed Hydrology, New Jersey: Prentice Hall, Englewood Cliffs, 1991.

[9] T. C. Sheng, "Challenges and Strategies in Integrated Watershed Management in Developing Countries," Guide to the Papers of Ted C. Sheng. Series 1: Publications. Finding aid: http://lib. colostate. edu/archives/findingaids/water/wtcs. html., 2016.

[10] B. Panigrahi and M. R. Goyal, Modeling Methods and Practices in Soil and Water Engineering; P:220, illustrated ed., CRC Press, 2017.

[11] P. Greminger, "Mountain watershed management: Lessons from the past–lessons for the future," Berne, 2003.

[12] B. Madurapperuma, P. Rozario, P. Oduor and L. Kotchman, "Land-use and land-cover change detection in Pipestem Creek watershed, North Dakota," *International Journal of Geomatics and Geosciences,* vol. 5, no. 3, pp. 416-426, 2015.

[13] K. L. Martin, T. Hwang, J. Vose, J. W. Coulston, D. N. Wear, B. Miles and L. E. Band, "Watershed impacts of climate and land use changes depend on magnitude and land use context," *Ecohydrology,* p. doi: 10.1002/eco.1870, 2017.

[14] A. Goettle, "Profit of watershed management for lowland areas - experiences made in Bavaria: Greminger: Mountain Watershed Management, Lessons from the Past - Lessons for the Future. Proceedings. Environmental Documentation No. 165: 97-114," Berne, 2002.

[15] W. Sidawi, "Report of the first phase of regional planning - Tartous Governorate. In arabic," Tartous, 2011.

[16] Y. Samoudi, "The martyr Bassel al-Assad dam in Safita is one of the most important projects in the region (in arabic)," *Al Wahda newspaper,* p. First page., 30 12 2015.

[17] H. Wattenbach, "Farming Systems of the Syrian Arab Republic," FAO Project GCP/SYR/006/ITA, the National Agricultural Policy Center (NAPC), Damascus, 2006.

[18] T. Rahmoun and W.-M. Zhao, "Restructuring of spatial urban-rural linkages in the Syrian coastal region.," in *Urban Planning and Architectural Design for Sustainable Development (UPADSD)* , Palermo, 2017.

[19] S. Akkoyunlu, "The potential of rural-urban linkages for sustainable development and trade," *International Journal of Sustainable Development & World Policy,* vol. 4, no. 2, pp. 20-40, 2015.

[20] A. E. Blacka, P. Couture, C. W. Coale, J. E. Dooley, A. G. Hankins, A. M. Lastovica, B. Mihalik, C. A. Reed and M. Uysal, "Agri-tourism," Virginia Polytechnic Institute and State University, Virginia State, 2009.

IOP Conf. Series: Earth and Environmental Science **151** (2018) 012034 doi:10.1088/1755-1315/151/1/012034

On the Relationship between Lean Practices and Environmental Performance

M Dieste[1] and R Panizzolo[2]

1 PhD researcher, Department of Management Engineering, University of Padua, Stradella San Nicola 3, 36100 Vicenza, Italy
2 Associate professor, Department of Management Engineering, University of Padua, Stradella San Nicola 3, 36100 Vicenza, Italy
E-mail: marcos.diestegracia@studenti.unipd.it, roberto.panizzolo@unipd.it

Abstract. Lean production has emerged in the past decades as one of the most popular topics in business and manufacturing literature and it is the most extended production paradigm currently applied in industry. Lean production is characterized by five principles (value, map the value stream, flow, pull and continuous improvement) and by the importance of reducing waste (*muda*). Alongside the Lean philosophy, the so-called green strategy has also gained importance in competition between firms. Many companies are trying to develop products that reduce environmental impacts throughout their life cycle. The aim is to reduce resource consumption, to replace hazardous substances, to increase recyclability, to enhance energy efficiency and to bring down CO_2 emissions. Lean and Green production paradigms are both focused on waste reduction and several authors have studied the relationship between Lean and Green practices and the synergic effects of joining these two management approaches. This research carries out a literature review in order to investigate if firms which have applied Lean principles and methods have improved their environmental measures. In particular, the work seeks to highlight which green indicators are more positively affected by Lean practices adoption. The results are synthesized in a final chart which illustrates the main green indicators cited in the literature and shows how these indicators have changed after a Lean transformation program. The research is to be understood as a work in progress and is part of a larger study that the authors are conducting on this topic.

1. Introduction

Lean production system [1] is the most influential production paradigm nowadays with its techniques and methods pioneered between the decades of 1950 and 1960 by Toyota (Toyota Production System) which have become the successors of the mass production practices initiated by Ford. This philosophy is based on the beliefs of "doing more with less", while coming closer and closer and providing costumers exactly what they want. In this effort to reduce non value added activities in the whole processes of the company Lean management establishes seven types of waste, also known as *muda* in Lean environments.

On the other hand, the sustainability concept arises as the "development which meets the needs of current generations without compromising the ability of future generations to meet their own needs" [2]. Reasons for adopting sustainability may vary from company to company but all have found that emphasising sustainability improves their profitability, generates greater loyalty and commitment from employees, and cements relationships with customers and suppliers. Consequently, sustainability is an important element of nowadays corporate strategies.

Limiting Sustainability and focusing this concept only in an environmental performance view, arises the notion of Green manufacturing [3] which "aims a continuous integration of environmental improvements of industrial processes and products to reduce pollution to air, water and land; to reduce waste at source; and to minimize risks to humans and other species". According to the US Department of Commerce, sustainable manufacturing is "the creation of manufactured products which use processes that minimize negative environmental impacts, conserve energy and natural resources, are safe for employees, communities, and consumers and are economically sound" [4]. Among the most important factors of the increasing importance of environmental performance is the rapidly depletion of natural resources that has compelled companies to continuously improve their manufacturing systems for environmental efficiency strategies. Consequently, central to the definition of green production is the theme of waste reduction management. In this perspective is seems that Lean production and Green production have many elements in common since both are focused on reducing waste and increasing efficiency of production processes.

In recent years, several scholars have analysed the possible relationships between the adoption of the Lean paradigm and the improvement of green performance. The results achieved so far are not always homogeneous, some studies strongly emphasize the existence of a positive relationship while others are more cautious by stating that not always adopting Lean practices entails improving green performances.

The objective of this paper is to conduct a literature review aimed primarily at understanding the phenomenon and secondly identify the relationships between the Lean and environmental performance measures. In particular, the literature review has made it possible to understand in detail the impact of adopting Lean practices for each of the environmental measures contained in the model developed by the USA Environmental Protection Agency (EPA). The results are of interest to academics in both the operations management and sustainability areas as well as to business managers designing Lean and sustainable operations and to policy advisors.

2. Literature Review

In order to analyze the relationship between Lean production and sustainability-environmental performance the web document database Scopus was consulted. Moreover, to find the documentation and articles needed for the literature review, the following keywords were searched: Lean and Green or sustainab* (using the asterisk in this word the database will take all the keywords which start with "sustainab", not discarding those referencing the same concept although they are not written equally as the keyword, for example "sustainability" or "sustainable"). These keywords were searched in the title, abstract and keywords of the article in Scopus. The time interval selected starts from 1993 and finishes in 2016, including in this way the most relevant preliminary publications about this topic.

The results obtained searching the keywords mentioned in the database Scopus are shown in Figure 1. In summary, 1897 articles resulted from the database research.

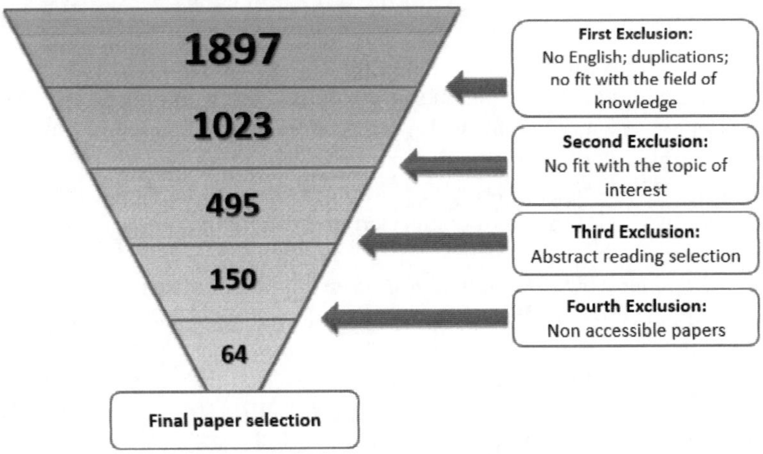

Figure 1. Article selection criteria

Following the steps detailed in Figure 1, from the initial results found in the database and after applying the filters indicated 64 papers were selected for the elaboration of this literature review study. Figure 2 outlines the date of publication of the papers emerging from the literature review. The papers are grouped in five-year periods starting from 1993.

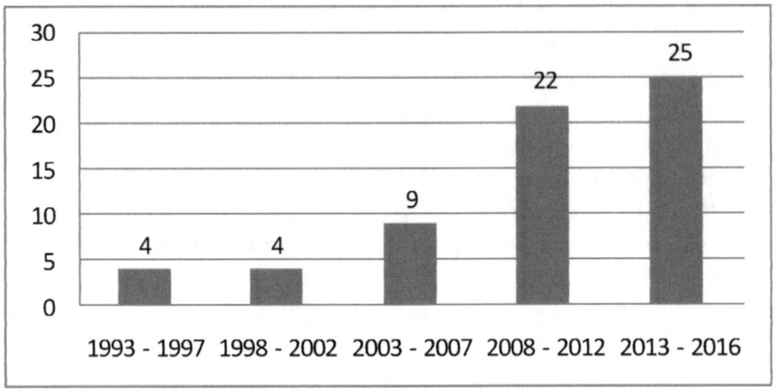

Figure 2. Publication dates of the documents

The data highlight the growing importance in literature of the topic regarding the relationship between Lean and environmental performance, in particular from 2008 onwards.

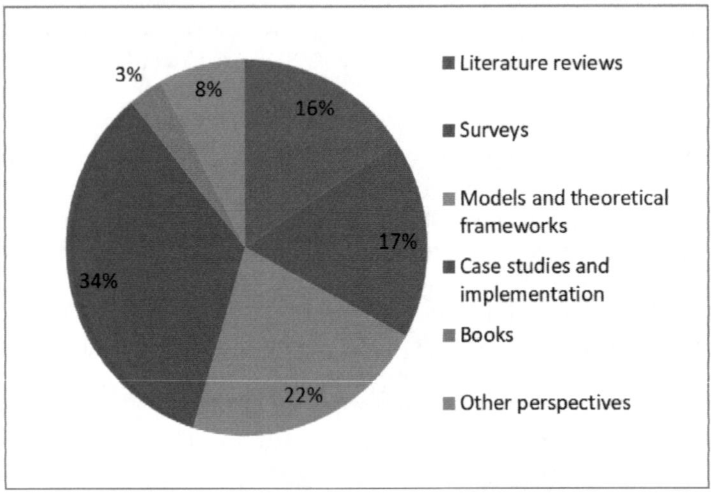

Figure 3. Research methodologies adopted

Figure 3 shows the breakdown of the papers considering the research methodology adopted. In particular, 22 papers out of 64 have employed a single or multiple case study methodology, 14 were models and theoretical frameworks, 11 papers out of 64 have investigated the topic through a survey approach, 10 papers out of 64 have developed a literature review analysis, the rest correspond to 5 manuscripts using other methodologies not comprised in these main categories and 2 books finally.

2.1. The Lean and Environmental Performance Relationship in Literature

The final papers selected have been carefully read and studied. Figure 4 summarizes the results of this analysis.

The overwhelming majority of the studies in literature sustains that Lean has positive effects in the environmental performance of the company [5], [6], [7], [8], [9], [10], [11], [12], [13], [14], [15], [16], [17], [18], [19], [20], [21], [22], [23], [24], [25], [26], [27], [28], [29], [30], [31], [32], [33], [34], [35], [36], [37], [38], [39], [40], [41], [42], [43], [44], [45], [46], [47], [48], [49], [50], [51], [52], [53]. A 12 percent of the papers analysed sustains that Lean may has positive and negative environmental effects; it depends on the practices applied and the indicators measured [54], [55], [56], [57], [58], [59], [60], [61]. Only the 3 percent of the studies declines the positive relationship between Lean and Green production [62], [63]. The remaining percentage specifies those papers that explain both concepts independently but are lacking the study of their relationships [64], [65], [66], [67], [68].

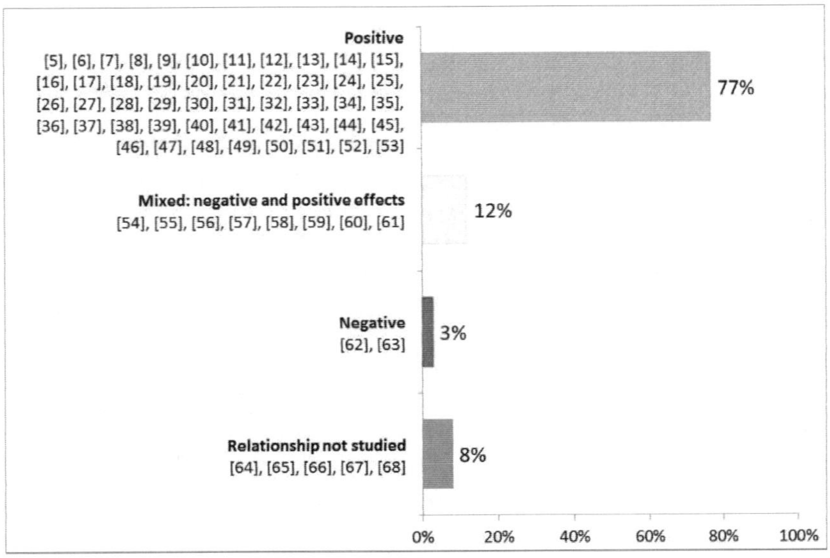

Figure 4. Relationships between Lean and Environmental performance

More in detail, according to the gross body of the literature both Lean and environmental performance activities have a waste reduction connotation. In one hand Lean manufacturing seeks the reduction of non-value added activities and the improvement of efficiency [1]. In the other hand, Sustainability and Green manufacturing have an environmental and ecologist point of view: banish waste in order to achieve pollution and emissions prevention as it is mentioned in one of the three pillars of the Triple Bottom Line (3BL) [2]. In particular, authors like Chiarini [13], King and Lenox [31] and Piercy and Rich [37] verify that the application of some practices composing the Lean transformation path go beyond economic results and Lean waste (*muda*) reduction: these techniques additionally enable the improvement of diverse environmental measures.

Anyway, not all authors agree that there is a positive relationship between Lean production practices and green indicators. As stated for example by Florida [23], Rothenberg et al. [58] and Venkat and Wakeland [63], Lean and Green production are in some cases divergent. Among the causes of this contrast, some authors suggest that the search for greater levels of productivity and efficiency is not coherent with sustainable manufacturing. Furthermore, business growth generally implies more production, transportation, deliveries and stock; these additional activities mean extra wastes in the normal functioning of the company. For example, integration with suppliers via *kanban* [69] requires more frequent deliveries and therefore greater fuel consumption for transport. Another point of reflection is the fact that the financial savings gained by environmental management techniques are not always significant or even non-existent.

However, applying Lean techniques some *blind spots* can arise during their implementation. This means that in addition to the Lean waste reduction and continuous improvement concepts, which frequently produce implicit environmental performance, there are further "hidden" opportunities to achieve this purpose [43]. Moreover, some practices as source reduction, production process improvements and facility downsizing which imply pollution prevention [23] can be enclosed in a Lean thinking view even if they are traditionally framed in the environmental management field.

3. Environmental Measures Presence and Performance in Literature

In this final chapter we focused the attention on the environmental performance measures considered in the literature. For this aim, the list of measures proposed by the EPA was used [45]. For each indicator Figure 5 shows the number of studies attesting that the adoption of Lean practices has improved its value (green color), worsened its value (red color), in some cases improved and in others worsened its value (yellow color). In addition, the gray color indicates the number of studies that claim that indicator among those that should benefit from the implementation of Lean practices but do not provide a measure of this improvement.

Within the sample of 64 papers analysed the EPA indicators most used to measure the environmental performance of companies and processes are air emissions, energy use, solid waste, money saved, water pollution, toxic/hazardous chemicals use, water use and materials use (see Figure 5, the indicators are listed in descending order of quotations).

As shown in Figure 5, air emissions [5], [12], [13], [15], [22], [25], [26], [27], [31], [43], [44], [45], [50], [56], [61]; energy use [5], [11], [12], [15], [22], [26], [27], [36], [37], [44], [45], [50], [53], [58], [60]; solid waste [11], [13], [15], [22], [26], [27], [31], [35], [36], [37], [44], [45], [50], [53], [61]; and money saved [5], [11], [12], [15], [25], [26], [33], [35], [36], [43], [44], [45], [50], [53]; are the indicators which are more benefited from the implementation of Lean practices. However, it is interesting to note that for the air emissions indicator there are four papers in which the relationship is not positive [58], [59], [62], [63]; money saved [59] and product impacts [62] both have one manuscript sustaining negative results.

Additionally, but to a lesser extent water pollution [11], [12], [15], [26], [27], [35], [36], [44], [45], [46], [53], [61]; materials use [11], [15], [22], [26], [35], [37], [43], [44], [45], [46], [53], [60]; toxic/hazardous chemicals use [11], [22], [25], [27], [35], [36], [43], [44], [45], [53], [61]; and water use [11], [12], [22], [26], [35], [36], [44], [45], [46], [53]; are also strongly related with Lean practices. Land use [5], [22], [43], [45]; and product impacts [22], [45]; are the indicators that have less positive evidence in the literature.

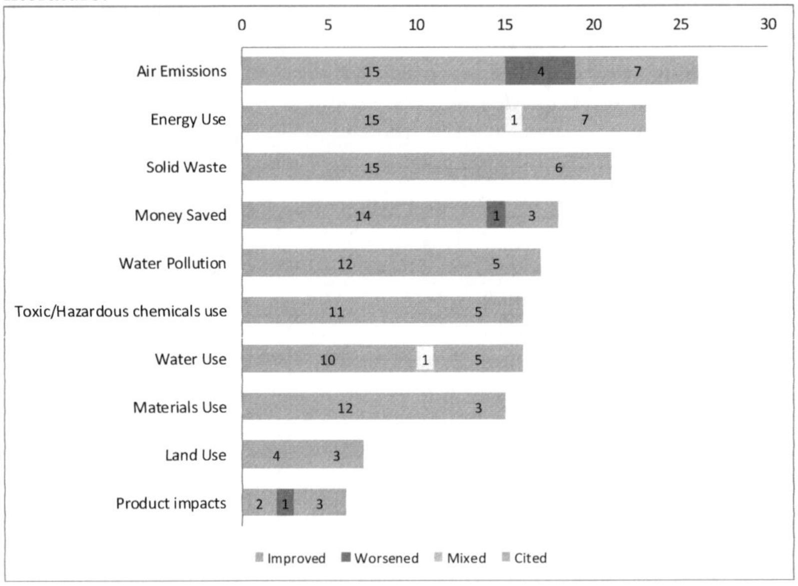

Figure 5. Distribution of the results of the environmental measures.

4. Conclusions and Future Research

In both academic and industrial-managerial fields it is increasing the interest in knowing the effects of Lean practices implementation on environmental performance. The analysis of the scientific literature showed that most scholars acknowledge that there is a positive correlation between adopting Lean production practices and improving environmental performance. Moreover, Lean facilitates a cultural background in the company that leads to green objectives, waste elimination and pollution prevention, which are mandatory for environmental performance [43]. There are now many empirical evidences that those businesses that are following a Lean transformation process will improve resource efficiency and therefore will increment their ecological outcomes [31].

The literature review conducted in this paper was also aimed at understanding which indicators are most cited in the literature and benefit most from the implementation of Lean practices. Using a model developed by USA Environmental Protection Agency the main environmental indicators were identified such as: air emissions, energy use, solid waste, money saved, water pollution, toxic/hazardous chemicals use, water use, materials use, land use and product impacts. For each

indicator, we counted the number of studies that found an improvement or a deterioration or both an improvement and a deterioration of its initial value in companies that have started a Lean transformation path.
The results are of interest because highlight which are the environmental indicators that are more benefited with the implementation of Lean practices. For example, the following indicators are the ones that perform best: air emissions, energy use and solid waste.
However, it should be emphasized that, regarding to air emissions, the results are controversial as there are several studies that state that the adoption of Lean practices does not improve this indicator. Companies have to manage well JIT activities since are a cause of air emissions increments [58], [59].
An important point emerging from the literature analysis concerns the need to involve customers and suppliers in the process of implementing Lean practices. As stated by Dhingra *et al.*, manufacturers should encourage suppliers and customers to collaborate in waste reduction activities in order to facilitate extra environmental performance and arrive to maximum sustainability levels [16].

5. References

[1] Womack JP, Jones DT. Lean thinking: Banish waste and create wealth in your organisation. *Simon and Shuster*, New York, NY 1996.
[2] Brundtland GH. Report of the World Commission on environment and development: "our common future". *United Nations*; 1987.
[3] Berkel R, Willems E, Lafleur M. The relationship between cleaner production and industrial ecology. *J Ind Ecol* 1997;1(1):51-66.
[4] Obama, B. Federal leadership in environmental, energy, and economic performance. *Executive Order* (13514) of October 2009;5.
[5] Aguado S, Alvarez R, Domingo R. Model of efficient and sustainable improvements in a lean production system through processes of environmental innovation. *J Clean Prod* 2013;47:141-148.
[6] Bergmiller GG, McCright PR. Are Lean and Green Programs Synergistic? *Proceedings of the 2009 Industrial Engineering Research Conference*, Miami, FL; 2009.
[7] Bergmiller GG, McCright PR. Lean manufacturers' transcendence to green manufacturing. *Proceedings of the 2009 Industrial Engineering Research Conference*, Miami, FL; 2009.
[8] Bergmiller GG, McCright PR. Parallel models for lean and green operations. *Proceedings of the 2009 Industrial Engineering Research Conference*, Miami, FL; 2009.
[9] Calia RC, Guerrini FM, de Castro M. The impact of Six Sigma in the performance of a Pollution Prevention program. *J Clean Prod* 2009;17(15):1303-1310.
[10] Campos LM, Vazquez-Brust DA. Lean and green synergies in supply chain management. *Supply Chain Management: An International Journal* 2016;21(5):627-641.
[11] Chapman CD. Leaning toward green. *Qual Prog* 2010;43(3):19.
[12] Cherrafi A, Elfezazi S, Govindan K, Garza-Reyes JA, Benhida K, Mokhlis A. A framework for the integration of Green and Lean Six Sigma for superior sustainability performance. *Int J Prod Res* 2017;55(15):4481-4515.
[13] Chiarini A. Sustainable manufacturing-greening processes using specific Lean Production tools: an empirical observation from European motorcycle component manufacturers. *J Clean Prod* 2014;85:226-233.
[14] Corbett CJ, Klassen RD. Extending the horizons: environmental excellence as key to improving operations. *Manufacturing & Service Operations Management* 2006;8(1):5-22.
[15] Deif AM. A system model for green manufacturing. *J Clean Prod* 2011;19(14):1553-1559.
[16] Dhingra R, Kress R, Upreti G. Does lean mean green? *J Clean Prod* 2014;85:1-7.
[17] Diaz-Elsayed N, Jondral A, Greinacher S, Dornfeld D, Lanza G. Assessment of lean and green strategies by simulation of manufacturing systems in discrete production environments. *CIRP Annals-Manufacturing Technology* 2013;62(1):475-478.
[18] Duarte S, Cabrita R, Machado, VC. Exploring lean and green supply chain performance using balanced scorecard perspective. *Proceedings of the 2011 International Conference on Industrial Engineering and Operations Management*; 2011.

[19] Duarte S, Cruz-Machado V. Modelling lean and green: a review from business models. *International Journal of Lean Six Sigma* 2013;4(3):228-250.
[20] Faulkner W, Badurdeen F. Sustainable Value Stream Mapping (Sus-VSM): methodology to visualize and assess manufacturing sustainability performance. *J Clean Prod* 2014;85:8-18.
[21] Fercoq A, Lamouri S, Carbone V. Lean/Green integration focused on waste reduction techniques. *J Clean Prod* 2016;137:567-578.
[22] Fliedner G. Sustainability: a new lean principle. *Proceedings of the 39th annual meeting of the decision sciences institute*, Baltimore, Maryland: Citeseer; 2008.
[23] Florida R. Lean and green: the move to environmentally conscious manufacturing. *Calif Manage Rev* 1996;39(1):80-105.
[24] Franchetti M, Bedal K, Ulloa J, Grodek S. Lean and Green: Industrial engineering methods are natural stepping stones to green engineering. *Industrial Engineer* 2009;41(9):24-30.
[25] Galeazzo A, Furlan A, Vinelli A. Lean and green in action: interdependencies and performance of pollution prevention projects. *J Clean Prod* 2014;85:191-200.
[26] Gustashaw D, Hall RW. From Lean to Green: Interface, Inc. *Target* 2008;24(5).
[27] Hajmohammad S, Vachon S, Klassen RD, Gavronski I. Lean management and supply management: their role in green practices and performance. *J Clean Prod* 2013;39:312-320.
[28] Ho SK. Integrated lean TQM model for global sustainability and competitiveness. *The TQM Journal* 2010;22(2):143.
[29] Ho SK. Integrated lean TQM model for sustainable development. *The TQM Journal* 2010;22(6):583-593.
[30] Hong P, Jungbae Roh J, Rawski G. Benchmarking sustainability practices: evidence from manufacturing firms. *Benchmarking: An International Journal* 2012;19(4/5):634-648.
[31] King AA, Lenox MJ. Lean and green? An empirical examination of the relationship between lean production and environmental performance. *Production and operations management* 2001;10(3):244-256.
[32] Kleindorfer PR, Singhal K, Wassenhove LN. Sustainable operations management. *Production and operations management* 2005;14(4):482-492.
[33] Miller G, Pawloski J, Standrigde CR. A case study of lean, sustainable manufacturing. *Journal of industrial engineering and management* 2010;3(1):11-32.
[34] O'Rourke D. The science of sustainable supply chains. *Science* 2014 Jun 6;344(6188):1124-1127.
[35] Pampanelli AB, Found P, Bernardes AM. A Lean & Green Model for a production cell. *J Clean Prod* 2014;85:19-30.
[36] Pampanelli AB, Found P, Bernardes AM. A lean and green Kaizen model. *Proceedings of the 22nd international conference of production and operations management*, Reno, Nevada, USA; 2011.
[37] Piercy N, Rich N. The relationship between lean operations and sustainable operations. International *Journal of Operations & Production Management* 2015;35(2):282-315.
[38] Porter M, Van der Linde C. Green and competitive: ending the stalemate. *Business and the Environment* 1996:61-77.
[39] Qureshi MI, Rasli AM, Jusoh A, Kowang TO, Bahru UJ. Sustainability: A new manufacturing paradigm. *Journal Teknologi* 2015;77(22):47-53.
[40] Sobral MC, Sousa Jabbour, Ana Beatriz Lopes de, Chiappetta Jabbour CJ. Green benefits from adopting lean manufacturing: a case study from the automotive sector. *Environ Qual Manage* 2013;22(3):65-72.
[41] Soltero C, Waldrip G. Using kaizen to reduce waste and prevent pollution. *Environ Qual Manage* 2002;11(3):23-38.
[42] Taubitz MA. Lean, green & safe. *Prof Saf* 2010;55(5):39.
[43] U.S. Environmental Protection Agency. *Lean Manufacturing and the Environment*. 2003.
[44] U.S. Environmental Protection Agency. *The Environmental Professional's Guide to Lean & Six Sigma*. 2013.
[45] U.S. Environmental Protection Agency. *The Lean and Environment Toolkit*. 2007.

[46] Vais A, Miron V, Pedersen M, Folke J. "Lean and Green" at a Romanian secondary tissue paper and board mill—putting theory into practice. *Resour Conserv Recycling* 2006;46(1):44-74.
[47] Verrier B, Rose B, Caillaud E, Remita H. Combining organizational performance with sustainable development issues: the Lean and Green project benchmarking repository. *J Clean Prod* 2014;85:83-93.
[48] Verrier B, Rose B, Caillaud E. Lean and Green strategy: the Lean and Green House and maturity deployment model. *J Clean Prod* 2016;116:150-156.
[49] Vinodh S, Arvind K, Somanaathan M. Tools and techniques for enabling sustainability through lean initiatives. *Clean Technologies and Environmental Policy* 2011;13(3):469-479.
[50] Wiese A, Luke R, Heyns GJ, Pisa NM. The integration of lean, green and best practice business principles: original research. *Journal of Transport and Supply Chain Management* 2015;9(1):1-10.
[51] Wong WP, Wong KY. Synergizing an ecosphere of lean for sustainable operations. *J Clean* Prod 2014;85:51-66.
[52] Yang MGM, Hong P, Modi SB. Impact of lean manufacturing and environmental management on business performance: An empirical study of manufacturing firms. *Int J Prod Econ* 2011;129(2):251-261.
[53] Zokaei K, Lovins H, Wood A, Hines P. Creating a lean and green business system: techniques for improving profits and sustainability: *CRC Press*; 2013.
[54] Corbett CJ, Van Wassenhove LN. The green fee: internalizing and operationalizing environmental issues. *Calif Manage Rev* 1993;36(1):116-135.
[55] Dües CM, Tan KH, Lim M. Green as the new Lean: how to use Lean practices as a catalyst to greening your supply chain. *J Clean Prod* 2013;40:93-100.
[56] Halldórsson Á, Kovács G, Mollenkopf D, Stolze H, Tate WL, Ueltschy M. Green, lean, and global supply chains. *International Journal of Physical Distribution & Logistics Management* 2010;40(1/2):14-41.
[57] Johansson G, Winroth M. Lean vs. Green manufacturing: Similarities and differences. *Proc. on the 16th International Annual EurOMA Conference, Implementation realizing Operations Management knowledge*, Göteborg, Sweden, June; 2009.
[58] Rothenberg S, Pil FK, Maxwell J. Lean, green, and the quest for superior environmental performance. *Production and Operations Management* 2001;10(3):228-243.
[59] Simons D, Mason R. Lean and green: 'doing more with less'. *International Commerce Review: ECR Journal* 2003;3(1):84.
[60] Torielli R, Abrahams R, Smillie R, Voigt R. Using lean methodologies for economically and environmentally sustainable foundries. *China Foundry* 2011;8(1):74-88.
[61] Zhu Q, Sarkis J. Relationships between operational practices and performance among early adopters of green supply chain management practices in Chinese manufacturing enterprises. *J Oper Manage* 2004;22(3):265-289.
[62] Cusumano MA. The limits of "Lean". *Sloan Manage Rev* 1994;35(4):27.
[63] Venkat K, Wakeland W. Is lean necessarily green? *Proceedings of the 50th Annual Meeting of the ISSS-2006*, Sonoma, CA, USA; 2006.
[64] Dornfeld D, Yuan C, Diaz N, Zhang T, Vijayaraghavan A. Introduction to green manufacturing. *Green Manufacturing*: Springer; 2013. p. 1-23.
[65] Garza-Reyes JA. Lean and green–a systematic review of the state of the art literature. *J Clean Prod* 2015;102:18-29.
[66] Ioppolo G, Cucurachi S, Salomone R, Saija G, Ciraolo L. Industrial ecology and environmental lean management: Lights and shadows. *Sustainability* 2014;6(9):6362-6376.
[67] Kainuma Y, Tawara N. A multiple attribute utility theory approach to lean and green supply chain management. *Int J Prod Econ* 2006;101(1):99-108.
[68] Klassen RD, Whybark DC. Environmental Management in Operations: The Selection of Environmental Technologies. *Decision sciences* 1999;30(3):601-631.
[69] Monden Y. Toyota Production System (Norcross, GA: *Industrial Engineering and Management Press*). 1983.

Effect of Aspect Ratio and Symmetrical Distribution on Urban Design in Baghdad City, and the Impact of Greenery Strategies on improving Outdoor Thermal Comfort

S J RIDHA
Civil Engineering Department, College of Engineering, Mustansiriyah University, Baghdad-Iraq
Suaad.ridha@gmail.com;dr.suaadalzubaydi@uomustansiriyah.edu.iq

Abstract. The concept of thermal comfort relies on the knowledge and content of the pedestrians, particularly in an arid climate. This work concentrates on the possible methods to improve the pedestrian thermal comfort. There are inadequate investigations researches carried out on the outdoor thermal comfort in a desert climate. The researches on enhancing outdoor thermal comfort are nearly nonexistent in cities like Baghdad. Baghdad has a sophisticated urban fabric with contemporary buildings, traditional houses. This work aims at investigating possible strategies to verify how pedestrian comfort is influenced by the constructions design choices, and how vegetation contributes to enhancing the outdoor thermal comfort. The evaluation was performed by using ENVI-met on the hottest day in summer. ENVI-met represents buildings design and calculates the impact on the vegetation on outdoor thermal comfort. We adopted PMV indices to evaluate the outdoor thermal comfort. This work aims to design cities in an arid climate and to maintain a convenient level of thermal comfort. Also, how to perform a proper climate of pedestrians during the daytime beneath the burning sun, particularly in cities with an extreme rise in summer temperatures and for numerous months, as in Baghdad. The research exhibits a proposal area design in an arid climate based on the architectural and implementation standards to increase the pedestrians thermal comfort requirements and how the urban factors such as the aspect ratio, symmetrical distribution, and the green strategies are essential factors that urban designer may be taken into account, mostly for design new urban area in an arid climate.

1. Introduction

Thermal comfort had been discussed since 1930 [1]. In dry climates, there are three crucial parameters influence the canopy layer climate: cooling methods, ventilation, evaporation and solar sheltering [2]. Solar access is a significant requirement for the effect of solar heating on the buildings. The exposure of solar radiation path is one of the fundamental control factors which effects microclimate condition in urban design. Open spaces such as parks are ordinarily more exposed to the solar radiation than the roads and semi-enclosed areas. Controlling on solar access could be defined summarily in two various elements as, solar access for buildings and solar access for pedestrians. The amount of solar radiation could directly influence the solar access. Consequently, have an influence on the outdoor thermal comfort. Hence, the effect of solar access in urban design canyon is needed to enhance urban microclimate [3]. Pedestrians are maintained from the direct impact of sunlight by buildings components such as arcades, overhangs, canopies or frameworks, also by trees and vegetation. Strategies for providing shade by limiting the width of the street, shade may also be produced by applying trees along the pavements or by using pedestrians arcades connected at the street level of the adjacent buildings [4]. Several investigations verify how the green impacts have a significant role in the method of sustainable cooling of the urban planning and in keeping energy and developing human thermal comfort. A study was conducted out in Cairo. There were satisfactory comfort levels and cooling reasonable for some orientations for the urban area due to the clustered form with cool green islands and the wind flow through the main canyons [5]. Shading surfaces contribute to reducing temperature and mean radiant temperature that are affecting the thermal comfort at any given moment in time. Shading surfaces can release the saved energy at night so that could be contributing to producing thermal comfort for the pedestrians in the next day [6]

1.1. Effect of Aspect Ratio and Sky View Factor

Urban canyon geometry described by Erell et al., [4] as the aspect ratio, also identified as height-width. Aspect ratio is determined as the proportion of the average height of neighbouring perpendicular parts like building facades and the average width of the space. As explain in Fig. (1). Bakarman and Chang [7] illustrated the superficial canyons that included H/W =0.42 encounter higher ambient temperatures through the daytime when contrasted to the deep traditional aspect ratio H/W=2.2. An investigation research in Morocco conducted by Alaoui et al., [8], the results confirmed that the aspect ratio more than 2 leads to growing the energy cooling demand due to the influence of low ventilation and the high thermal of the recession.

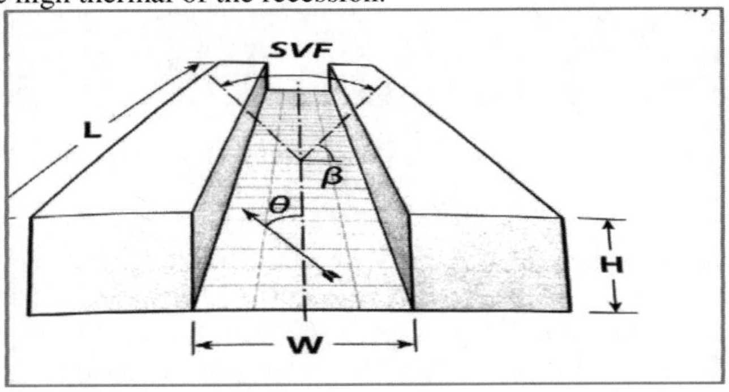

Figure 1. Cross-section of a symmetrical pattern of canyon [4].

Sky view factor (SVF) is almost correlated to its aspect ratio (H /W), as well represents the cross-sectional ratio of the canyon. Figure 2 presents the sky view factor as a function of canyon aspect ratio (H/W) [4]. Brown et al., [9], indicated that for the case where buildings cover 50% of the area of the sky, the sky view factor would be more significant than 5% because sky view factor is measured by the prevalence of the solar radiation overhead the surface of the area.

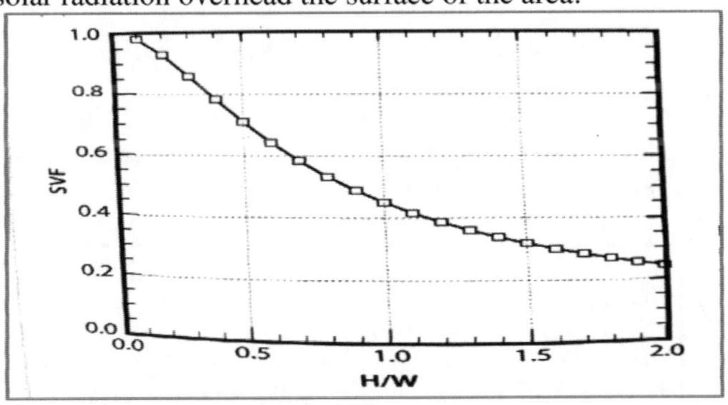

Figure 2. SVF as a function of (H/W) [4].

There is a correlation between the efficacious emissivity and SVF. The high amount of the buildings in the urban area meant that there is a small SVF and increased trapping of radiance [10]. According to a comparison study of the environmental circumstances of urban street canyons in the conventional area and other one contemporary neighbourhoods of Aleppo, Syria the study showed that the effect of shading on the Tmrt decreases significantly with the solar radiation decrease in canyons with low SVF (high aspect ratio) [11].

1.2. Effect of Orientation
Street oriented to [NE-SW] or [NW-SE] in an arid climates means that the shading impacts on the fences and walls are more effective than for an E-W orientation, accordingly, the deficiency of the direct exposure to sunlight due to the availability of shade leads to better comfort conditions [12]. According to Polservice [13], the typical street trend in Hot-Zone should be revolved towards the axis [Northeast-Southwest] and [Northwest - Southeast] to obtain the better insulation in the winter and insignificant heat and adequate shading in the summer season. Figure (3) depicts the buildings orientations and ideal streets in the hot climate in Baghdad.

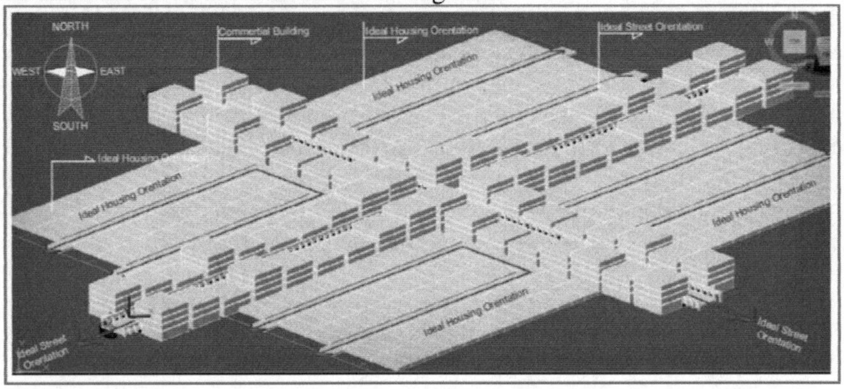

Figure 3. Typical streets and buildings orientation in arid climates [14].

1.3. Effect of Vegetation
Numerous investigations show how the greenery impacts have a significant role in the method of sustainable cooling of the urban planning and in keeping energy and enhancing human thermal comfort. A research was conducted in Egypt; Cairo explained that the satisfactory comfort levels and mitigating reasonable for some orientations for the urban area due to the collected form with cool green islands and wind flow through the main canyons [5]. In an investigation study on the effect of greenery strategies in an arid climate in Dubai, Rajabi and Hijleh [15] confirmed that green rooftop contributed to performing inadequately in decreasing the surface temperatures in urban areas; this is because of the cooling impacts of green roofs lessen by distance and consequently this influence is negligible on the overall temperature decrease in urban areas. Also, concerning the distribution of greenery, trees have the best participation to the decrease in surface temperatures.

2. Methodology

2.1. Study Area
Baghdad is positioned in the middle portion of Iraq on both sides of Tigris River Fig (4). It lies on latitude 43 east and longitude 34 north. The climate of the Baghdad region is described as a semi-arid, with hot dry in summer, and cold winter. Baghdad area comprises $4555Km^2$, which described 1.047% of the whole area of Iraq.

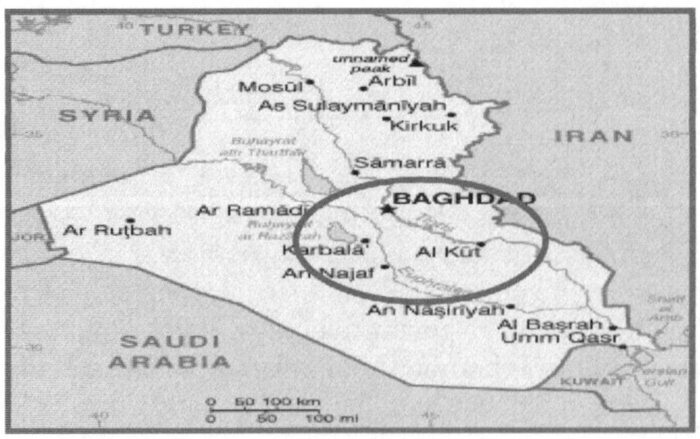

Figure 4. location of Baghdad-Iraq Map (www.googlemap.com).

2.2. Micro-Scale Numerical Modeling ENVI-met

ENVI-met software simulates micro scale interplays urban surfaces, the atmosphere, and vegetation. The reliability of ENVI-met results for simulating the outdoor thermal spaces validated in many investigations. These investigations showed that the data measured at local meteorological locations seemed to accord with the simulated results [16]. Monam and Ruckert [17] validated that concerning to several investigations [18], [19] and [20] ENVI-met results could be recognized as accurate and reliable.

2.3. Data for the Proposed Model

Data adopted to simulate the proposed model was presented by Iraqi Meteorological Organization and Seismology. The microclimate features describe weather conditions of the hottest day of summer in Baghdad that was 12th of July in 2010 [21]. Accordingly, the original climatology environments for the primary conditions were 315 deg. for wind direction and 5 m/s for wind speed. Simple forcing for relative humidity and air temperature were applied along one day, which explains that the maximum temperature was 50 °C at 4 pm, and the minimum air temperature was 35 °C at 6 am. The minimum relative humidity at 4 pm was 24 %; the maximum relative humidity at 7 am was 36%. Simulation period is 24 hours. The assumed area has the model dimensions (180×180). The area for the mode has been performed with grid size x= 60, y= 60 and z= 20, this grid size is represented in a grid cell, the size of the grid cell is dx= 3 m, dy=3m and dz=5m. The model has been rotated of 45° depending on the buildings orientations and ideal streets in an arid climate [14].

2.4. Model Configuration

The work focuses on imposing an area 32400 m^2 to represent the specific urban area in Baghdad city. This area is an approach to an area that was adopted by the researcher in a previous study 48750 m^2on improving the thermal comfort in Baghdad, where the area was taken from the real district in Baghdad city. The researcher relied on the recommendations and results for the previous researches and studies, which were listed in this work to design the area that has been imposed in Baghdad in order to improve the outdoor thermal comfort for the pedestrians at the hottest day in summer. Also, depended on the specifications of building and the construction requirements from the Ministry of Construction, Housing and Public Works in Iraq. The principles adopted to construct the proposed model for the suggested urban area are as follows:

1. The height of the residential buildings is 15 m (the new requirements of design the residential buildings in Baghdad), see Fig (5).
2. Spacing between the buildings based on the typical Aspect Ratio in an arid climate is ($H/W \geq 1$), so we choose H/W=2.5.
3. Depending on the Standards of Housing Technical and the Code of Practice of Iraq prepared by (Polservice) in 1982, also the outcome in this work, the orientation is (NW-SE).
4. The research focuses on the concept of courtyard design within the building to give proper ventilation within the courtyard of one building as shown in Fig (6), thereby contributing to adequate ventilation of the outer perimeter, and therefore creating appropriate thermal comfort for pedestrians between the corridors.
5. Concerning the role of albedo for enhancing the outdoor comfort, we choose light concrete pavement for the entrance and for the main street grey concrete pavement, see Fig (7).
6. To heighten the function of vegetation in outdoor comfort, we concentrate on utilizing trees and grass. Also, the role of afforestation and ventilation within the courtyards was also enhanced through the use of evergreen and shady trees (Sophora tree) see Fig (8), Thick tree branches contribute to obstructing the sunlight and reducing, and reduce its impact on the surfaces thus contributing to the enhancement of thermal comfort.
7. To enhance the active role of shading, which mainly affects the thermal comfort of the pedestrians, it has been proposed to design the buildings symmetrically in order to take advantage of the amount of shade contributed by the buildings each to other to protect pedestrian.

Figure 5. Perspective view of the proposed urban design simulated by ENVI-met.

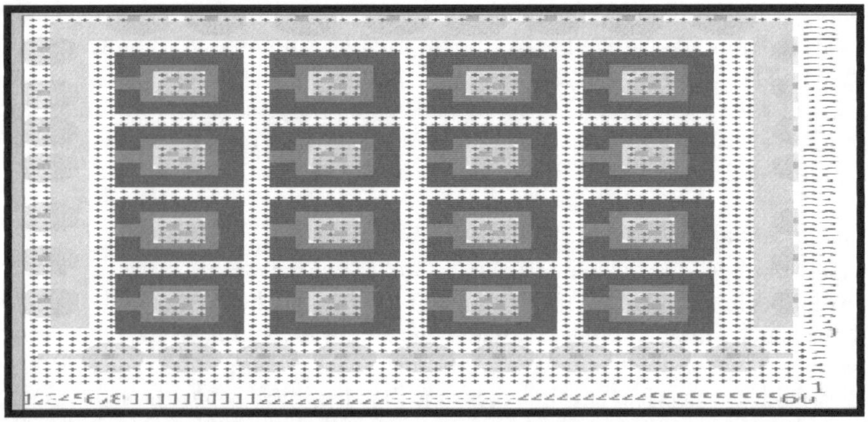

Figure 6. Top view of the proposed urban design simulated by ENVI-met.

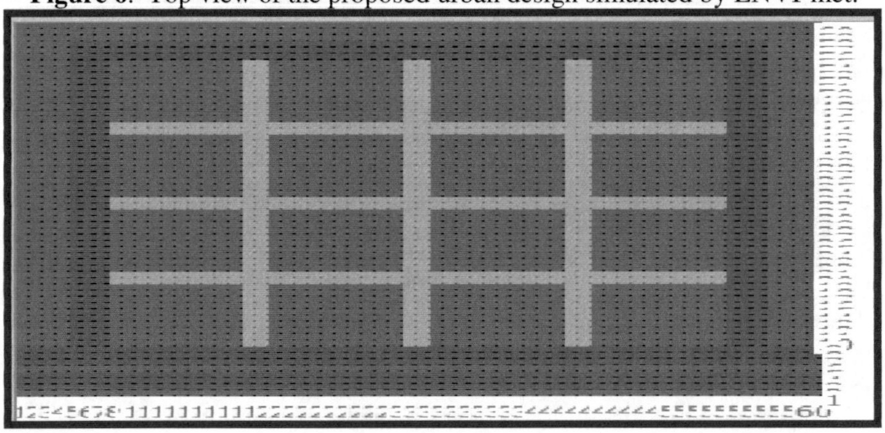

Figure 7. Distribution of roads and the main streets of the proposed urban design simulated by ENVI-met.

Figure 8. Sophora tree (https://www.pinterest.com/).

3. Results Analysis

The real temperature at noon time was measured 47°C according to the Iraqi Meteorological Organization and Seismology. We observe a decree 2 °C in air temperature, the concentrated air temperature in the simulated urban area ranges between 43°C and 45 °C, as shown in Fig (9). Figure (10) depicts the distribution of temperature at height 1.5 m over the ground; this level indicates the human thermal sensation. We observe that the minimum air temperatures are recorded in the courtyards because the courtyards play an essential role in the cooling system in the simulated urban area. The existences of courtyards contribute to minimizing air temperature.

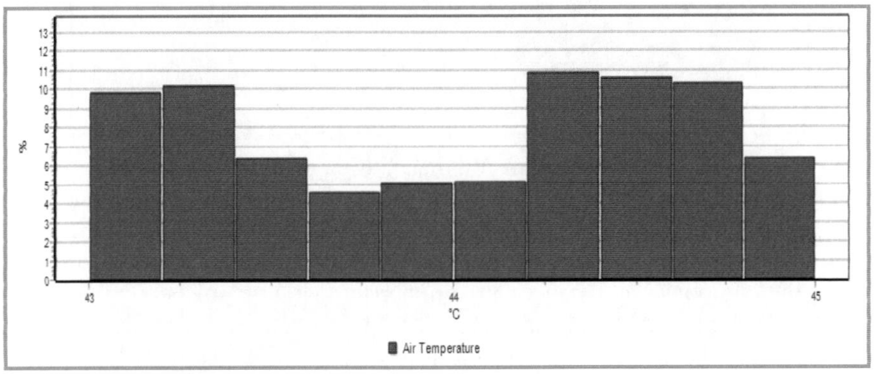

Figure 9. Percentage value of air temperature for the simulated urban area.

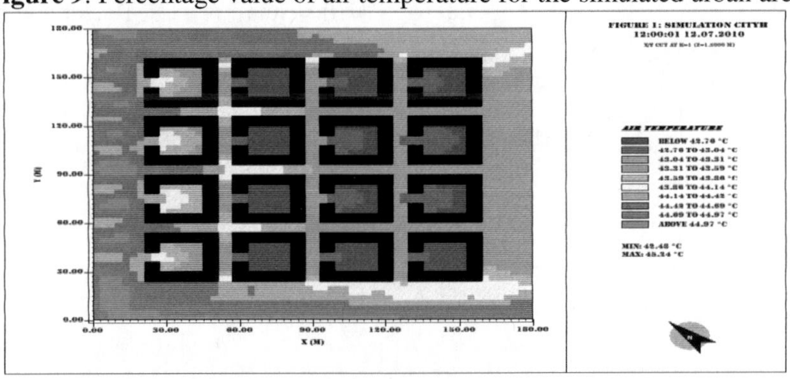

Figure 10. Air temperature distribution for the simulated urban area.

The real wind speed that was used in the simulation work was 5 m/s, we observe from Fig. (11) that the wind speed is around (0-4), this is attributed to the existence of vegetation, which leads to blocking the air flow.

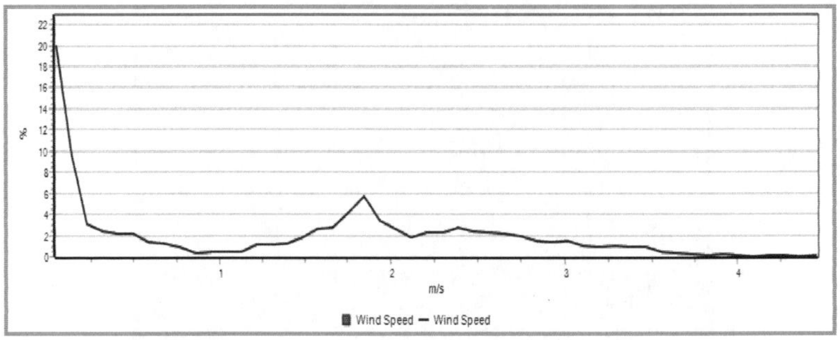

Figure 11. Wind speed Percentage value for the proposed design at midday noon.

Sky view factor also decreased, the existence of vegetation minimizes the open spacing in the urban area, Fig. (12) clarifies that the minimum values of sky view factor concentrated in the courtyards which are around (0.19-0.28), while the maximum values 0.65 and above are concentrated outside the distribution of the buildings.

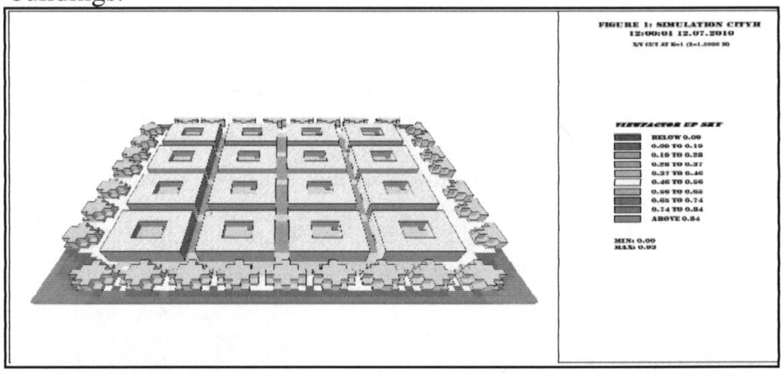

Figure 12. Three-dimension distribution of Sky view factor.

Typically, PMV range is described within [-4] very cool and [+4] very hot, 0 is the thermal comfort value (neutral), see Fig (13). Although PMV values are the mathematical function of the local environment, in most utilization, it can also give values above [-4] or below [+4]. Implementing of PMV equation to the outdoor circumstances in summer temperature stress conditions can efficiently provide values of PMV high over +4 [+8 and more]. While these results are mathematically accurate, it infringes the scale of the actual PMV method (www.ENVI-met.com). Figure (14) depicts the percentage value of PMV at noon; we observe that the proposed design achieves an improvement in the thermal comfort PMV values range (5 - 6.5). The distribution of PMV for the proposed urban design at noon illustrated in Fig (15).

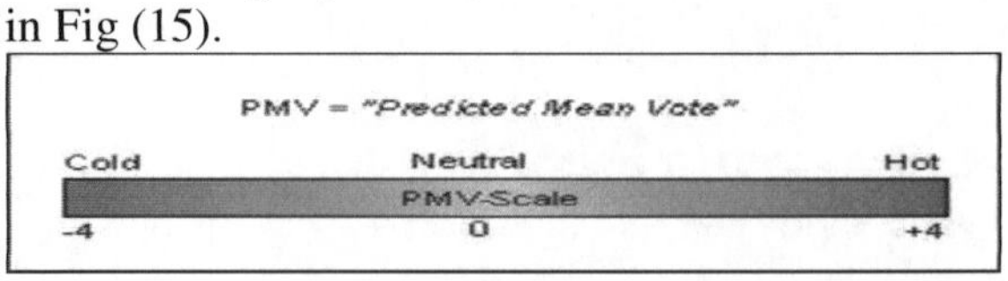

Figure 13. PMV scale (www.ENVI-met. com).

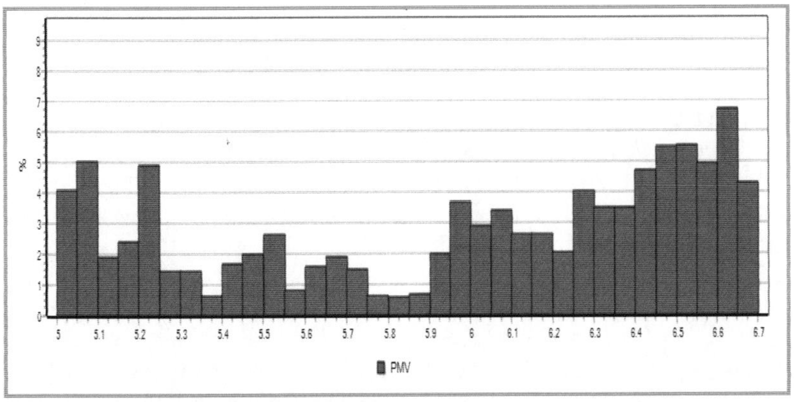

Figure 14. PMV percentage value for the proposed design at midday noon.

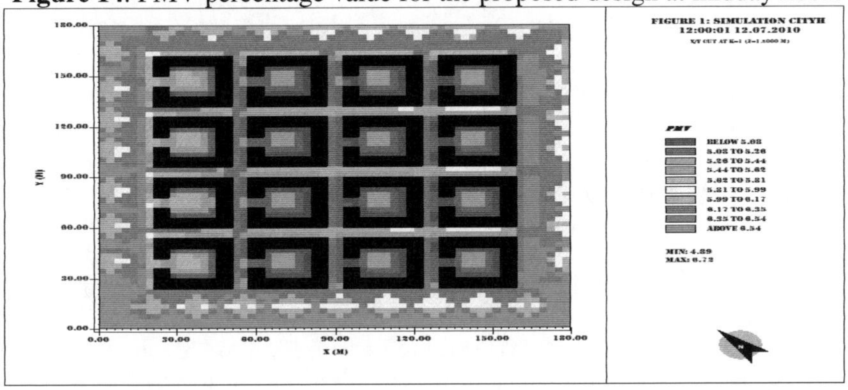

Figure 15. Distribution of PMV for the proposed urban design at noon.

4. Conclusion

The increase in desertification in the hot and dry areas in recent years, especially in Baghdad city, and lack of interest in the optimal design of cities, which simulates the environment and improves the pedestrians thermal comfort all these factors have motivated the researcher to do this research to achieving the urban environmental design under increasingly changing climate conditions. The results showed a decrease in temperature more than 2 °C. Concerning PMV outcomes in hottest circumstances day, the outcomes showed that the proposed design had adequate hot conditions. The characteristics of vegetation covering influence the results of thermal sensation directly such the leaves absorb and convey solar radiation and evapotranspiration provides for improving thermal sensation. The results of this work may help for enhancing cities design in an arid climate.

5. References

[1] Taleghani M., Tenpierik M., and Dobbelsteen A., V., D., (2012) Environmental Impact of Courtyards-A review and Comparison of residential Courtyards buildings in Different Climates. Journal of Green Building 7, 113-136.

[2] Fahmy M., A., R., (2010) Interactive Urban Form Design of Local Climate Scale in Hot Semi-Arid Zone, PhD thesis, The University of Sheffield School of architecture, UK.

[3] Shishegar N., (2013) Street Design and Urban Microclimate: Analyzing the Effects of Street Geometry and Orientation on Airflow and Solar Access in Urban Canyons. Journal of Clean Energy Technologies 1,52-56.

[4] Erell E., Pearlmutter D., Williamson T., (2011)Urban Microclimate Designing the Spaces between Buildings, New York.

[5] Fahmy M., and Sharples S., (2009) On the Development of an Urban Passive Thermal Comfort System in Cairo, Egypt, Building and Environment 44. 1907-1916.

[6] Rosheidat A., (2014)Optimizing the Effect of Vegetation for Pedestrian Thermal Comfort and urban Heat Island Mitigation in a Hot Arid Urban Environment, PhD thesis, Arizona State University.

[7] Bakarman M. and Chang J., (2015) the Influence of Height/Width Ratio on Urban Heat Island in Hot-Arid Climates. Procedia Engineering 118, 101-108.

[8] Alaoui J., and Tahiri M., (2016) Analysis of Canyon Aspect Ratio Impact on Urban Heat Island and Buildings Energy Construction in FEZ Climatic Zone, Morocco, ARPN Journal of Engineering and Applied Sciences 11, 3059-3073.

[9] Brown M., J., Grimmond S., and Ratti C., (2001)Comparison of Methodologies for Computing Sky View Factor in Urban Environments, International Society of Environnemental Conf. Tempe, AZ.

[10] Yang J., Wong M., S., Menenti M., and Nichol J., (2015) Modeling the Effective Emissive of the Urban Canopy Using Sky View Factor, ISPRS Journal of Photogrammetry and Remote Sensing 105, 211-219.

[11] Zakhour S., 2015) The Impact of Urban Geometry on Outdoor Thermal Comfort Conditions in Hot-Arid Region. Journal of Civil Engineering and Architecture Research 2, (862-875.

[12] Ali-Toudert F. Mayer H., (2007) Thermal Comfort in an East-West Oriented Street Canyon in Freiburg (Germany) Under Hot Summer Conditions, Theoretical and Applied Climatology. 87, 223-237.

[13] Polservice, (1982)Housing Technical Standards and Codes of Practice. Report Two, printed by design office MIASTOPROJEKT-KRAKOW; Ministry of Housing and Construction-State Organization for Housing, Baghdad, Iraq.

[14] Al-jumaili S., (2014) Street orientations: The first step towards a sustainable place. American Journal of Environmental Protection 3, 305-317.

[15] Rajabi T., and Abu-Hijleh B., (2014)The Study of Vegetation Effects on Reduction of Urban Heat Island in Dubai, World SB4, Sustainable Building Conference Series, Barcelona.

[16] Ghaffarianhoseini A., Berardi U., and Ghaffarianhoseini A., (2015) Thermal Performance Characteristics of Unshaded Courtyards in Hot and Humid Climates. Building and Environment 87, 154-168.

[17] Monam A., Ruckert K., (2013)The Dependence of Outdoor Thermal Comfort on Urban Layouts, German-Iranian Research Project. Young Cities. Developing Energy-Efficient Urban Fabric in the Tehran-Karaji Region. 10623 Berlin-Germany.

[18] Ali-Toudert F.," (2005) Dependence of Outdoor Thermal Comfort on Street Design in Hot and Dry Climate "Ph.D. Thesis, Freiburg, Germany, November.

[19] Ozkeresteci I., Crewe K. and Bruse M., (2003) Use and Evaluation of the ENVI-met Model for Environmental Design and Planning: an Experiment on Linear Parks. The 21st International Conference (ICC). Durban, South Africa 10-16.

[20] Lahme E. and Bruse M., (2002) Microclimatic effects of a small urban park in densely built–up areas: Measurements and model simulations. The European Commission in the 5th Framework Program under the contract.

[21] Hassoon, A., F., (2015) Assessment of Air pollution Elements Concentrations in Baghdad City from Periods (May-December) 2010, International Journal of Energy and Environment 6, 191-200.

IOP Conf. Series: Earth and Environmental Science **151** (2018) 012036 doi:10.1088/1755-1315/151/1/012036

Strategy to Improve Outdoor Thermal Comfort in Open Public Space of a Desert City, *Ouargla,Algeria*

H GHERRAZ[1] I GUECHI[1] and A BENZAOUI[1]
1 Department of architecture, university of Oum El Bouaghi, 04000, Algeria,
Email: halima.gherraz@gmail.com.

Abstract: The control of outdoor thermal comfort became a question of extreme importance in hot desert cities, knowing that people of these cities seek to shelter solar rays by all means especially during the hot season. The aim of this research is to study and analyze the effect of urban vegetation and shade on the microclimate and thermal comfort within an open public space in a hot desert city (the city of Ouargla in Algeria). As an element that minimizes and intercepts the sun's rays, it generates solar energy shade and absorbs the fluxes of radiation, modifies the temperature, the relative humidity and minimizes the wind velocity and modifies its directions. For this, we have used Ray Man which is simulation software to calculate the indices of thermal comfort. The work consists in calculating these indices before and after the insertion of two types of vegetation within our case study (the martyr's plaza in the traditional neighborhood of the city of Ouargla). The results obtained show the main role of vegetation by the improvement of thermal comfort.

1. Introduction

Any urban project must take into account climatic factors; in fact the climatic conditions of an urban agglomeration have an impact on the lifestyle of its inhabitants, in particular the use of outdoor spaces, which are considered as supports of urban life.

Public spaces are fundamental elements of the urban environment and the architectural and aesthetic aspect of the built environment of cities. These spaces contribute to develop the social relations. These are spaces that offer a certain comfort to the inhabitants who often feel assaulted in urban environments [1].

Outdoor thermal comfort is a complex function of atmospheric conditions and physical, physiological, psychological, and behavioral factors [2]. According to S Reiter user comfort in the outdoor space have several positive effects such as improving the quality of living in an outside environment, valorization the image of the city, increasing the rate of the outdoor spaces use for meetings, relaxation and leisure, the reduction of energy consumption related to the use of inside spaces [3].

The microclimate of these spaces is influenced by several parameters such as geometry of space, vegetation, water and the thermal properties of surfaces. It is possible to improve the comfort conditions of an outdoor space taking into account the urban design. Vegetation as an element of urban design that influences the thermal comfort of outdoor spaces has three main characteristics that influence the climate: shading, humidification, and windbreak.

By its shade, the vegetation planted in urban environment modifies the physical environments (light, heat, structure, wind, sound, humidity). It influences the perception and appropriation of urban spaces by users [4]. In terms of protection against nuisances, the interests of vegetation are several: reduction of atmospheric pollution and improvement of air quality by fixing certain dust, reduction of the effect of run off by rain interception, protection against erosion due to wind and water [5].

Therefore the climatic advantages of urban vegetation are assessed either by their effects on meteorological factors or by their energy savings induced in buildings. It has been suggested that several small areas with adequate intervals are more efficient in cooling than a large green space [6].
The density of foliage determines a more or less large permeability to the light, and radiation of wavelength. This density varies significantly depending on the species considered, the configuration of its development (pruning, size ...) and the season. Taking into account these different settings, we can consider that the urban trees can be three types of Screen: an opaque screen (coniferous), a screen semi-transparent (deciduous trees) and a transparent screen (some fruit trees or species of PINE) [7].
Several studies have investigated the impact of parks and green spaces on urban micro-climate and found that temperature reduction in a treed urban environment can reach up to 4°C. However, many studies suggested that urban green spaces can be warmer than the surrounding built-up area and can exacerbate unpleasant micro-climatic conditions [8]. Measurements indicate that the surface temperature of the walls is oriented west and concrete protected from direct solar radiation by trees, are weaker in average of more than 13 °C,Figure1(a) [9]. A.Chatzidimitriou revealed the importance of shading paved surfaces. Shaded surfaces have surface temperatures 21% lower than the air temperature, and are 40% cooler than the same surfaces exposed to the sun's rays[10][11], Figure 1(b).

(a) (b)

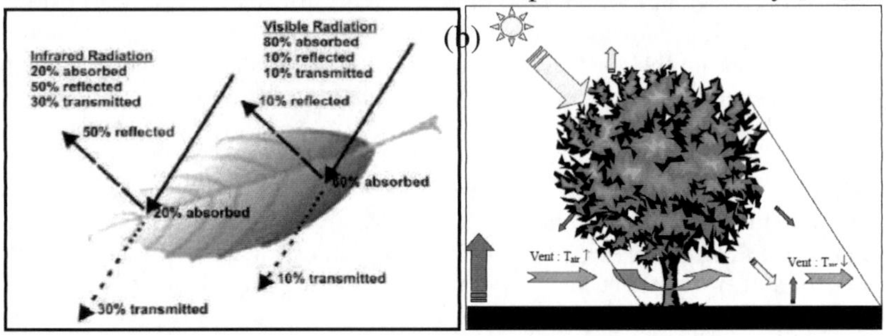

Figure 1 (a). Absorption, transmission and reflection of solar radiation by foliage [11]

Figure 1 (b).Modification of the transfers of heat around a tree [5]

Bernatzky's study in Fraukfurt (Germany) has shown that an urban square lowers the air temperature by 3 to 3.5 °C and increases the relative humidity by 5 to 10%. It purifies polluted air and turns it into fresh air in the city center [12]. The study of S.Louafi on the esplanade of the University of Constantine, which is characterized by a large expanse of concrete totally devoid of vegetation and another vegetated space intended for the creation of students, the study shows that the outdoor space can be affected by the presence of vegetation that plays an important role in its environmental quality. In the Saharan cities, it is noticed that people seek to avoid the sun's rays, especially during the hot season.
The main object of our study is the verification of outdoor thermal comfort to find the good type of vegetation and the pertinent strategy to introduce it into an open public space in a hot desert city.

2. Assessment of Outdoor Thermal Comfort

The thermal comfort could be defined as a feeling of well-being vis-à-vis a thermal environment[13]. The assessment of outdoor comfort does not only allow improving the Thermal Conditions of these spaces but also the protection of people who use those [14].

2.1. Thermal Indices of Outdoor Comfort

Several indices integrating thermal environmental factors and heat balance of the human body are applied for accessing thermal comfort, these indices are PMV (predict mean vote), SET (standard affective temperature), PET (physiologically equivalent temperature) and PPD (Predictable Percentage of dissatisfied) [15].

2.1.1.PMV. Today, one of the best-known thermal indices is the Fanger comfort equation, which allows the calculation of the Predicted Mean Vote (PMV). The equation connects the classical parameters which are the temperature of the air, the average radiant temperature, the hygrometry, the speed of the air, the metabolism and the thermal resistance of the clothing. In addition, two physiological parameters are added, the average temperature of the skin and the evacuation of the heat by sweating. Fanger arrived at the average assessment for the thermal environment by the PMV which varies on a scale of 7 degrees between -3 and +3, but Michael Bruse, in its work performed during the updating of its simulation program, has adopted nine scales of own comfort to the outdoor spaces, where the theoretical area of comfort outside is between -3 and +3, Figure 2 (a) [16].

2.1.2.PPD. it is the percentage of unsatisfied people in relation to this situation. The PPD is a formula derived from statistical analyzes. The people are considered as dissatisfied if they vote ± 2 or ± 3 on the scale of 7 points of ASHRAE[17], Figure 2 (b).

(a)

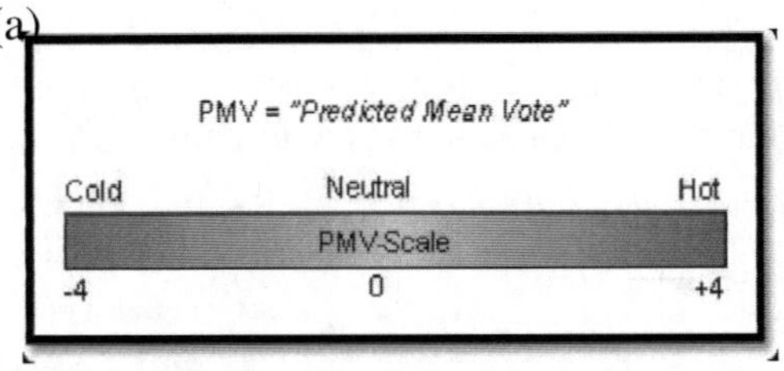

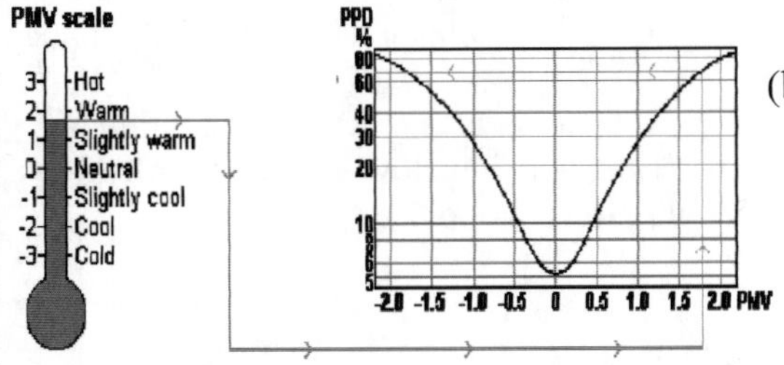

(b)

Figure 2 (a). Classification of values of the PMV according to nine scales of comfort [16].

Figure 2(b).Correspondence between the PMV and the PPD[17].

2.1.3.PET. The PET index is defined as the temperature of the air which, in a typical local indoor, without wind and solar radiation, the energy balance of the human body is balanced with the same internal temperatures and skin than those obtained in the conditions to assess, according to T.Lin, PET is defined as the air temperature at which, in a typical indoor setting (air temperature =mean radiant temperature, relative humidity=50 %, wind speed= 0.1 m/s)[15].

*2.1.4.SET.*the standard temperature effective is defined as the temperature of a reference isotherm environment whose surface temperatures and air temperature are equivalent, where the relative humidity is 50% and the speed of the air is (0.12 m/s), The reference environment is defined by a Metabolism (M)=1.2met and the resistance of clothing (R clo)=0.9clo [12]. Table 1 show the correspondence between SET sensation and physiological state [13].

Table 1. Correspondence between SET sensation and physiological state[13].

SET (C°)	sensation	Physiological state of a sedentary individual
>37,5	Very hot, uncomfortable	Failure of regulation
34,5-37,5	Hot, very unacceptable	Abundant sweat
30,0-34.5	Hot, uncomfortable, unacceptable	Sweat
25,6-30,0	Slightly warm, slightly unacceptable	Low sweat, vasodilatation
22,2-25,6	Comfortable and acceptable	Neutrality
17,5-22,2	Slightly cool, slightly unacceptable	vasoconstriction
14,5-17,5	Cool, unacceptable	Slow body cooling
10,0-14,5	Cold, very unacceptable	Thrill

3. Area of the Study

Our Area of study is situated in the city of Ouargla which is located at 128m altitude, 190 km East of Ghardaia, 388 km South of Biskra, 160 km South-West of Touggourt, it extends between 28 ° and 32 ° north latitude and meridians 4 ° and 8 ° east, Figure 3.

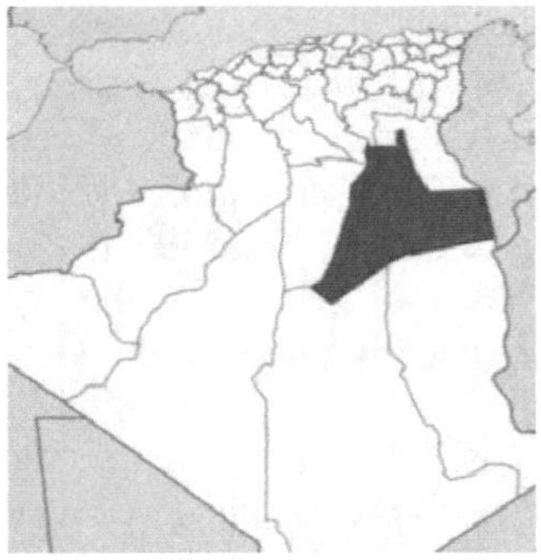

Figure 3.geographical situation of Ouargla, The city is located at an altitude of 135 m, its territory is characterized by dominance of the massifs dunes [1].

The climate of Ouargla is characterized by a quasi-permanent marked aridity which is expressed by a high temperature, a serious low precipitation and evaporation, (a large hot summer contrasting with a cold winter)Figure 4, 5 and 6.

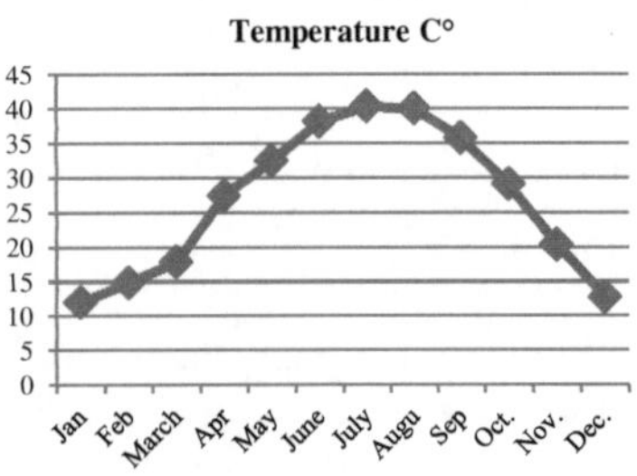

Figure 4.air temperature.

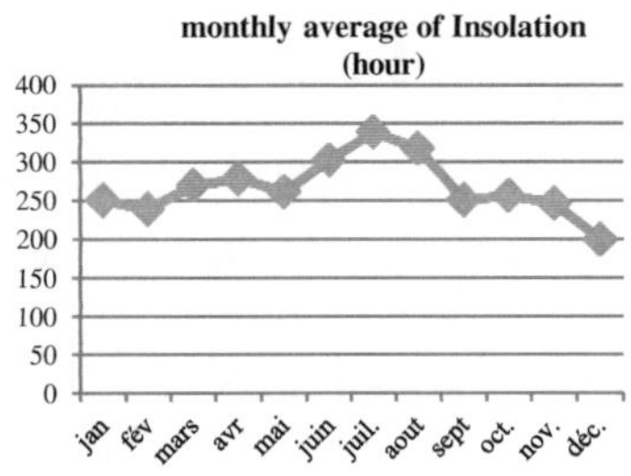

Figure 5.monthly average of insulation.

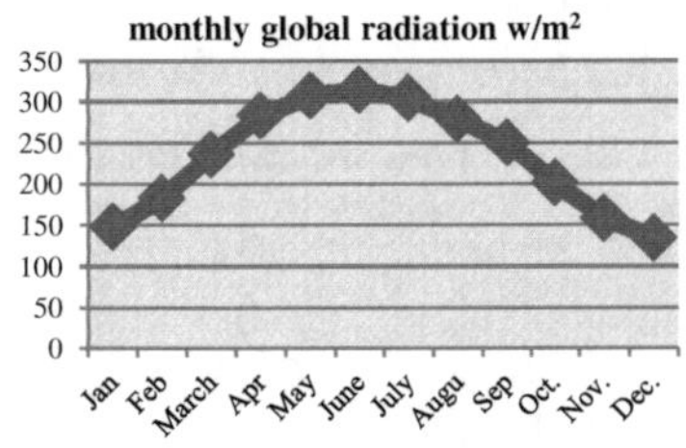

Figure 6. monthly global radiation

4. Methodology

Our case of study is the martyrs plaza, which occupies the south western part of the traditional quarter playing a very important role of identification and orientation, it constitutes a particular space, especially by the inclusion of different types of daily, weekly and occasional activities, this plaza is poor in vegetation except the existence of some palms that offer shade in some times of the day. This plaza was selected for according to its historical past. The plaza is very crowded during almost the whole year, Figure 7 (a), (b) and (c).

In order to evaluate the effect of vegetation on outdoor thermal comfort into this plaza, a series of measurements were collected; the measurements were performed in summer during the very hot season in the month of July 2016. The simulation will be done by the software RayMan Pro; the software can analyze complex urban structures and other environments (building and tree). It requires only basic meteorological data (air temperature, air humidity, type of sky and wind speed) for the calculation of radiation fluxes, soil temperature and common thermal indices (PMV, PET, SET, and SVF). It is taken into consideration the geographical coordinates of the site studied [18] [19].

(a) (b) (c)

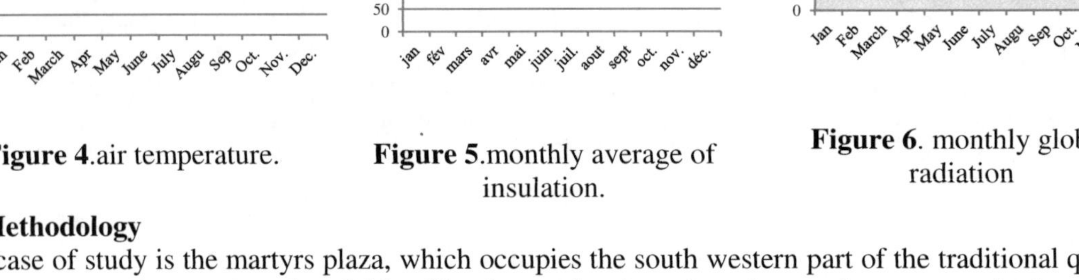

Figure 7. (a), (b) and (c) the open public space (martyrs plaza)

The first step is to simulate the current state of the plaza in other words to calculate the global radiation, soil temperature, PMV, PET, SET, and SVF and the shadow during the day. The second step is to propose a type of vegetation useful for our case; the choice is based on the maturity of the tree, its height, the size of crown, the density of foliage and its adaptability to the plaza (soil temperature and humidity). Note that the large crown size of the tree ensures more shade, especially at noon when the sun is at the azimuth state [20].

5. Results and Discussion

5.1. Before the Intervention

Figure 8 (a) show an aerial view of the plaza, and Figure 8 (b) shows the plaza with the software RayMan pro.

(a)
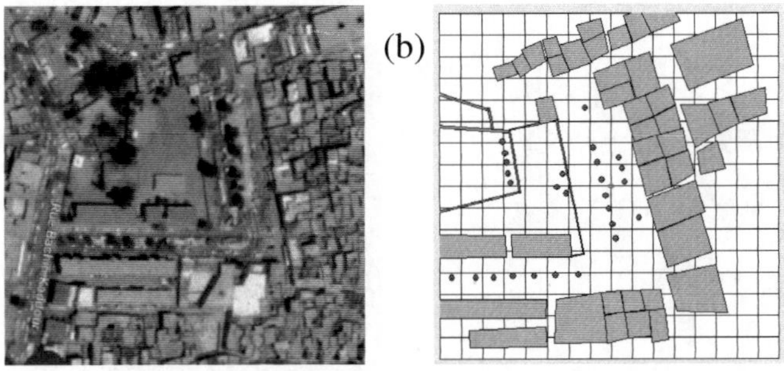

Figure 8. (a) and (b): the area of study

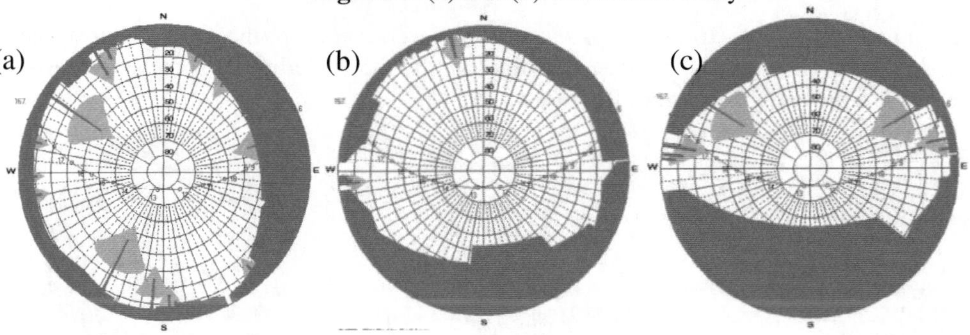

Figure 9. (a) (b)and (c): sky view factor in different point of the plaza before intervention

The SVF of the plazavaries between 0.637 and 0.677 which is so high; this is why people in this plaza always feeluncomfortable. In some other words it is due tothe high rate of solar radiation (see Figure 9 a, b and c). The plaza is exposed to sunlight almost all day except in the evening by blessing the shade provided by the houses (see Figure10 a, b and c).

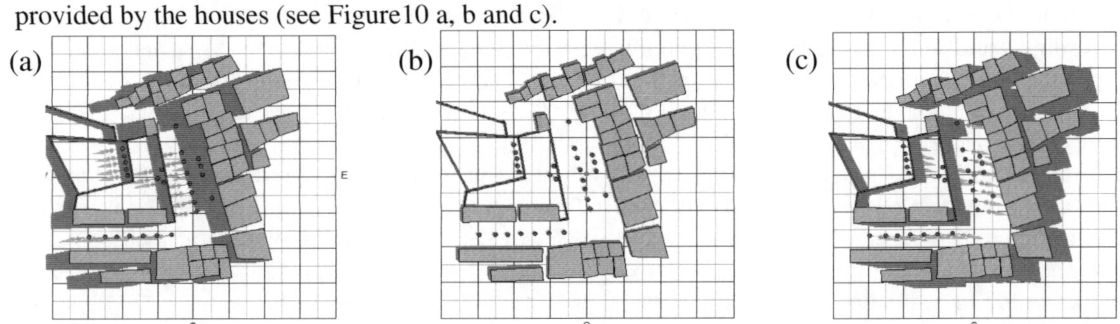

Figure 10. (a) shadow at 8.00 AM (b) shadow at 12.00 (c) shadow at 17.00 PM before intervention

5.2. After the Intervention

Figure 11 (a) shows the optimum proposal of design of the vegetation, which it gave the best results compared to current situations of the plaza. The conception of the vegetation was based on two essential parameters: first, do not plant the entire surface of the plaza except the parts of sitting most used by the inhabitants of this plaza to protect them from solar radiation. The second to use two types of vegetation; Acacia Raddiana; a deciduous tree, very used in Algerian Sahara that resists the aridity of our area, (Figure 11 b). The second type is coniferous tree (Ficus Retusa) very used too in Algeria (Figure11 c), It is of category tree of tropical origin of the persistent type of dense port shape erected, rapid growth. In addition it has been in the city of Ouargla for a long time and it resists its climate [18]. These two types were chosen in order to protect inhabitants in summer from solar radiation and allow the penetration of solar rays through the deciduous trees in winter.

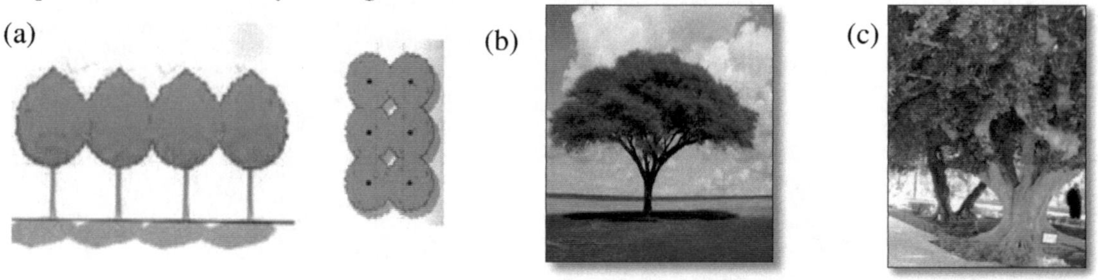

Figure 11. (a) a true mask with solar rays (b) acacia raddiana tree (c) Ficus Retusa tree

Figure 12 (a) and (b) show the variance after the insertion of vegetation, the area of study has shade almost all day long. This increase of the shadow will facilitate the unfolding of some activities like the weekly trade, and the grouping before or after the prayer.

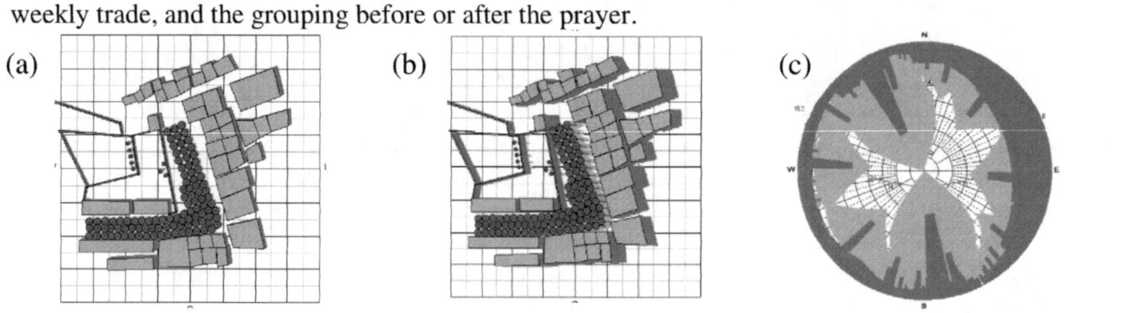

Figure 12. (a) shadow at noon (b) shadow at 17.00 pm (c) SVF after the insertion of vegetation

It is noticed that the SVF is decreased after the insertion of the vegetation (SVF=0.117) which will mitigate the penetration of radiation, especially in the hottest times of the day (Figure12c).

Concerning Tmrt, in the first situation without vegetation, we measure 42.9°C at 8.00am and it reaches 52.9°C at noon but it decreases at 18.00 pm until 45.8°C. in the second situation with the vegetation we measure 35.7 ° C at 8.00 am, it reaches 43.5 ° C at noon and decreases to 36.5 at 18.00 pm, Figure13 (a).

Figure 13(b) shows the global radiation before and after the insertion of the vegetation, for the first situation we measure 526w/m² at 8.00 am, 789 w/m² at 2.00 pm and 649w/m² at 18.00 pm, for the second situation we measure 322w/m² at 8.00 am, 587 w/m² at 2.00 pm and 439 w/m² at 18.00 pm.

For the PMV obtained in the second situation, we observe a significant reduction compared to those obtained in the first situation, the maximum PMV is 4.8 at 2.00 pm, Figure14 (a).

For the PET, the results obtained for the second situation, we observe a significant attenuation compared to the resultsobtained in the first situations, the maximum PET is at 2.00pm with 47.1 °C very hot, the minimum PET is at 8.00am we measure 35°C hot layer, Figure 14 (b).

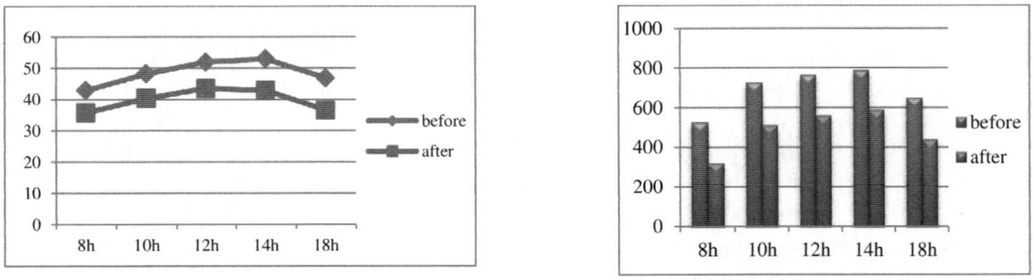

Figure 13. (a) Tmrt before and after the intervention

(b) Global radiation before and after the intervention

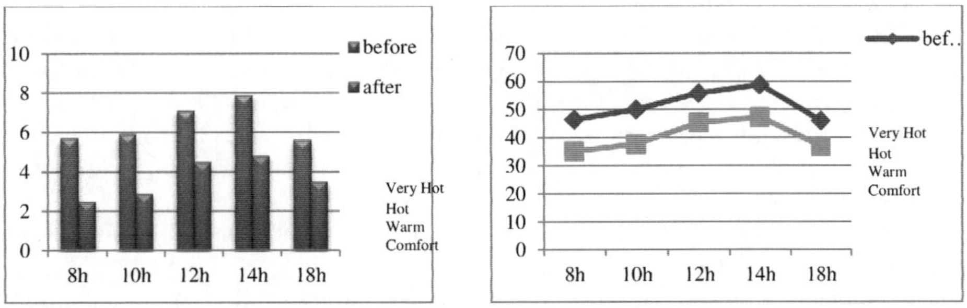

Figure 14. (a) PMV before and before the intervention, (b) PET before and before the intervention

For the standard temperature effective (SET), the results obtained for the second situation, we observe a significant attenuation compared to the results obtained in the first situations, the maximum SET at 2.00pm is 41.2°C but in the first situation is 46.2°C, Figure15.

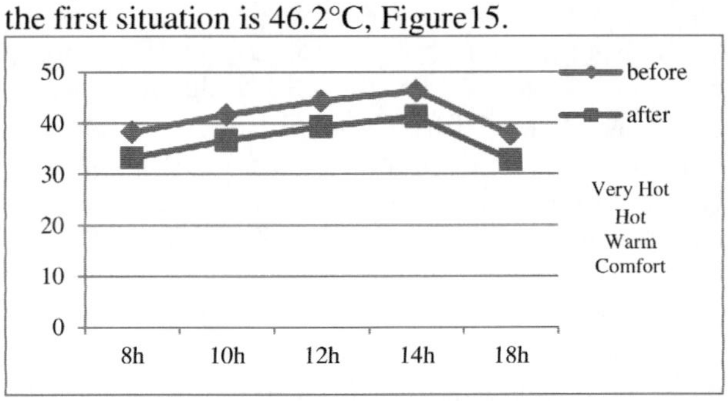

Figure 15.SET before and after the intervention

6. Conclusion

Vegetation can affect the microclimate in many ways; it reduces air temperature, while providing shade. Vegetation brings aesthetic improvements to an environment otherwise dominated by asphalt and concrete. The main object of this research is to improve the impact of vegetation on outdoor thermal comfort, the two types of trees introduced into the area of study has given good results and wide variations in climatic parameters and indices. The Sky view factor (SVF) was changed from 0.677 in the situation without vegetation to 0.117 after the insertion of vegetation. For the global radiation we passed from 789w/m2 in the first situation as maximum level to 587w/m2 after the insertion of vegetation. For the PMV we passed from 7.9 to 4.8 as maximum level in the second situation, the mean radiant temperature was changed from 52.9°C to 42.9°C as maximum temperature in the second situation. So vegetation is often an important component of the open outdoor space, which contributes greatly to the quality of its development. Generally trees provide significant improvements in thermal comfort mainly at noon and in the early afternoon as they provide shading, and attenuate solar radiation. In addition, trees increase the environmental quality of outdoor space and the thermal and visual comfort felt in a semi-arid Mediterranean climate.

7. References

[1] Gherraz H 2013 Les espaces publics entre forme et pratiquedans les villesarides et semi arides (Cas des places publiques de la ville de Ouargla), *mémoire de magister*, université Mohamed Khider ,Biskra, Algérie p 16.

[2] Middel A Selover N Hagen B and Chhetri N 2016 Impact of shade on outdoor thermal comfort—a seasonal field study in Tempe Arizona *Int J Biometeorol* (2016) 60 p 1849–1861.

[3] Reiter S 2007 Elaboration d'outils méthodologiques et techniques d'aide à la conception d'ambiances urbaines de qualité pour favoriser le développement durable des villes »- *thèse de doctorat,* UCL Université catholique de Louva

[4] Louafi Bellara S Abdou S 2012 Effet de l'ombrage sur le confort thermique et visuel dans les espaces extérieurs : cas de l'esplanade de l'Université Mentouri de Constantine, Est de l'Algérie, *Revue Nature & Technologie.* n° 07/Juin 2012 p 26–37

[5] Vinet J2000 Contribution à la modélisation thermo-aéraulique du microclimat urbain. Caractérisation de l'impact de l'eau et de la végétation sur les conditions de confort en espaces extérieurs. *Sciences de l'ingénieur [physics]*. Université de Nantes, 2000. Français.

[6] El Hozayen A 2013 Le traitement des espaces urbains dans les villes au climat chaud et sec pour diminuer leurs problèmes climatiques étude de cas : la ville du Caire Sciences agricoles. *<dumas-00907165>*.

[7] Muret J P Allain Y M and Sabrié M L 1987 Les espaces urbains, concevoir, réaliser, gérer, Paris, édition du *Moniteur*, p 364.

[8] Cohen P Potchter O and Matzarakis A 2012 Daily and seasonal climatic conditions of green urban open spaces in the Mediterranean climate and their impact on human comfort, *Building and Environment* 51 (2012) p 285-295

[9] Parker J H 1989 the impact of vegetation on air conditioning consumption. In: proceedings of workshop on saving energy and reducing atmospheric pollution by controlling Summer heat islands. Berkeley: *Applied Science Division LBL*, pp45 – 52.

[10] Chatzidimitriou A Chrissoumallidou N and Yannas S 2006 Ground surface materials and microclimates in urban open spaces, *PLEA,* the 23rd conference, Genève, Suisse.

[11] Shahidan M F Mustafa K M Sand Elias S 2007 Effects of tree canopies on solar radiation filtration in a tropical microclimatic environment *PLEA 2007 conference* Singapore.

[12] MERROUCHE M 2011 Microclimat végétation urbaine dans les espaces publics cas : jardin Bennaceur (Constantine) *sciences et technologie D-* N° 33p 37-45

[13] Boukhalkhal I 2014 Impact de la morphologie urbaine Sur le confort thermique extérieur Dans les climats chauds et arides, Cas de la nouvelle ville de Tafilelt à Ghardaïa, *mémoire de magister*, université de Constantine, Algérie p 34–36.

[14] Tebbani H 2006 Impact de la végétation sur le microclimat et le confort thermique des espaces extérieurs, cas d'étude cour de la révolution d'Annaba, *mémoire the magister*, université de Constantine p 86–123.

[15] Lin T Matzarakis A and Hwang R 2010 shading effect on long-term outdoor thermal comfort, *Building and Environment* 45 *(2010)* p 213–221.

[16] Bruse M 1999 Modeling and Strategies for Improved Urban Climates proceedings,International conferences on urbanism.

[17] Bellout A 2010 Le rôle de la végétation et l'eau dans la création d'un microclimat urbain, Cas de la place d'Ain El Fouara à Sétif, *mémoire de magister*, université de Constantine.

[18] Matzarakis A Rutz F and Mayer H 2007 Modeling Radiation fluxes in simple and complex environments, Application of the RayMan model. *Int Jof Biometeorolog.*51p 323–334

[19] Matzarakis A Rutz F and Mayer H 2010 Modelling Radiation fluxes in simple and complex environments Basics of the RayMan model. *Int J of Biometeorology 54*p 131–139.

[20] HANAFI A 2017 Role of the urban vegetal in improving the thermal comfort of a public place of a contemporary Saharan city *International Conference on Technologies and Materials for Renewable Energy, Environment and Sustainability*17, Beirut Lebanon p142-148.

Research on Treatment Status of Inner Rivers in Chinese Cities

Xiuhua Wu and Ruizhao Shao

Professor, Information and Electrical College of Shenyang Agricultural University, Shenyang, China
E-mail: wuxih70@126.com

Abstract. The pollution of the Chinese inner rivers was more and more seriously in nearly 20~30 years. The people had already pay widespread attention to this status. The researchers had done a great deal of works on the treatment of these rivers and it had obtained much achievement and gained many effective treating methods. At same time the government had done some policy support to this situation. This paper summarized the treatment status of Chinese inner rivers. It included treatment methods and their advantages and disadvantages. It discussed the government's relevant policies and the implementing of the local governments. It also gave some improved directions in the future.

1. Current Situation

Since the 1980s, the city construct of China was changing day by day. The industry enterprises developed rapidly and many people moved in and began to live in the cities. The population of Chinese cities began increasing annually. The pollution water from the factories and thousands of rooms mostly was been rowing into the city inner rivers directly or indirectly. The living rubbish from the residents and visitors of all over the country was poured out at the river banks. The city inner rivers became drainage ditches year by year. At the same time, the management on the rivers was lack of and treatment had no any. These inner rivers gradually turned into the situations that water body heavily polluted, the ecological function declined, water-carrying capacity seriously insufficient[1], some rivers' beds dried up, most rivers pollution dirt stacking on the beds and the water became black and odorous. These inner rivers were not the beauty signs of the cities any more. They were the scars of the city development.

In recent years, Chinese people and the government already seen the rivers pollution situations and realized the ponderance of this problem. They began to treat the water pollution and had already gained some good effect. The researchers also had done more works on treatment methods in order to change the river status rapidly and enduringly and reasonably.

2. Treatment Methods

The treatment methods[2] of the black and odorous rivers are generally included of physical methods, chemical methods and biological methods. The physical methods are included of artificial aeration, sediment releasing and sweeping pollutant through drawing water, etc. The chemical methods are mainly to use chemical agents to remover the target pollutants in polluted rivers in order to improve the water clarity. But it is obviously that the chemical methods will produce the secondary pollution. The biological methods are new and effective methods in recent years. It is a better and more optimum treatment route when they combined with the aeration technique.

2.1. Artificial Aeration

Artificial aeration is a kind of technical method[3] which injects oxygen to water body through the artificial controlling to recover the aquatic ecology. It has been used to the river treatments since 1960s. The actual experiences showed that the artificial aeration not only could change the black and odorous status of the water body, but also could oxidize and degrade the reducing matters in the upper layer sediments of the water[4]. The aeration method can product a mainly amphimicrobe ecological environment in the surface layer of the water sediments so that there possessed aerobic bacteria growing and exciting probability. It can reduce the organic pollutants during a shorter time, raise the dissolved oxygen content, enhance the water self purification to improve the water quality. The method was generally effective so many researchers had taken lager time to study it. It also should be one of necessarily water treatment methods.

2.2. Drawing Water to Sweep Pollutants

Drawing water to sweep pollutants is a treatment method through drawing lager flow, high velocity clear water or the release flood water to sweep the polluted black and odorous rivers in order to reach the aims of clean up the pollutants and improve the water quality. Generally, after several hours the water quality will be changed. If it could be swept a certain time every day the water should have been gotten a fundamental improvement[5] and cleared away the black and odorous water.

But it has the same effect with the other physical method, sediment releasing, the cost of the method is huge. For example, if it took 19 hours to sweep the river every day, the cost would be 25,000,000 Yuan per year. At the same time, its purifying effect is not durable. After 12 hours stop sweeping the water quality of the inner rivers began to recover. This method can hardly treat those so called cecum rivers and beheaded rivers.

The method is a way that easy to operate, easy to realize, instant[6] to effect and no negative effects. So if the conditions allowed, it would be worth to do more discusses, researches and experiences[7].

2.3. Biological Treatments

Biological methods can recover the biological function of the rivers from the roots. It is a rapidly developing method in the world of recent years. It utilized cultivated plants or culture and inoculated microorganic metabolic activity to transfer, change or degrade the pollutants in the rivers and rebuilt a perfect trophic level aquatic ecosystem. It can self repair river system health and improve the water quality safely and durably. Its project cost and running cost are both lower and it can mix together with the greening environments and sceneries[8].

The ecological floating bed is one of the typical and successful applications. The ecological floating bed is a new, low cost, good effect, high biosecurity technique. It cultivated the terrestrial plants in the beds floating on rivers and utilized the plants' roots to absorb the nutritive salts, mineral substances or other pollutants in water aiming to improve water quality. When the plants were reaped these pollutants were also removed away. Moreover, the root systems of these plants could adhere to those suspended solids in the rivers. The rhizospheric microorganism could degrade the organic matters efficiently. These two facts both could clear up the river water. Furthermore, some plants were able to secret the algal inhibition materials to inhibite the growing of the red tide planktons to reduce the rates of the red tide. At last, the floating beds formed perfect aquatic ecological landscapes or called water gardens. They supplied the urban residents a clean, beauty and comfortable living circumstance. So, the researches about this technique had been done a great deal of and it had achieved much progress inland or outland in recent years.

But the effect slower was the weakness of the ecological method. The purifying function began to work always after the ecosystem was built completely. Simultaneously, the ecological systems, in most cases, would spare several or ten more years to be established because they depended on a great many biological factors and environmental factors[9].

The progress of the ecological floating bed technique still had many problems to be solved. Firstly, in different areas, of different rivers, there would be different types of pollutants. So it needed to study and find different aquatic plants or microorganism to adapt it. It was not only able to treat the polluting water but also able to prevent the biotic intrusion destroy the ecosystem. This must take a long period to research and experience. Secondly, the water body circumstance of the same river would be changed by seasons, by various situations of its upstream and downstream, etc. But the plants in the beds were hardly adjusted in time. It made the treatment efficiency not enough. Thirdly, the most plants in the beds could not smoothly go through the winter. It made the application to be restricted in the cold regions. Lastly, yielding process of the floating beds needed to be standardized and factory-like to shorten the product periods. The problems about the joint, fixation, stability of the beds in heavy waves, high flow rate river regions must be solved. At the same time, the less and more scientific daily maintenance were worth to be done more researches in order to make the maintenance-free floating beds be possible in the future.

2.4. Combination Treatment Methods

2.4.1. Artificial aeration-aquatic plants combined treatment technique. This technique was that the aquatic plants played the main parts and the artificial aeration played the auxiliary parts in the purification process. It was a technology combined the artificial treatment with the nature ecological treatment[10]. It originated from aeration biology pools and artificial wetlands using in purification of sewage. It was largely used to treat the urban and village sewage in Europe, America and Japan for the past few years. For example, this system had been used about 20,000 series and occupied 90% of all the sewage treatment equipments in Japanese villages.

2.4.2. Aquatic-microorganism combined treatment technique. This technique was according to the biology and ecology principles, took full use of the synergistic effects and actions between the aquatic plants and various types of micro-organism. The aquatic plants supplied growing surfaces and intermedias for the microorganism. The aquatic plants' roots and floating beds formed a safe and thick enough living environment for microorganism to degrade the pollutants. The degradation products might help the growing of the aquatic plants. In this complex ecology system, there were germs, fungus, mycetes, algas, protozoa, metazoa[11], benthonic animals, aquatic animals and plants. They got rid of the pollutants, purified the water together by interaction, physical absorption, biological absorption, biological degradation etc.

2.4.3. Aquatic-microorganism- aeration combined treatment technique. This technique took the aquatic and microorganism as the core, the artificial aeration as assisted. Its effect was rapid as well as durable. Its treatment efficiency was far better than any one only or any two. When the rivers were occurred such events as large amount of pollution incoming, abrupt changes of the weather, this technique would not be seriously impacted. So it had prominent advantages.

3. The Government Actions

The pollution of the urban inner rivers was not only the natural environment destructive from the human beings but also the problems of the unsound and inexhaustive government management. It had been over 30 years from the Chinese economic reform since 1980s. The environmental loads became more and more heavy as the economic energetically developing, industry and handicraft industry rapidly stronger, agricultural production greatly growth, urban population intensive expansion etc. The pollution of the inner rivers was only the one of all kinds of cases. The attention of the government for urban river water pollution problem was beginning at nearly ten more years. The local government had taken some measures, but they were mostly superficial, short-term and one-sided. There were no long-term, effective, prompt, applicable method. So there was still a long way to go through for us to treat the pollution and there were many severe problems to solve.

The general problems of the government in river water quality controlling were as follows:

3.1. The One-sidedness of Governance.

The treatment was not comprehensive enough. The black and odor situations on unfocused areas were more serious. Those rivers which related with the local governments "face" were generally abundant quantity, good quality, and becoming beautiful landscapes. But the other rivers which did not concern the governments "face" were dirty smelly, full of garbage, pollution. The different areas of a same river may be different.

3.2. Management Discontinuity
Once the government changed or the leader changed, it was easy to be a no-end project or the no-management project. Urban water environment management was a systematic project, which included not only the technical engineering of improving the water environment, but also the management measures to maintain the quality of water environment. The technical engineering which was spent a huge sum of money, because of the lack of follow-up supervision and management, might make the water pollution repeatedly. It was not only becoming huge waste, but also harmed the engineers' faith.

3.3. Lack of Management Coordination
Many environmental projects must have regular inspection, supervision, durable operation, cleaning the rivers, transport the rubbish, so that the effect can be kept. It needed the coordination within the different government departments to treat all kinds of problems together. But the reality was hard to be done. For example, Living garbage was poured optionally, for example, was the main cause to lead to the city water body black and odorous. But the environmental protection and health departments could not form linkage mechanism to solve this problem for a long time. So the Garbage clearance and treatment problems become the deep ache of many cities.

3.4. The Method of Governance Was Not Scientific
In some areas, the artificial landscape construction in the process of governance was excessively pursued. Too many artificial lakes were too broad and needed too much water to replenish a large amount of evaporation loss because they were often lack of natural runoff recharge. Those lakes not only can't accumulation rainwater for the city, but become large "water -users" in the city water cycle. They perennially needed abundant water diversion to supplement water quantity, maintain landscape use from the major rivers. They took up a lot of water resources and those small inner river were becoming black and odorous because lack of water supply.
Again, for instance, construction of riparian zone also exist many problems. The landscape requirements were only considered to cement, marble, stone. Using hard revetment was mostly from the perspective of cost clean flood etc, but the water permeability of riparian zone of the water and the ecosystem protection was far less than the ecological slope protection or natural slope protection. These will lead to poor urban rainwater collection, worse inner river ecological condition and aggravating rivers pollution.
There are many problems in the government's work, but the problem of water pollution in the inner rivers of cities was also a common problem in the developing countries. It was unrealistic to expect a quick solution. In the past, some measures are feasible and worthy of promotion and extension. Such as:
1) River Chief System.
It was the rules that the leaders of various levels of the government became the managers of all the inner rivers. They were responsible for river management and protection work within their respective jurisdictions. River Chief System avoided the problem of property right of public rivers. It cleared the main body of management, giving river manager responsibility, permissions and rewards and punishments measures. The chief's main work included strengthening the protection of water resources, the full implementation of the most stringent water management system, strengthen the management and protection of the coastal areas of the river and water areas, and control the ecological environment of water space and water line. It also included improving the prevention and control of water pollution, coordinate pollution control over the water and on the bank, searching for the pollution sources of the rivers, optimizing the distribution of sewage outlets in the rivers, strengthening water environmental management, ensuring the safety of drinking water, increasing the management of black and odorous water, and realizing the clean and beautiful water of the river lake. It would strengthen water ecological restoration, delineate river and lake management scopes

according to laws, strengthen the management of the lake system, and strengthen the supervision of law enforcement and a crackdown on violations involving of lakes.

River Chief System provided a new path to solve the problems of public waters pollution and ecological destruction. In operating one more years, it played a positive role to reduce the river sewage, keep the river clean, reasonable development and protection of river resources. It had gained some significant progress.

2) Public-Private Partnership

It refers to the government made public bidding to purchased water quality maintenance services. The social capital participated in pollution control. It was to realize the water transparency increased significantly, the comprehensive pollution index decline year by year. The inner river water was no smell, no impurity. At the same time gradually restore the ecosystem of the river itself, river self-purification ability of ascension.

4. Conclusion

Urban river management is a comprehensive systematic project for a long time. It needs human beings protect, pay attention to and inherit. The inner water is not only the blood of a city, the life of a city, but also the eyes and face of a city. It is worth to be protected, to be paid more hard work and effort for her. Both the scientists and the government managers are worth doing more researches, more endeavors and more attention for our future generations.

5. References

[1] Zhu Qingping, HuangLiqun, Integrated treatment of urban rivers in the water ecological civilization construction. China Water Resources. 2013,(15):47-50+35

[2] Deng Zhiqiang, Research progress of application of artificial floating beds technology. Environmental Pollution and Prevention, 2013.5 .35(5)

[3] Zhang Jiang-shan. Urban River Restoration and Ecological City Construction in Fuzhou City. Journal of Fujian Normal University (Natural Science Edition).2005.9.21(3):99-103

[4] Burke V, Richter D. Redox-dependent removal of 27 organic trace pollutants: compilation of results from tank aeration experiments. Environ Earth Sci 2014.71(8):3685–3695CrossRefGoogle Scholar

[5] Chua L, Tan SBK, Sim CH.Treatment of baseflow from an urban catchment by a floating wetland system. Ecol Eng .2012.49:170–180

[6] Doyle MW, Patterson LA, Chen Y. Optimizing the scale of markets for water quality trading. Water Resour Res.2014. 50(9):7231–7244

[7] Li Yunxi, Experience and Enlightment of Cheonggyecheon Restoration Project in Seoul，Korea，Chinese Public Administration,2012,3(321):96-100

[8] Wang Jun, Enlightenment of Cheonggyecheon Restoration Project for China, ChinaDevelopment,2009,6(9): 15-18

[9] Lai Jinmei, Research Process on the Treatment of odorous Gas in City River, Science & Technology Vision, 210-211

[10] Zhang Ying, The Application and Research Status of Deodorization Techniques for Wastewater, Environmental Pollution and Prevention and Control,2013,36(1):81-85

[11] Eatrada J M,A Comparative Analysis of odour treatment technologies in wastewater treatment plants[J].Environ.Sci.Technol.,2011,45(3):1100-1106

IOP Conf. Series: Earth and Environmental Science **151** (2018) 012038 doi:10.1088/1755-1315/151/1/012038

Prediction of Optimal C: N Ratio in Different Palm Oil Mill Waste Mixtures and Its Evaluation of Earthworm Biomass

P F Rupani[1*], A Embrandiri[2], A F M Alkarkhi[3], M H Ibrahim[4] and M Abbaspour[1]

*1Faculty of Environment and Energy, Science and research branch, Islamic Azad University, Tehran, Iran

2 Faculty of Agro-Based Industry, Universiti Malaysia Kelantan, Kota Bharu, Malaysia

3 Universiti Kuala Lumpur - Malaysian Institute Chemical and Bioengineering Technology (MICET), 78000 Alor Gajah, Melaka, Malaysia

4 School of Industrial Technology, University Sains Malaysia, 11800, Penang, Malaysia

Email: p-rupani@srbiau.ac.ir; ashanty66@gmail.com;abbas@unikl.edu.my; mhakimi@usm.my; m-abbaspour@jamejam.net

Abstract. The oil palm industry has been recognized for its contribution towards economic growth and rapid development, it has also contributed to environmental pollution due to the production of huge quantities of by-products from the oil extraction process. Current research reports prediction of optimal CN ratio of different mixture percentages of palm oil mill effluent (POME) and palm pressed fiber (PPF) namely 100% (only POME), 90%, 80%, 70%, 60%, 50%, 40%, 30%, 20%, 10% and 0% (only PPF). Different mixture percentages with varying initial characters results in different C:N ratio which could affect the earthworm growth and the vermicomposting stability. Therefore, the present research aims to predict the optimal mixture by establishing a 3D base model. The models were generated having an average unified formula using mathematical software Matlab. The results obtained in this study indicates that the 3D polynomial graph can explain the relationship between different POME-PPF concentration with respect to earthworm growth and time showing ideal R^2 value of 0.99. The regression analysis showed positive correlation between different mixture percentages and earthworm growth in 50% ($r = 0.412$), 60% ($r = 0.509$) and 70% ($r = 0.441$). Therefore, from the model it can conclude that 60% mixture of POME-PPF is an optimal mixture for the vermicomposting process.

1. Introduction

This experiment therefore, was carried out to test how the different mixtures of POME-PPF namely 100% (only POME), 90%, 80%, 70%, 60%, 50%, 40%, 30%, 20%, 10% and 0% (only PPF) with having different starting CN ratios can affect the resultant CN ratio and earthworm growth. Carbon and nitrogen are two primary nutrients required for cell growth; therefore an optimal carbon to nitrogen ratio is necessary for vermicomposting process [1]. Technological application is to make the best use of epigeic group of earthworms for their ability to feed on partially degraded organic wastes. Earthworms that can survive on different organic substances, tolerate a wide range of temperature and pH are best suited for culturing [2].

POME is one of the by-products which produces in abundant from the palm oil mills. Palm oil mill board of Malaysia reported discharge of 44 million ton in 2014 [3]. POME is acidic in nature and releasing it naturally is harmful for the environment and causes depletion in water bodies. Rupani et al [4], studied on the applicability of acidic POME for vermicomposting technology and could successfully achieve matured compost (CN-14) in 40 days of experiment. Yet, no further study conducted on evaluating the optimal mixture for vermicomposting process of palm oil mill wastes. This study aims to evaluate the kinetic study of different mixtures of POME-PPF and generate a mathematical model in order to predict the optimal mixtures.

2. Material and Methods

2.1. Material

The fresh palm oil mil effluent (POME) and palm pressed fiber (PPF) were obtained from MALPOM industry Sdn. Bhd, Nibong Tebal, Malaysia. The earthworms belong to E. eugeniae species were achieved from the vermiculture unit (Penang, Malaysia). The culture of the earthworm was maintained in plastic containers with partially decomposed mixture of bio-waste (as a growth medium) under laboratory conditions (26±2°C) for further use.

2.2. Experimental Setup

A fixed amount of 50 g of well mixed POME and PPF of different combination (namely: 100%, 90%, 80%, 70%, 60%, 50%, 40%, 30%, 20%, 10% and 0% POME - PPF) on a wet weight basis were placed in each reactor. All the vermireactors were set in triplicates, with a total of thirty three experimental units. As this experiment focuses on the efficiency of the earthworm in different mixtures, hence, each reactor offered as a comparison study for the other. Therefore, after 5 days of pre-composting, 10 g sample from each mixture was placed in a petri dish to conduct the vermicomposting experiment. pH were analysed using pH Meter (Hach Sension 3); total organic carbon (TOC) and total nitrogen (N) were measured by CHN analyser (Shimadzu Model 2400), [5]

2.3. Statistical Analysis and Modelling

The data was subjected to ANOVA, Post-Hoc Tukey's test to determine the significant differences between the parameters at 5% confidence level. Data were analysed using the regression analysis to find the model that describes the relationship between independent variables (pH and C: N ratio) and the dependent variable (earthworm biomass). Polynomial model was used to fit the data for different POME-PPF mixtures to predict the optimal changes in carbon to nitrogen (C: N) ratio in relation to different POME-PPF mixtures by using Matlab software [6].

3. Results and Discussion

3.1. pH Changes in Different POME-PPF Mixtures

The pH of different POME-PPF mixtures during the vermicomposting process through 15 days is shown in Figure 1. The initial pH was in the range of 5.3 - 5.8 in different mixtures which increased to 6.3 to 7 in 15 days of vermicomposting. The highest pH value (7.1- 7.13) on the 15th day was observed in 50 and 60% mixtures of POME-PPF. The results of ANOVA for pH during vermicomposting of different POME-PPF mixtures show significant difference ($P < 0.05$) on 15th day, while on 0, 5th and 10th day no significant difference ($P > 0.05$) was observed . The pH may be due to the rapid metabolic degradation of organic acids, as well as intense proteolysis of alkaline ammonia due to protein degradation [7], [8]. Earthworms in 100% mixture of POME-PPF could not survived, therefore, data pertaining 100% eliminated from the results.

3.2. CN Changes in Different POME-PPF Mixtures

Figure 2 shows the CN ratio of different POME percentages mixed with PPF using *E. eugeniae*. Initially CN was in the range of 50.26 - 25.2 in POME-PPF mixtures. The figure revealed that all the treatments showed a reduction in the CN ratio throughout the vermicomposting process, ranging between 35.27 and 9.6 in different mixtures using *E. eugeniae*. Standard deviation from 40 to 90% was minimal and is not feasible to illustrate graphically in bar charts. The results of ANOVA for CN

ratio of different POME-PPF mixtures show significant difference ($P < 0.05$) in all the mixtures during the vermicomposting process.
In order to normalize the CN ratio, the CN ratio at any time of measurement (CNt) was divided to CN at time zero (CNo). Thus, at starting time, t=0 the CNt /CNo is equal to 1.0 to measure the efficiency of CN stabilization (Figure 3). The normalized CN ratio which starts from 1 could clearly demonstrate the significant decreasing pattern in all the mixtures ($P < 0.05$). The results show that the maximum normalized value was achieved in 50% mixture of POME-PPF due to the equal ratio of POME and PPF.

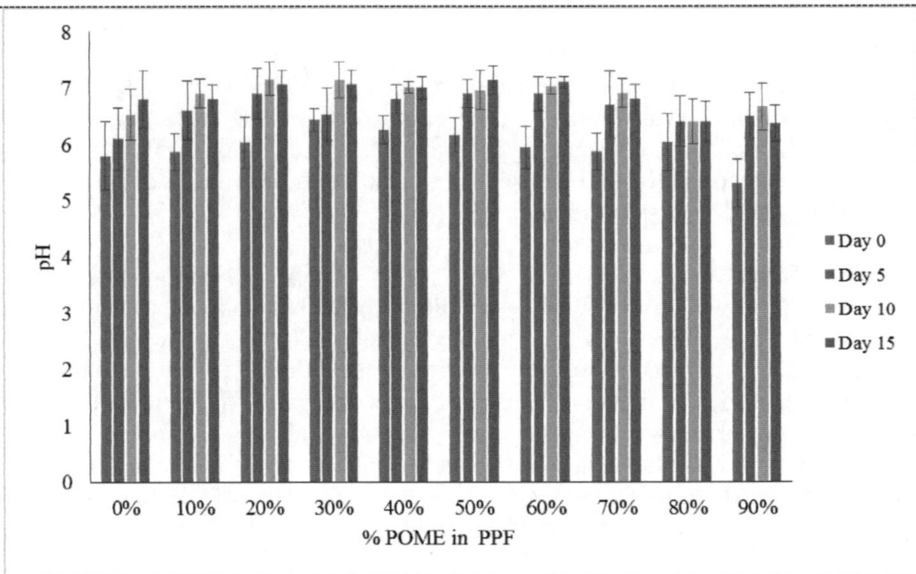

Figure 1. pH values during the vermicomposting of different POME percentages in PPF (*E. eugeniae*)

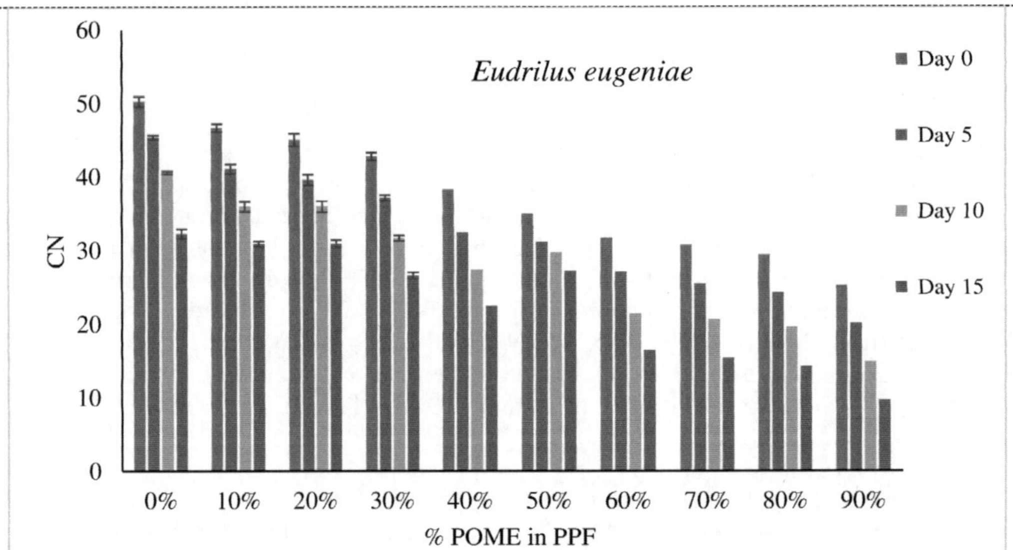

Figure 2. Changes in the C: N ratio of different POME-PPF mixtures using *E. eugeniae* from day 0-15

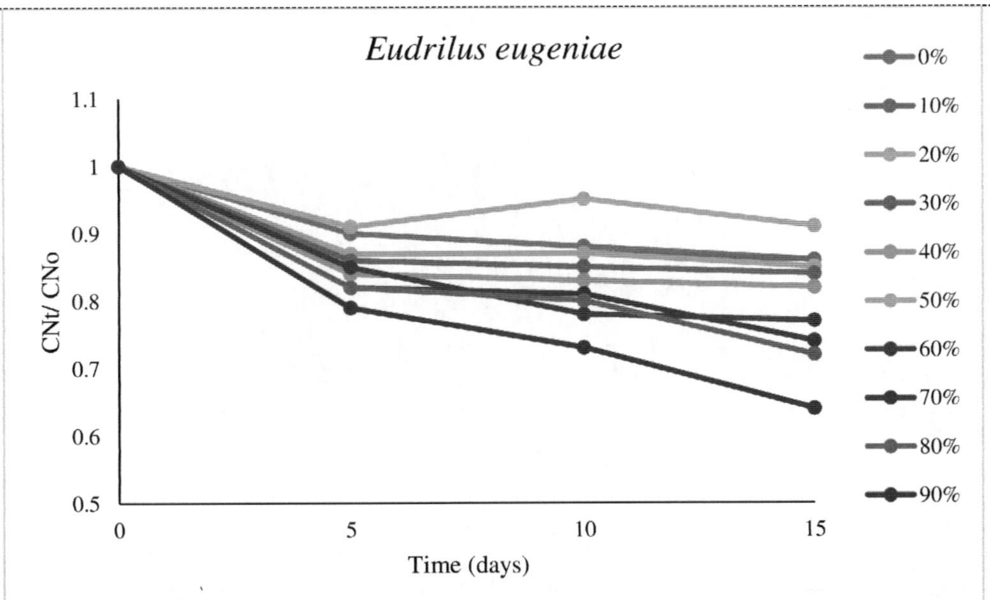

Figure 3. CNt /CNo obtained through vermicomposting process using *E. eugeniae* in different POME percentages mixed with PPF

3.3. The Relationship between POME Concentration and CN Ratio with Time

In order to relate dynamic CN ratio of different POME concentrations, a surface fitting method was applied. A 3D polynomial graph of normalized CN ratio obtained from different POME-PPF percentages as shown in Figure 4. The equations represents the fitted polynomial surface (Eq) which can be very useful in order to identify the corresponding values lying in between the collected data. In other words, the equation gives the fraction of the reduction in CNt /CNo with respect to time and make possible predictions and comparison.

$$Y = 1.008 - 0.0426\,x - 0.001247\,z + 0.004784\,x^2 + 0.0004697\,xz + 0.0000378\,z^2 - 0.0001743\,x^3 + 0.000003133\,x^2z \quad \text{(Eq.1)}$$

The goodness of the fitting techniques for CNt /CNo and the square root of coefficient (R2) and the sum of squares of errors (SSE) were calculated. The values demonstrated as the goodness of the surface fitting is very high (0.027) which is very close to the ideal value (zero). It indicates that the model has a smaller random error component, and the fit can be more useful for the prediction of CNt /CNo. The square root of coefficient showed above 0.9 and very close to the ideal value (1). Therefore, it can be concluded that the fittings are very close to the actual data and it can serve as an ideal equation to predict the CN ratio at any time given.

The regression analysis were performed to evaluate the effect of pH and CN ratio on earthworm biomass. Regression models for earthworm biomass of different POME-PPF mixtures are given in Table 1. The results shows that the relationship between earthworm biomass and pH was highly positively related ($r = 0.784$) at 20% mixtures of POME-PPF and moderately positive related at 10% ($r = 0.539$), 50% ($r = 0.412$), 60% ($r = 0.509$) and 70% ($r = 0.441$) mixtures of POME-PPF, while in other percentages of POME-PPF, earthworm biomass were negatively correlated with pH. However, C: N ratio at 80% ($r = 0.312$) and 90% ($r = 0.685$) mixture of POME-PPF were positively related to earthworm biomass and other percentages were negatively related.

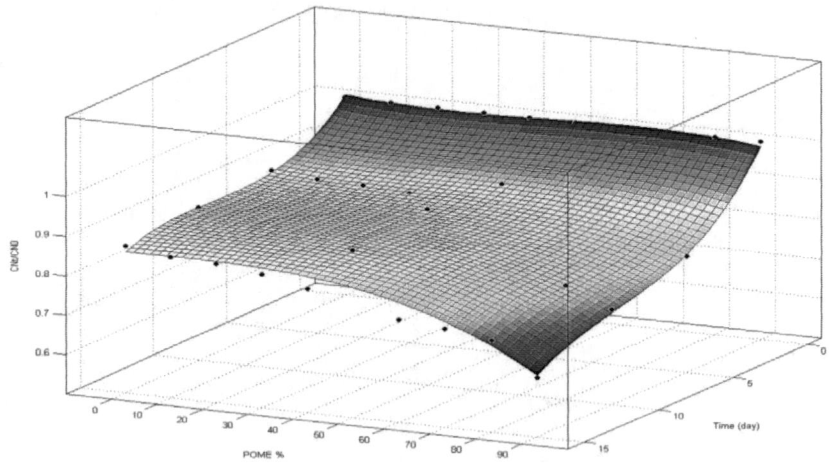

Figure 4. Relationship between POME concentration and normalized C: N ratio (CN_t /CN_o) in vermicomposting process using *E. eugeniae*

Table 1. Regression analysis between earthworm biomass on pH and CN in different POME % concentrations

Earthworms as dependent variable in POME-PPF	Beta Coefficient (pH)	Beta Coefficient (C: N)	Regression models
0%	- 0.045	- 0.642	Ew_0 = 0.954 – 0.10 pH – 0.16 CN
10%	0.539	- 0.420	Ew_{10} = 0.306 + 0.125 pH – 0.08 CN
20%	0.784	- 0.081	Ew_{20} = 1.128 + 0.214 pH – 0.002 CN
30%	- 0.086	- 0.643	Ew_{30} = 0.940 -0.029 pH – 0.015 CN
40%	0.429	- 0.576	Ew_{40} = 0.938 + 0.309 pH – 0.024 CN
50%	0.412	- 0.651	Ew_{50} = 0.784 + 0.238 pH – 0.065 CN
60%	0.509	- 0.474	Ew_{60} = 0.724 + 0.238 pH – 0.020 CN
70%	0.441	- 0.572	Ew_{70} = 0.307 + 0.166 pH – 0.020 CN
80%	- 0.695	0.312	Ew_{80} = 0.088 – 0.017 pH + 0.108 CN
90%	- 0.096	0.685	Ew_{90} = 0.624 – 0.002 pH + 0.139 CN

4. Conclusion

The results of this experiment showed that pH value showed significant differences through vermicomposting process except for 20% and 80%. However, CN value decreased significantly in all mixtures of POME-PPF. The results of regression analysis indicates that 60% mixture of POME-PPF can be select as an optimal mixture for vermicomposting of POME-PPF. This experiment can serve as a novel findings with regards to kinetic studies during vermicomposting process of palm oil mill by products. Further studies are recommended to generate a mathematical prediction model for any biowaste products.

5. References

[1] Khan, A.A., *Vermicomposting of poultry litter using Eisenia foetida*. 2006, Oklahoma State University.

[2] Kale, R.D., *Earthworm Biotechnology Source of Sustainable Agriculture. In Biology and Ecology of Tropical Earthworms*. Vol. First Edition. 2014, New Delhi- India.

[3] MPOB, *Malaysian Oil Palm Statistics.* *http://bepi.mpob.gov.my/index.php/statistics/export/138-export-2015/759-monthly-export-ofoil-palm products-2015.html* 2015.

[4] Rupani, P.F., et al., *Review of Current Palm Oil Mill Effluent (POME) Treatment Methods: Vermicomposting as a Sustainable Practice.* World Applied Sciences Journal, 2010. 10(10): p. 1190-1201.

[5] Benner, R., et al., *Bulk chemical characteristics of dissolved organic matter in the ocean.* Science, 1992. 255(5051): p. 1561-1564.

[6] MATLAB, U.s.G., *The mathworks.* Inc., Natick, MA, 1998. 5: p. 333.

[7] Batham, M., Arya, R. and Tiwari, A., 2014. Time Efficient Co-composting of Water Hyacinth and Industrial Wastes by Microbial Degradation and Subsequent Vermicomposting. *Journal of Bioremediation & Biodegredation*, *5*(3), p.1.

[8] Satisha, G. and L. Devarajan, *Effect of amendments on windrow composting of sugar industry pressmud.* Waste management, 2007. 27(9): p. 1083-1091.

Acknowledgments
The author (Parveen Fatemeh Rupani) acknowledges Universiti Sains Malaysia (USM) for providing research facilities to conduct this research.

IOP Conf. Series: Earth and Environmental Science **151** (2018) 012039 doi:10.1088/1755-1315/151/1/012039

Selective Isolation and Phenotypic Characterization of Bacteria and Actinomycetes from Oil-contaminated Soils

C A Rodríguez[1], T C Gavilánez[1,2], J P Chamorro[1], A G Vinueza[1,3], D M Salazar[1] and M Y Arancibia[1]

1 G+ BioFood & Engineering Group, Faculty of Food Science and Engineering, Technical University of Ambato. Av. Los Chasquis & Río Payamino, Z.C. 180150, Ambato-Ecuador
2 Department of Agricultural Sciences and Natural Resources, Technical University of Cotopaxi. Z.C. 050202, La Maná - Ecuador
3 Bioseborganics Cia. Ltda. Av. Indoamérica Km 6. Z.C. 180156, Ambato-Ecuador
E-mail: ca.rodriguez@uta.edu.ec

Abstract. Large numbers of bacteria and actinomycete strains were isolated from two contaminated soils with high total petroleum hydrocarbons (TPH) content. From five selective media formulations used, nutrient agar and glucose yeast extract agar showed the best results. Bacterial colony forming units per dry gram soil (ufc/g) ranged from $5.14 x 10^3$ to $4.20 x 10^{11}$, whereas for actinomycetes was from $8.99 x 10^4$ to $9.2 x 10^7$ ufc/g. Seventy-three bacteria and ninety-three actinomycete isolates were recovered from the two soils. Numerical taxonomy analysis of phenotypic data shows that at 90% similarity, the bacteria could be divided into sixteen multimembered and fifteen single-membered phenetic-groups, while the actinomycetes were separated in twenty multimembered and four single-membered phenetic-groups. Out of the one hundred and sixty-six strains, only ten were unable to use gasoline or oil as sole carbon and nitrogen sources. Most of the isolates could degrade both hydrocarbons. The results of this study show that bacteria and actinomycetes are present in large numbers in high TPH contaminated soils, with an extraordinary degree of diversity and that they can use oil or its derivatives as nutrients.

1. Introduction

Oil production around the world is a process that involves many steps where contamination risks to the environment are potentially high [1]–[3]. Efforts to clean the environment of pollutants that come from the petroleum industry have been implemented in oil-producing countries [4]. Modern techniques are used to clean water or soils where oil spillages have occurred. These combine chemical, physical and biological methods with varied rates of decontamination success [5]–[7]. Many biological techniques rely on microorganisms that come from different bacterial genera such as *Acinetobacter*, *Alcanivorax*, *Gordonia, Pseudomonas* or *Rhodococcus* [8],[9]. These bacteria have been isolated from diverse environmental sources, where soils with some content of hydrocarbon contaminants from oil have been the preferred choice [10],[11].

Many microbial groups are better suited to adapt to petroleum contaminants and its many derivatives in the soil [12],[13]. Dependant culture methods have been looking for that type of microbes, which are very common in contaminated soils. These actinomycete, bacteria and fungi grow very quickly on rich nutrient media. Members of Proteobacteria, and a few from the class Actinobacteria have been isolated and characterised, with some of them used commercially in bioremediation programs [11],[13]–[15]. On the other hand, molecular techniques have shown that there is a sizeable microbial population that has not been studied. These microbes belong to bacterial groups that are challenging to isolate in pure culture. One alternative to get growing these types of microorganisms could be by developing new approaches to look for potentially novel bacterial species capable of transforming oil pollutants into harmless compounds [10],[15],[16]. This study aimed to isolate and characterise bacteria and actinomycetes present in oil-contaminated soils and to assess their potential to use hydrocarbons sources as nutrients.

2. Materials and Methods

2.1. Soil Samples

Two soil samples were provided by LAPSU S.A. (an Ecuadorian private laboratory that performs chemical analysis for the oil industry) which had been collected from the Sacha Oil Fields in the Ecuadorian Amazon Basin. Soil sample A contained 31000 ppm of total polycyclic hydrocarbons (TPHs) and a high content of clay. Sample B was a sandy soil with 46000 ppm TPH content. Both samples were used to carry out the process of selective isolation of any bacteria and actinomycete strains that had been present in the soils. Once the samples arrived at our facilities, they were homogenised with the help of a sterile mortar and pestle and kept at a temperature of 10°C in sterile plastic flasks.

2.2. Moisture and Organic Matter Contents.

Five grams of each soil sample was added to a porcelain crucible and dried at 105°C for 24 hours. This procedure was performed in triplicates and the moisture content recorded as the average of weight loss to the initial weighed soil. A known sample of the dried soil was weighed and placed into a new crucible and placed in a muffle furnace at 400°C for 24 hours, to burn any organic matter present in the samples. Before the crucibles were weighed again, they were cooled down to room temperature. The organic matter content was determined as the percentage of the weight loss in each of the triplicates.

2.3. Electrical Conductivity and pH of the Soil Samples

Ten grams of each soil were placed in 100 ml beakers where 25 ml of distilled water was added. The resultant soil suspension was stirred for 30 minutes. A thin aqueous layer was visible on top of the soil suspensions after the samples were left to equilibrate for one hour. The electrical conductivity and pH values were obtained using a glass electrode pH meter (Fisher Scientific, Acumet AB200), once the reading in the equipment stabilised. The final results were reported as the average of three recordings per sample.

2.4. Selective Isolation of Microorganisms

Before the selective isolation of any bacterial and actinomycete strains present in the samples, the colony forming units per dry weight soil and colony diversity were determined. Ten grams of each soil was placed in a glass bottle that contained 90 ml of sterile tap water. For the bacterial studies, serial dilutions were prepared down to 10^{-7} and 100 uL from dilutions 10^{-2} to 10^{-4} were spread over the surface of two sets of sterile media that had been added either gasoline or crude oil. The basal medium contained 1ml of trace salts solution ($MgSO_4 7H_2O$, $FeSO_4 4H_2O$, $CuSO_4 4H_2O$, each at a concentration of 0,1g/100mL), K_2HPO_4 (4 g/L), KNO_3 (2 g/L) and agar (15 g/L). Additionally, 100 uL from dilutions 10^{-4} to 10^{-7} were inoculated over the surface of nutrient agar (NA), glucose yeast extract agar (GYM) and glucose yeast extract malt extract agar (GYEA). For the actinomycete selective isolation, the dilution 10^{-2} was prepared separately and placed in a water bath at 55°C for 6 minutes. The dilutions 10^{-3} and 10^{-4} were prepared, and 100 uL of these serial dilutions were spread over the surface of plates that contained starch casein agar (SCA) and glucose yeast extract agar (GYM). All plates containing the different media formulations were supplemented with nystatin (75 ug/mL) and adjusted to a pH of 6.5. After the inoculation procedure was completed, the plates were left to dry at room temperature for half an hour and then incubated at 26°C, one week for bacteria and two weeks for the actinomycetes. The colonies growing on the surface of the many media were expressed as colony forming units per dry weight soil and as the number of different colonies based on their macro-morphology.

2.5. Isolation, Purification, and Preservation of the Strains

Seventy-three bacterial and ninety-three actinomycete colonies were isolated using sterile toothpicks on NA or GYM, respectively and incubated at 26°C. The resultant microbial cultures were examined by eye to discard any contamination. All pure cultures were used to inoculate fresh agar plates and to prepare cell or spore suspensions in 20% glycerol (w/v) for long-term preservation and 1 ml sterile tap water tubes for the phenotypic tests.

2.6. Colour Grouping of the Isolates

Microbial cultures that were grown on AN and GYM plates were used to determine the colony pigmentation of the bacterial cultures whereas for actinomycete strains the colour of the aerial spore mass, substrate mycelium and any diffusible pigments. In both cases, a standard colour chart from the British Standard Specification for Colours for Identification Coding and Special Purposes was used. Strains that shared the same pigments were grouped to form preliminary bacterial and actinomycete colour groups.

2.7. Microscopic Characterization

Bacterial smears on microscopic slides were prepared to determine cell shape, Gram stain, and any capsule or endospore formation. Gram stain was achieved by covering a fixed bacterial smear with crystal violet for 1 minute then rinsed with distilled water. Then, an iodine colourant was used for 1 minute and washed with 100% ethanol. Finally, safranin was used as a counterstaining dye for 30 seconds. Capsule determination was performed by using a non-fixed bacterial smear which was stained with crystal violet for 1 minute and washed with a 20% copper sulphate solution (w/v). Finally, any bacterial culture that presented gram-positive bacilli characteristics was used to determine the presence of endospores inside its cell. To that aim, a fixed bacterial smear was stained with malachite green for 10 minutes, in a hot water vapour atmosphere. Safranin was used a counterstaining dye. At the end of all these techniques, the glass slides were left to dry at room temperature prior observation under a binocular microscope. For actinomycete cultures, sterile coverslips were inserted at an angle of approximately 45°, in GYM plates that had been previously streaked with a fresh actinomycete inoculum. After a 10 – 12 days incubation period, the coverslips were taken out of each plate and fixed on the surface of a glass slide. With the help of a light microscope, the spore chain morphology was determined for each actinomycete strain.

2.8. Physiological Tests for Growth

Five microliters of the cell or spore aqueous suspensions were used to inoculate plates with different media formulations and incubation conditions. Temperature, pH, and NaCl tolerance range growth were determined for the whole bacterial and actinomycete culture collection. As otherwise stated in each test, bacterial strains were inoculated on nutrient agar whereas actinomycetes were on GYM agar. After the incubation period, the presence of actinomycete or bacterial growth was recorded as 1, and absence as 0. Table 1 shows the details of each test.

Table 1. Conditions studied to determine the physiological conditions for growth.

Test	Conditions
Temperature	4°C, 37°C and 50°C
pH (buffered)	4.5, 5.5, 6.5, 7.5 and 8,5
NaCl (w/v)	1%, 5% and 10%

2.9. Determination of the Ability to Use Hydrocarbon Compounds as Sole Nitrogen and Carbon Sources

A basal medium (1ml of trace salts solution, K_2HPO_4, 4 g/L, and agar, 15g/L) supplemented with crude oil or gasoline (5 mL per litre), was used to determine the ability of the bacterial and actinomycete strains to grow on these two hydrocarbon compounds. The pH of the media was adjusted to pH 6.5, Five microliters of the bacterial or spore aqueous suspensions were inoculated over the surface of the sterile culture media. Plates were incubated at 26°C for two weeks, and the presence of growth was recorded.

2.10. Numerical Taxonomy of Phenotypic Data

All the results from the phenotypic characterisation were used to construct a data matrix in binary code. The data were analysed with the NTSYS programme suite [17]. Similarities were calculated using the simple matching coefficient (Ssm) and Jaccard's coefficient (S_J). Clustering was achieved by the UPGMA (Unweighted Pair Group Method with Arithmetic Means) algorithm. Phenetic clusters were identified by analysing the results from the two resulting dendrograms.

3. Results and Discussion

3.1. Physicochemical Properties of the Samples

Table 2 shows the results for the physicochemical characteristics of the soil samples. The high amount of TPHs in soil sample B is probably responsible for the higher counts on electrical conductivity, organic matter content, and pH.

Table 2. Physicochemical parameters of the soil samples.

Parameter	Sample A	Sample B
Main soil particle	Clay	Sand
TPHs (ppm)	31000	46000
pH	4,99	6,37
Electrical conductivity (uS/cm)	127	5190
Moisture Content (%)	29,96	4,11
Organic Matter Content (%)	18,10	35,02

3.2. Number and Colony Diversity of Bacteria and Actinomycetes

The colony forming units per dry weight soil and colony diversity are shown in Table 3. The results from the isolation medium that contained crude oil are not shown because it was challenging to differentiate the colony growing on the dark surface of the medium and fungal growth from strains that apparently are resistant to the nistatyn concentration used in the medium.

Table 3. Colony forming units per dry weight soil and colony diversity of bacteria.

Soil Sample	Selective medium	Colony Forming Units per Dry Weight Soil	Different Colony Number
A	AN	1,55E+09	14
	AG	2,47E+07	5
	GYM	5,14E+03	3
	GYEA	1,35E+04	6
B	AN	4,20E+11	6
	AG	1,36E+09	11
	GYM	1,29E+08	5
	GYEA	1,50E+08	4

In general, more significant numbers of bacteria were found in sandy soil B, despite that the TPH value is close to 1.5 times higher than in sample A. It seems that the sandy soil texture makes more oxygen available, giving an opportunity to the microbes present to adapt better to the conditions [18]. The results also show that the electrical conductivity and TPHs content do not affect the microbial population. Another factor for lower cfu counts found in soil A could be attributed to the pH of de soils and the pH of the selective media. The pH in sample B is very close to the one used in this experiment whereas the pH in soil sample A is acid (4.99). There is a chance that the microbes present

in soil A were not able to grow due to the almost neutral pH on the selective isolation plates. Regarding the different bacterial colonies growing on the plates, there are no significant differences when comparing the two soils. On the other hand, when analysing if there is any correlation between the cfu counts and the selective media used, it is evident that there are higher numbers related to nutrient agar than the rest of the media. It is well known that NA is a nutrient rich medium that has significant amounts of compounds quickly available for microbial use. In the other selective media, it is probable that many of their components are not available unless the microbes possess the enzymatic machinery to use them.
For the actinomycete count, Table 4 shows the colony forming units per dry weight soil and colony diversity. The results agree to what was found in the bacterial population. Soil sample B presents higher actinomycete cfu/g numbers than soil sample A, however, the differences in numbers are not as significant than in the bacteria. What it is remarkable it is that regardless of soil sample or isolation media, the actinomycete counts are very high for this type of contaminated soils. This kind of numbers is typically found in non-contaminated soils [19],[20]. For us, this is the first report of such findings, probably meaning that actinomycetes could play a significant role in bioremediation programmes. It is widely known that nocardioform actinomycetes can degrade different hydrocarbon sources. However, not much is known about what other non-nocardioform actinomycetes can do in such circumstances.

Table 4. Colony forming units per dry weight soil and colony diversity of actinomycetes.

Soil Sample	Selective medium	Colony Forming Units per Dry Weight Soil	Different Colony Number
A	GYM	7,04E+07	16
	SCA	9,20E+07	11
B	GYM	8,99E+04	6
	SCA	2,70E+06	5

3.3. *Colour Grouping of Bacterial and Actinomycete Strains*

Seventy-three bacteria and ninety-three actinomycete strains were isolated from both soils samples. Table 5 and 6 show the corresponding grouping by colour similarities between the different bacterial and actinomycete isolates. Sixteen multimembered and six single membered colour groups were recovered for the bacterial collection, whereas fifteen multimembered and thirteen single membered colour groups for the actinomycete isolates. These differences could be because of more actinomycete strains were isolated from the contaminated soils, and in any case, higher colony diversity was found on the media plates that corresponded to the actinomycetes.

3.4. *Phenotypic Characterisation of the Isolates*

Figures 1 and 2 show the dendrograms from the analyses of the different phenotypic tests carried out for the bacterial and actinomycete isolates. The dendrograms were drawn based on similarity matrices calculated from datasets consisting of 40 and 53 results reported as binary code, from the phenotypic characterisation of the bacterial and actinomycete isolates. The bacteria could be divided into sixteen multimembered and fifteen single-membered phenetic-groups, while the actinomycetes were separated in twenty multimembered and four single-membered phenetic-groups.

Table 5. Assignment of bacterial isolates to multimembered and single-membered colour-groups.

Colour Group	Colony Colour	ID Actinomycete Isolates
1	Yellow	TG17, TG24, TG25, TG26, TG27, TG61, TG62, TG82, TG83, TG84, TG89, TG92, TG119, TG120, TG16, TG73, TG53
2	Light Yellow	TG63, TG64, TG65, TG66, TG67, TG80, TG50, TG8
3	Light Brown	TG103, TG91, TG112, TG122, TG125, TG18
4	Cream	TG7, TG107, TG108, TG130, TG131, TG110
5	Orange	TG69, TG70, TG71, TG21, TG54, TG22
6	White	TG31, TG117, TG118, TG95, TG126
7	Grey	TG104, TG105, TG106, TG87
8	Ivory	TG93, TG88, TG89, TG97
9	Dark Yellow	TG23, TG72, TG113
10	Beige	TG47, TG114, TG51
11	Light Purple	TG4, TG6, TG5
12	Light Red	TG2, TG52, TG60
13	Light Green	TG14, TG15, TG46
14	Pink	TG20, TG123

Table 6. Assignment of actinomycete isolates to multimembered and single-membered colour-groups.

Colour Group	Colony Colour	ID Actinomycete Isolates
1	Cream	BG007, BG008, BG011, BG016, BG037, BG065(1), BG071, BG088, BG090, BG104, BG105, BG107, BG126, BG142 (1), BG169
2	Dark Red	BG004, BG042, BG064, BG066, BG075, BG082, BG089, BG115, BG152, BG153, BG160
3	Beige	BG019, BG034, BG072, BG110, BG112, BG132, BG182, BG187, BG189
4	Pale Yellow	BG010, BG020, BG036, BG085, BG094, BG143, BG155, BG165
5	Light Brownish	BG002, BG015, BG035, BG057, BG109, BG127, BG149, BG161
6	Brown	BG032, BG087, BG101, BG118, BG123, BG163, BG171, BG180
7	Yellow	BG058, BG065(2), BG097, BG124, BG140, BG145, BG146, BG159
8	Red	BG009, BG051, BG120, BG128, BG133, BG150
9	Orange	BG061, BG086, BG092, BG154, BG174
10	Dark Green	BG060, BG125, BG158, BG177
11	White	BG014, BG083, BG114, BG137
12	Light Brown	BG038, BG095, BG136
13	Dark Yellow	BG043, BG080
14	Orange	BG151
15	Dark Brown	BG111

Based on the analyses of the dendrograms, there are thirteen bacterial phenetic multimembered groups recovered partially or completely by using both algorithms (S_{SM} and S_J). When the comparison was made between the colour and phenetic groups, there is a congruence of fourteen out of the sixteen phenetic groups. Regarding the analysis of the dendrogram for the actinomycete strains, there are nineteen phenetic groups that are found clustered together, either partially or completely, when both algorithms are used. About the coherence of the colour and phenetic multimembered groups, there is less agreement since only fourteen out of the twenty phenetic groups were found in both characterisations. In any case, these results show that there is great coherence between the colour

IOP Conf. Series: Earth and Environmental Science **151** (2018) 012039 doi:10.1088/1755-1315/151/1/012039

grouping and the results from the numerical taxonomy analyses from the phenotypic data, thus validating the colour determination and formation of preliminary colour groups of isolates when there is a large collection of microbes.

These results also show that the procedures used in this study allowed the cultivation of many bacteria and actinomycetes present in the soil samples contaminated with TPHs. Regarding the bacteria, this fact is relatively normal since other studies, which have used either culture dependant or molecular detection techniques, have demonstrated that there are numerous bacterial groups present in this type of soils [10]. However, for the actinomycetes, this is the opposite since studies about their presence in TPH contaminated soils is limited. In any case, this study opens the door to carry out more experiments to elucidate the role of the actinomycetes in oil-contaminated soils and to explore the true extent of their diversity in such habitats.

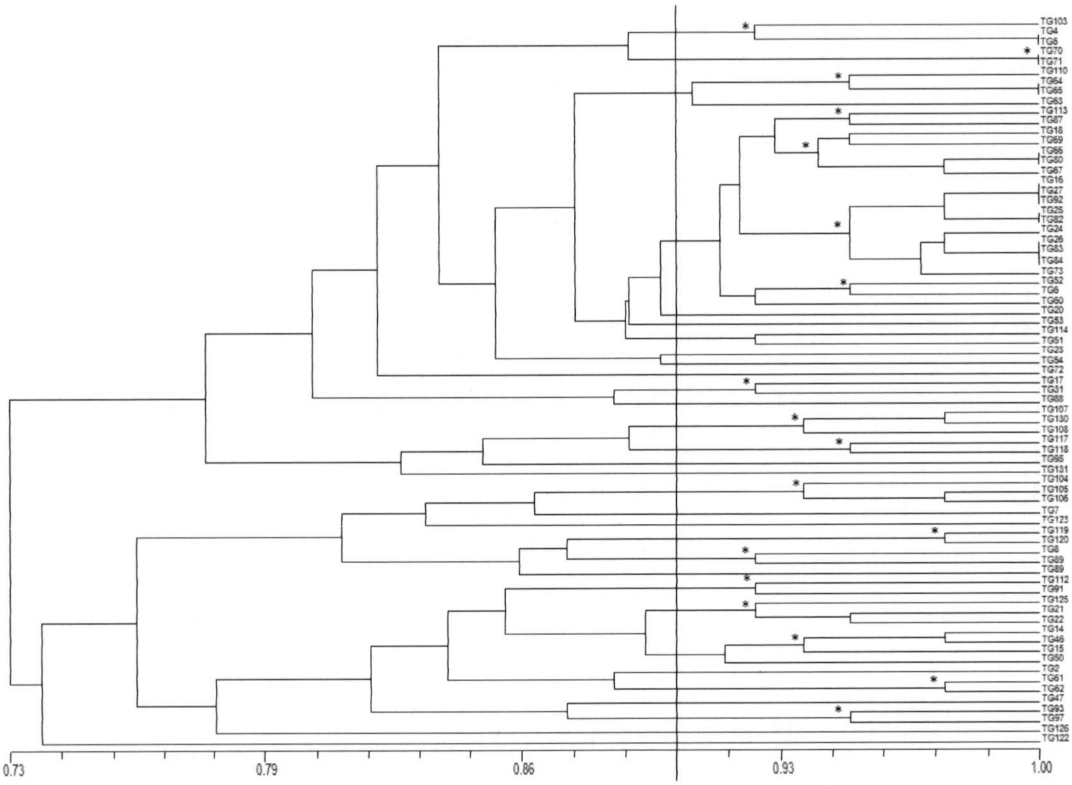

Figure 1. Dendrogram based on the analysis of phenotypic data of seventy-three bacterial strains. Clustering was achieved by the UPGMA (Unweighted Pair Group Method with Arithmetic Means) algorithm from a similarity matrix calculated using the single matching coefficient (S_{SM}). Phenetic groups were obtained at 90% similarity. The asterisks denote phenetic groups that were also recovered using the Jaccard's coefficient (SJ).

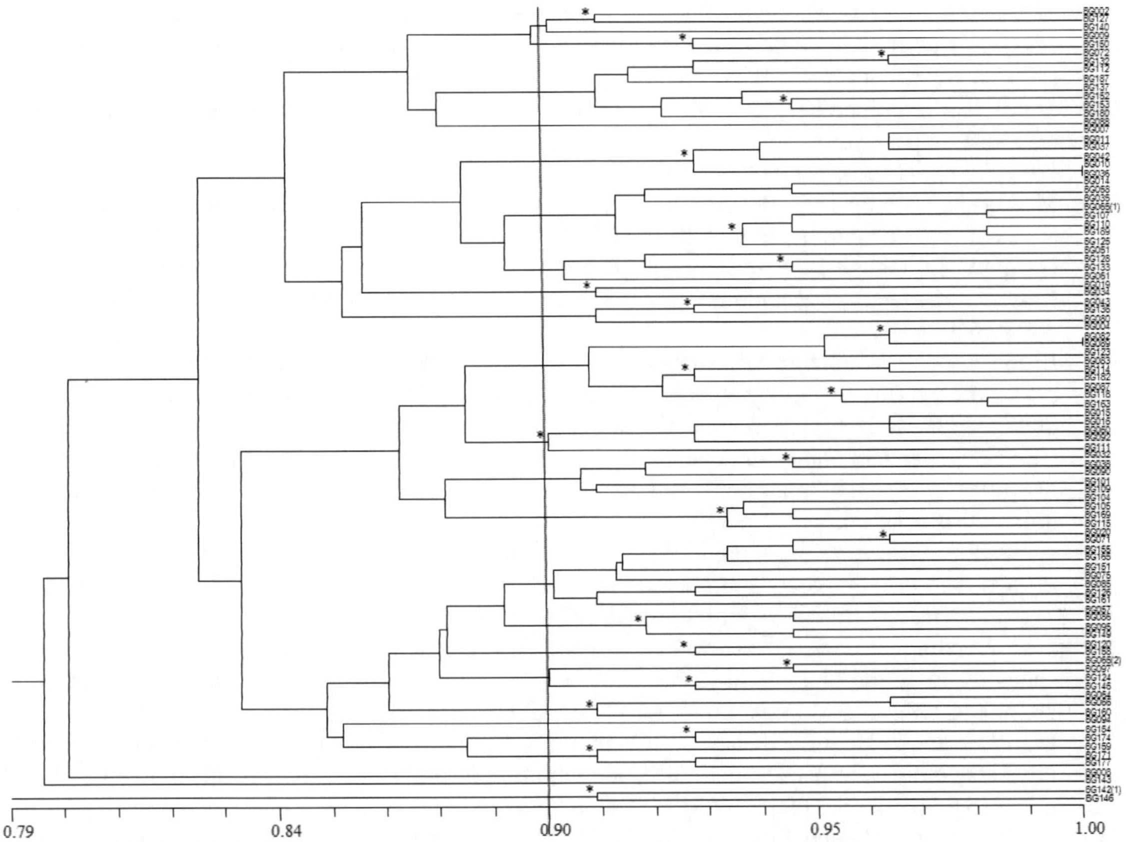

Figure 2. Dendrogram based on the analysis of phenotypic data of ninety-three actinomycete strains. Clustering was achieved by the UPGMA (Unweighted Pair Group Method with Arithmetic Means) algorithm from a similarity matrix calculated using the single matching coefficient (S_{SM}). Phenetic groups were obtained at 90% similarity. The asterisks denote phenetic groups that were also recovered using the Jaccard's coefficient (S_J).

4. Conclusions

A large diversity of bacteria and actinomycete strains was found in two contaminated soil with TPHs. This result is supported by the numerical taxonomy analyses of phenotypic data which divided the microbial culture collection in thirty-one and twenty-four bacterial and actinomycete phenetic groups. Besides the microbial diversity findings, it is also important to note that more than 90% of the isolates could use gasoline or oil as sole carbon and nitrogen source. This result could be of great importance later in the process of developing a bioproduct for bioremediation purposes.

5. References

[1] Finer M, Jenkins CN, Pimm SL, Keane B, Ross C. Oil and Gas Projects in the Western Amazon: Threats to Wilderness, Biodiversity, and Indigenous Peoples. PLoS One 2008;3. doi:10.1371/journal.pone.0002932.

[2] Abdol Hamid HR, Kassim WMS, El Hishir A, El-jawashi SAS. Risk assessment and remediation suggestion of impacted soil by produced water associated with oil production. Environ Monit Assess 2008;145:95–102. doi:10.1007/s10661-007-0018-3.

[3] Jernelöv A. The Threats from Oil Spills: Now, Then, and in the Future. Ambio 2010;39:353–66. doi:10.1007/s13280-010-0085-5.

[4] Paulauskien T, Jucik I. Aquatic oil spill cleanup using natural sorbents. Environ Sci Pollut Res Int 2015;22:14874–81. doi:10.1007/s11356-015-4725-y.

[5] Mohsenzadeh F, Chehregani Rad A, Akbari M. Evaluation of oil removal efficiency and enzymatic activity in some fungal strains for bioremediation of petroleum-polluted soils. Iran J Environ Health Sci Eng 2012;9:26. doi:10.1186/1735-2746-9-26.

[6] Tamis JE, Jongbloed RH, Karman CC, Koops W, Murk AJ. Rational application of chemicals in response to oil spills may reduce environmental damage. Integr Environ Assess Manag 2012;8:231–41. doi:10.1002/ieam.273.

[7] Mercer K, Trevors JT. Remediation of oil spills in temperate and tropical coastal marine environments. Environmentalist 2011;31:338–47. doi:10.1007/s10669-011-9335-8.

[8] Dasgupta D, Ghosh R, Sengupta TK. Biofilm-Mediated Enhanced Crude Oil Degradation by Newly Isolated Pseudomonas Species. ISRN Biotechnol 2013;2013:250749. doi:10.5402/2013/250749.

[9] Brooijmans RJW, Pastink MI, Siezen RJ. Hydrocarbon-degrading bacteria: the oil-spill clean-up crew. Microb Biotechnol 2009;2:587–94. doi:10.1111/j.1751-7915.2009.00151.x.

[10] Stefani FOP, Bell TH, Marchand C, de la Providencia IE, El Yassimi A, St-Arnaud M, et al. Culture-Dependent and -Independent Methods Capture Different Microbial Community Fractions in Hydrocarbon-Contaminated Soils. PLoS ONE 2015;10:e0128272. doi:10.1371/journal.pone.0128272.

[11] Augustine A, Prathap P, Kamarudheen N, Bhaskara Rao KV. Isolation, Screening and partial identification of Indigenous Biosurfactant Bacteria from Oil Contaminated Soil. Res J Pharm Technol 2017;10:1697–702. doi:10.5958/0974-360X.2017.00300.6.

[12] Liao J, Wang J, Jiang D, Wang MC, Huang Y. Long-term oil contamination causes similar changes in microbial communities of two distinct soils. Appl Microbiol Biotechnol 2015;99:10299–310. doi:10.1007/s00253-015-6880-y.

[13] Liao J, Wang J, Huang Y. Bacterial Community Features Are Shaped by Geographic Location, Physicochemical Properties, and Oil Contamination of Soil in Main Oil Fields of China. Microb Ecol 2015;70:380–9. doi:10.1007/s00248-015-0572-0.

[14] Al-Wasify RS, Hamed SR. Bacterial Biodegradation of Crude Oil Using Local Isolates. Int J Bacteriol 2014;2014:863272. doi:10.1155/2014/863272.

[15] Salam LB, Obayori OS, Olatoye NO. Biodegradation of anthracene by a novel actinomycete, Microbacterium sp. isolated from tropical hydrocarbon-contaminated soil. World J Microbiol Biotechnol 2014;30:335–41. doi:10.1007/s11274-013-1437-7.

[16] Mueller JG, Devereux R, Santavy DL, Lantz SE, Willis SG, Pritchard PH. Phylogenetic and physiological comparisons of PAH-degrading bacteria from geographically diverse soils. Antonie Van Leeuwenhoek 1997;71:329–43. doi:10.1023/A:1000277008064.

[17] Rolf F. NTSYSpc Numerical Taxonomy and Multivariate Analysis System Version 2.0 User Guide. New York: Applied Biostatistics Inc., Setauket; 1998.

[18] Liu Q, Tang J, Gao K, Gurav R, Giesy JP. Aerobic degradation of crude oil by microorganisms in soils from four geographic regions of China. Sci Rep 2017;7:14856. doi:10.1038/s41598-017-14032-5.

[19] Lee JY, Hwang BK. Diversity of antifungal actinomycetes in various vegetative soils of Korea. Can J Microbiol 2002;48:407–17.

[20] Poomthongdee N, Duangmal K, Pathom-aree W. Acidophilic actinomycetes from rhizosphere soil: diversity and properties beneficial to plants. J Antibiot (Tokyo) 2015;68:106–14. doi:10.1038/ja.2014.117.

Acknowledgements
C A Rodríguez is grateful for support from the Technical University of Ambato as a recipient of a research grant.

Antimicrobial potential of *Chlorella* algae isolated from stacked waters of the Andean Region of Ecuador

L P Acurio[1], D M Salazar[1], A F Valencia[1], D R Robalino[1], A C Barona[1], F C Alvarez[1], C A Rodriguez[1]

1 G+ BioFood & Engineering Group, Department of Food Science and Engineering, Technical University of Ambato. Av. Los Chasquis & Río Payamino, Z.C. 180150, Ambato
Email: dm.salazar@uta.edu.ec

Abstract. In this study, the focus will be primarily on antimicrobials extracted from Microalgae *Chlorella* from stacked water from three different places. The results provide dates for possible applications in the pharmaceutical industry for developing antiseptic and antimicrobial products to guarantee the health of the people. The study initially outlines the isolation of microalgae and secondly was the extraction of chlorellin as an antimicrobial metabolite. To replicate the environmental conditions which are useful for growth colonies of *Chlorella* it was maintained at room temperature, oscillating between 10 °C to 18 °C in Bold Basal medium enriched with NH4Cl. After that, chlorellin was extracted from aqueous supernatant and sediment with ethanol, isopropyl alcohol, and water. The results of this study show that the chlorellin extracted from *Chlorella* present an essential antimicrobial capacity against bacteria isolated from the hand. The antimicrobial capacity was equal with ampicillin and oxacillin to inhibit *Staphylococcus spp.*

1. Introduction

The increasing development of resistance of microorganisms to drugs and treatments, to combat specific diseases, has made scientists look for new sources of antimicrobial substances. Nowadays, several of those natural sources are being studying because they have antimicrobial activity, such as exotic Plantae, plant root and microalgae [1, 2, 3, 4]. Microalgae are microorganisms which are represented by a wide variety of species, living in diverse ecological habitats [5, 6, 7]. Likewise, these organisms are a rich source of structurally novel and biologically active metabolites. Primary and secondary metabolites which are produced by microalgae could be unusual for chemical and pharmaceutical industry [8, 9, 10, 11]. The industrial applications of macro and microalgae have been focused on the farming of edible species, or on the production of agar, carrageenan, and alginate [12, 13, 14]. Microalgae can support extreme environments for that reason they have developed defensive and adaptive strategies, including the synthesis of a diversity of compounds from different metabolic pathways [15]. Many of their metabolites have been shown to possess antibacterial, antiviral, antifungal, enzyme inhibiting, immunostimulant, cytotoxic and antiplasmodial activities [16, 17, 18, 19]. Although the potential that microalgae could have, the attention on it has recently been focused on its biotechnological potential for obtaining antimicrobial metabolites [20]. One of the antibacterial substances, named "chlorellin", extracted from the microalgae *Chlorella* was found to exhibit inhibitory activity against both Gram-positive and Gram-negative bacteria [21, 22, 23, 24]. In this sense, this research aimed to evaluate the antimicrobial potential of the *chlorella* algae extracted from the stacked waters from three different irrigation ditch in Ambato-Ecuador.

2. Materials and Methods

2.1. *Sample Collection*

Stacked water was collected from three different irrigations ditches of Ambato-Ecuador. The places for collecting the samples were Atocha, Tilulum and El Socavón. Taking account the green color which indicated the presence of algae, 100 ml was collected from each point in sterile plastic containers; the samples were collected in triplicate. Once the samples arrived at our facilities, they were kept at a temperature of 18 °C in sterile plastic flasks without covering them.

2.2. *Selective Isolation of Chlorella*

Before the particular isolation, a drop of each sample was placed on a slide and observed under the microscope. Identification of *Chlorella* was based on its spherical morphology, about 2-10 microns in diameter, due to its absence of flagella so it is immobile, due to its lack of hairs, to be unicellular, and in its pigmentation, since at containing high levels of chlorophyll a and b the medium in which it grew was green. In a Bold Basal medium which is a nutrient medium for multiplication of microalgae that contained 1ml of trace salts solution (MgSO4.7H2O, FeSO4.4H2O, CuSO4.4H2O, each at a concentration of 0,1g/100mL), K2HPO4 (4 g/L), KNO3 (2 g/L) and agar (15 g/L). The dilution 10-3 were prepared, and 100 uL of this dilution were spread over the surface of plates. To replicate the environmental conditions of Ambato, the colonies growing at room temperature, oscillating between 10 °C to 18 °C. Finally, the microalgae were isolated several times to obtain pure cultures.

2.3. *Growth and Purification in Liquid Medium*

For growth and purification, 250ml of Bold Basal medium enriched with NH_4Cl for the massification microalgal was prepared. Samples were placed in sterilized bottles, aerated each of them and exposed to sunlight to help to the cells to be adapted to the medium. Each bottle had continuous aeration because the microalgae are aerobic, avoiding their sedimentation and allowing the homogenization of the nutrients significantly. This process improves the assimilation of nutrients because the movement of the water decreases the layer laminar around the microalga [25, 26, 27]. This procedure was repeated until obtaining pure cultures of *Chlorella* based mainly on its microscopic round morphology. Finally, the samples were allowed to stand for seven days for their overcrowding microalgal.

2.4. *Cell Counting*

Using 1 liter of bold basal medium and 10% of pure cultures per liter samples from Atocha, Tilulum and El Socavón were prepared. Formula (1) was used to calculate the concentration of the microalga by the hemocytometer method.

$$Concentration\ (cell/ml) = Cells\ number/Volume\ (ml) \quad (1)$$

One drop of each sample was placed in the Hemocytometer to be observed; each sample was covered with a coverslip to avoid air bubbles that could affect the visualization. Formula (1) was used taking account the four external squares were counted, the volume was calculated by multiplying the height, width and length of the hemocytometer (0.1cm×0.1cm×0.01cm). The data were recorded during ten days; each day cell counts were made to obtain the growth curve with the number of *Chlorella* cells.

2.5. *Chlorellin Extraction*

Taking into account that cell growth did not vary significantly (cells of the tests are in an exponential state) for three days, 10 ml of each sample was collected in test tubes, which were centrifuged at 5000 rpm for 20 minutes. To determine the antimicrobial compound the measurements were developed in aqueous supernatant and sediment after centrifugation. This procedure was repeated three times to obtain three tubes of sediment and three tubes of supernatant for each test; then different solvents were placed in each one; 5 ml of ethanol, 5 ml of isopropyl alcohol and 5 ml of water to extract the antimicrobial compound

2.6. *Antimicrobial Capacity*

2.6.1. Bacterial Identification. With the purpose to establish the antimicrobial potential, bacteria which are in hands were isolated. Tests were carried out on blood agar (Gram-positive) and on Mac Conkey agar (Gram-negative) to identify the type of bacteria[28]. Likewise, the catalase test [29, 30] and gram test [31] were performed. When it was observed growth on blood agar, the coagulase test [31, 32] was performed to establish the presence of *Staphylococcus spp.*

2.6.2. Antimicrobial Potential. The antimicrobial potential of chlorellin as a metabolite of *Chlorella* was evaluated on hands bacteria because chlorellin could be used as an antiseptic gel for hands. An antimicrobial sensitivity test with five discs, four at the ends and one in the center, was placed with 300 μ l of chlorellin. The discs were placed manually with sterile tweezers ensuring that they had perfect contact with the surface of the agar, so the discs were pressed lightly on the surface. The discs were placed less than 15 mm from the edge of the plate, and they were distributed so that no overlapping of the inhibition halos occurred. After this process, the plates were incubated in groups of no more than five at 37 ° C for 24 hours. The inhibition halos were measured with a ruler in a light-illuminated box on a black background to facilitate measurement.

2.7. *Positive Control*

The positive control to establish the susceptibility of microorganisms to a variety of microbial agents was developed through an antibiogram. The diameter of the area of inhibition around the disk was converted to the categories of sensitive, intermediate or resistant (S, I, or R) according to tables of the parameters of minimum established inhibitory concentration [33, 34]. An antibiogram in Petri dishes with Mueller-Hinton agar (specific for bacteria) with isolated colonies of gram-positive bacteria was developed. The inoculum was adjusted to the MacFarlane turbidity pattern No. 0.5, antibiotic sensitives such as 10 μ g ampicillin, 10 μ g oxacillin were used for having a predominant gram-positive spectrum including *Staphylococcus spp*, and the inhibition halos were measured after 24 hours of incubation.

2.8. *Statistical Analysis*

Statistical analysis was performed with the GraphPad Prism 5.0 program (GraphPad Software, San Diego, California, USA) with a bi-directional analysis of variance. The test of comparisons was carried out with the Tukey test with a significance level of $P \leq 0.05$.

3. Results and Discussion

3.1. *Identification of Chlorella*

The samples collected from the stacked water from three different irrigations ditches (Atocha, Tilulum and El Socavón) of Ambato-Ecuador were observed under a microscope at 40 x. Taking into account that the size of colonies were 2 to 10 microns in diameter, green color attributable to the presence of chlorophyll, irregularly shaped colonies retaining their spherical shape and mainly the absence of flagellum allowed to establish the presence of the *Chlorella* microalgae in all samples [25, 35].

3.2. *Growth and Purification*

The average initial population of each sample was calculated using equation 1. Socavón sample had an initial population of $2.60 \pm 0.57 \times 10^5$ cell×ml^{-1} lower than Atocha sample ($4.8 \pm 0.54 \times 10^5$ cell×ml^{-1}) and Tilulum sample ($13.60 \pm 0.51 \times 10^5$ cell×ml^{-1}). From the growth curve (Figure. 1) can be concluded that induction phase lasted for 24 hours, for this reason color of the culture was almost transparent. In this phase cells were treating to adapt to the new environment and there was not cellular replication. After 24 hours, the exponential phase appears and cell density increased as a function of time according to an exponential function ($r^2 = 0.99$) until 72 hours. The color of the culture was light green and significant difference in specific growth rate was showed between Atocha and Tilulum ditches ($p \leq 0.05$) (Table 1). The specific growth rate in this phase mainly depend on algal species, light intensity, and temperature [36]. Next, of this phase, a linear phase was showed since 72 hours until 192 hours (r^2

= 0.99). The color of the cultures changed to dark green without significant differences in specific growth rate between ditches. Finally, in the stationary phase cell division slows down as a function of time according to an exponential function ($r^2 = 0.98$) until 240 hours, without significant differences in specific growth rate between ditches. At this point, there was no production of new cells because nutrients, light, pH, carbon dioxide or others physical and chemical factors limit growth, and there were not significant change in coloration.
The extraction process of the chlorellin was performed at 240 hours, in this point the final population was $54.23\pm0.57\times10^5$ cell×ml^{-1} in Tilulum sample followed by Atocha sample ($45.76\pm 0.57\times10^5$ cell×ml^{-1}) and Socavón sample ($42.86\pm0.57\times10^5$ cell×ml^{-1}).

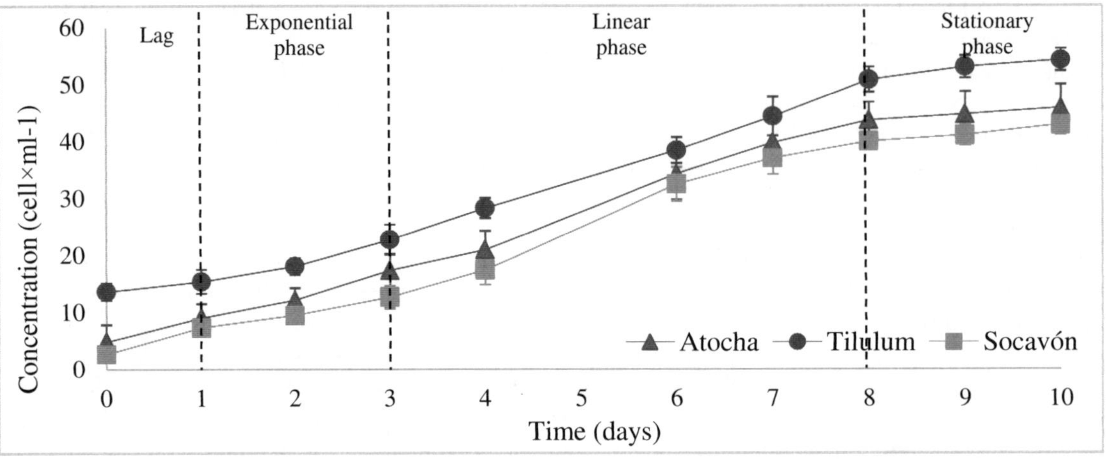

Figure 1. The growth of algae culture.

Table 1. Specific growth rate.

Ditches	Exponential phase		Linear phase		Stationary phase	
	$\times10^5$ cell×ml^{-1}	r^2	$\times10^5$ cell×ml^{-1}	r^2	$\times10^5$ cell×ml^{-1}	r^2
Atocha	0.328 ± 0.06^a	0.997	5.560 ± 0.137^a	0.992	0.024 ± 0.013^a	0.999
Tilulum	0.197 ± 0.023^b	0.989	5.524 ± 0.508^a	0.998	0.033 ± 0.006^a	0.966
Socavón	0.276 ± 0.033^{ab}	0.999	5.815 ± 0.036^a	0.982	0.035 ± 0.009^a	0.981

Results are the mean ± standard deviation. One-way ANOVA: different letters (a,b) in the same column indicate significant differences between samples ($p\leq 0.05$).

3.3. *Antimicrobial Capacity*

3.3.1. Chlorellin extraction. Tubes with sediment and supernatant were used to determine the antimicrobial compound of *Chlorella*. Different solvents such as 5 ml of ethanol, 5 ml of isopropyl alcohol and 5 ml of water to extract chlorellin were placed in each one. It was observed that the tubes with sediment-solvent were dark green, which indicated the predominance of microalgae, while the tubes with supernatant-solvent were almost transparent.

3.3.2. Bacterial identification tests. The bacterial identification was performed 24 hours after the samples were inoculated on blood agar with cultures obtained from the hands. The results showed growth of whitish colonies of circular shape and smooth surface with yellow and orange coloration. The catalase test to establish the presence of Enterococcus was performed, the presence of bubbles was observed, it indicates a splitting of the peroxide in water and gaseous oxygen, which allowed discarding the presence of Enterococcus. The gram test was positive. The results showed the presence of gram-positive bacteria. The growth of microorganisms on blood agar, which is a standard medium where Staphylococcus spp grows. Likewise, the coagulase test was positive; the results indicate the presence of Staphylococcus aureus because it is the only species that produces the coagulase enzyme and yellow pigmentation [37, 38, 39]. This isolated microorganism was used for the halo inhibition test.

3.3.3. Antimicrobial Potential. The sample from Atocha in which isopropyl alcohol was applied to extract chlorellin had an inhibition halo of 0.22 ± 0.01 mm, higher value than the Atocha sample that was obtained with ethanol whose inhibition halo value it was 0.17 ± 0.02 mm. In the Tilulum sample whose extraction solvent was isopropyl alcohol, the inhibition halo was 0.28 ± 0.01mm while the halo of inhibition of the sample in which ethanol was used was 0.19 ± 0.01 mm. In the sample from El Socavón in which Isopropyl Alcohol was used shown an inhibition halo of 0.19 ± 0.01 mm, whereas the sample of El Socavón in which ethanol was used shown an inhibition halo of 0.12 ± 0.01 mm. The samples from Atocha, Tilulum, and Socavón with isopropyl alcohol had higher values of inhibition compared to the treatments in which ethanol was used for extraction; however, the difference was not significant $p < 0.05$, which indicated that both solvents could be used to extract Chlorellin. The results show that the *extract of Chlorella* microalgae can inhibit bacterial growth when compared to the positive control [1, 2]. The positive control showed that the gram-positive bacterium was sensitive to ampicillin with an inhibition halo of 5 ± 0.01 mm and the oxacillin with a halo of 1 ± 0.01 mm, and these values are within the sensitivity range according to the Minimum Inhibitory Concentration table. The samples from Tilulum had greater inhibition, followed by the Atocha sample and finally the El Socavón samples. In this study it was possible to determine antibacterial activity against the bacteria *Staphylococcus aureus* using the active principle of *Chlorella*, thus proving that the *Chlorella* microalga extract has activity against gram-positive bacteria. The results were similar to other obtained by other researchers [9, 17, 20].

4. Conclusion

The study was carried out with the purpose of determining the antibacterial power of the *Chlorella* microalgae. The growth kinetics was established for the identification of the exponential phase in which the samples were taken for extraction with the different solvents. The results show the antibacterial potential of the *Chlorella* microalgae extracted from the stacked waters of the ditches of various sectors of the city of Ambato-Ecuador. The results obtained from the antibiogram method show that the best treatment corresponds to the samples from the Tilulum area in which isopropyl alcohol was used for its extraction, however, with ethanol the inhibition halos did not have significant differences, whereas with the water solvent no inhibition was observed with both the sediment and the supernatant. The results of the antibacterial activity of the isolated microalgae show promising possibilities for obtaining secondary metabolites that could be exploited by the pharmaceutical industry.

5. References

[1] Pina-Pérez MC, Rivas A, Martínez A, Rodrigo D. Antimicrobial potential of macro and microalgae against pathogenic and spoilage microorganisms in food. Food Chemistry. 2017;235(Supplement C):34-44.

[2] Najdenski HM, Gigova LG, Iliev II, Pilarski PS, Lukavský J, Tsvetkova IV, et al. Antibacterial and antifungal activities of selected microalgae and cyanobacteria. International Journal of Food Science & Technology. 2013;48(7):1533-40.

[3] Pathak VV, Kothari R, Chopra AK, Singh DP. Experimental and kinetic studies for phycoremediation and dye removal by Chlorella pyrenoidosa from textile wastewater. Journal of Environmental Management. 2015;163:270-7.

[4] Manivannan K, Anantharaman P, Balasubramanian T. Evaluation of antioxidant properties of marine microalga Chlorella marina (Butcher, 1952). Asian Pacific Journal of Tropical Biomedicine. 2012;2(1):S342-S6.

[5] Aravantinou AF, Theodorakopoulos MA, Manariotis ID. Selection of microalgae for wastewater treatment and potential lipids production. Bioresource Technology. 2013;147(Supplement C):130-4.

[6] Wehr JD, Sheath RG, Kociolek JP. Freshwater algae of North America: ecology and classification: Elsevier; 2015.
[7] Ramanan R, Kang Z, Kim B-H, Cho D-H, Jin L, Oh H-M, et al. Phycosphere bacterial diversity in green algae reveals an apparent similarity across habitats. Algal Research. 2015;8:140-4.
[8] Tan X-B, Zhang Y-L, Yang L-B, Chu H-Q, Guo J. Outdoor cultures of Chlorella pyrenoidosa in the effluent of anaerobically digested activated sludge: The effects of pH and free ammonia. Bioresource Technology. 2016;200:606-15.
[9] Syed S, Arasu A, Ponnuswamy I. The uses of Chlorella vulgaris as antimicrobial agent and as a diet: the presence of bio-active compounds which caters the vitamins, minerals in general. Int J Biosci Biotechnol. 2015;7(1):185-90.
[10] Michalak I, Chojnacka K. Algae as production systems of bioactive compounds. Engineering in Life Sciences. 2015;15(2):160-76.
[11] Abedin RM, Taha HM. Antibacterial and antifungal activity of cyanobacteria and green microalgae. Evaluation of medium components by Plackett-Burman design for antimicrobial activity of Spirulina platensis. Global Journal of Biotechnology and Biochemistry. 2008;3(1):22-31.
[12] Abu-Ghannam N, Rajauria G. 8 - Antimicrobial activity of compounds isolated from algae A2 - Domínguez, Herminia. Functional Ingredients from Algae for Foods and Nutraceuticals: Woodhead Publishing; 2013. p. 287-306.
[13] M Sanghvi A, Martin Lo Y. Present and potential industrial applications of macro-and microalgae. Recent patents on food, nutrition & agriculture. 2010;2(3):187-94.
[14] Ibañez E, Herrero M, Mendiola JA, Castro-Puyana M. Extraction and characterization of bioactive compounds with health benefits from marine resources: macro and micro algae, cyanobacteria, and invertebrates. Marine Bioactive Compounds: Springer; 2012. p. 55-98.
[15] Varshney P, Mikulic P, Vonshak A, Beardall J, Wangikar PP. Extremophilic micro-algae and their potential contribution in biotechnology. Bioresource technology. 2015;184:363-72.
[16] Gallardo-Rodríguez JJ, López-Rosales L, Sánchez-Mirón A, García-Camacho F, Molina-Grima E, Chalmers JJ. New insights into shear-sensitivity in dinoflagellate microalgae. Bioresource Technology. 2016;200:699-705.
[17] Ghasemi Y, Moradian A, Mohagheghzadeh A, Shokravi S, Morowvat MH. Antifungal and antibacterial activity of the microalgae collected from paddy fields of Iran: characterization of antimicrobial activity of Chroococcus dispersus. 2007.
[18] Águila-Ramírez RN, Arenas-González A, Hernández-Guerrero CJ, González-Acosta B, Borges-Souza JM, Véron B, et al. Antimicrobial and antifouling activities achieved by extracts of seaweeds from Gulf of California, Mexico. Hidrobiológica. 2017;22(1):8-15.
[19] El Shafay SM, Ali SS, El-Sheekh MM. Antimicrobial activity of some seaweeds species from Red sea, against multidrug resistant bacteria. The Egyptian Journal of Aquatic Research. 2016;42(1):65-74.
[20] Kokou F, Makridis P, Kentouri M, Divanach P. Antibacterial activity in microalgae cultures. Aquaculture Research. 2012;43(10):1520-7.
[21] Pratt R, Daniels TC, Eiler JJ, Gunnison JB, Kumler WD, Oneto JF, et al. Chlorellin, an Antibacterial Substance from Chlorella. American Association for the Advancement of Science; 1944. p. 351-2.
[22] Yenkar AW. Bio-Diversity of fresh water algae of dahegaon gondi reservoir of Wardha district of Maharashtra, India. Bulletin of Pure & Applied Sciences-Botany. 2015.
[23] Asadi Y, Doudi M, Darki BZ. An In-Vitro Investigation of the Antibacterial Effects of the Acetone and Ethanol Extracts and the Supernatant of the Algae Chlorella vulgaris CCATM-210-1 on Some of the Gram-Negative Bacterial Foodborne Pathogens. International Journal Of Advanced Biotechnology And Research. 2016;7:278-87.
[24] Athbi AM. Atimicrobial Bioactive Compound Isolated From Cyanobacterium Nostoc linkia. Al-Qadisiyah Medical Journal. 2017;10(17):239-47.

[25] González González LM. Influence of nitrogen and phosphorus deficiency on competitive interactions between Chlorella vulgaris and Scenedesmus acutus: National University of Colombia; 2010.

[26] Gómez-Guzmán A, Jiménez-Magaña S, Guerra-Rentería AS, Gómez-Hermosillo C, Parra-Rodríguez FJ, Velázquez S, et al. Evaluation of nutrients removal (NO 3-N, NH 3-N and PO 4-P) with Chlorella vulgaris, Pseudomonas putida, Bacillus cereus and the consortium of these microorganisms in the treatment of wastewater effluents. Water Science and Technology. 2017:wst2017175.

[27] Tandon P, Jin Q. Microalgae culture enhancement through key microbial approaches. Renewable and Sustainable Energy Reviews. 2017;80(Supplement C):1089-99.

[28] Turnidge JD. Susceptibility test methods: general considerations. Manual of Clinical Microbiology, Eleventh Edition: American Society of Microbiology; 2015. p. 1246-52.

[29] Taylor WI, Achanzar D. Catalase test as an aid to the identification of Enterobacteriaceae. Applied microbiology. 1972;24(1):58-61.

[30] Harrigan WF, McCance ME. Laboratory methods in microbiology: Academic press; 2014.

[31] Hemraj V, Diksha S, Avneet G. A review on commonly used biochemical test for bacteria. Innovare Journal of Life Science. 2013;1(1):1-7.

[32] Sperber Wz, Tatini S. Interpretation of the tube coagulase test for identification of Staphylococcus aureus. Applied microbiology. 1975;29(4):502-5.

[33] Andrews J. Susceptibility testing and antibiotic assay. Medical Microbiology. 2014.

[34] Jorgensen JH, Turnidge JD. Susceptibility test methods: dilution and disk diffusion methods. Manual of Clinical Microbiology, Eleventh Edition: American Society of Microbiology; 2015. p. 1253-73.

[35] Adhoni SA, Shivasharana C, Kaliwal B. Identification and characterisation of chlorella vulgaris for biodiesel production. Int J Scient Res Eng Stud. 2016;3(1).

[36] Singh SP, Singh P. Effect of temperature and light on the growth of algae species: A review. Renewable and Sustainable Energy Reviews. 2015;50(Supplement C):431-44.

[37] Bradley P, Gordon NC, Walker TM, Dunn L, Heys S, Huang B, et al. Rapid antibiotic-resistance predictions from genome sequence data for Staphylococcus aureus and Mycobacterium tuberculosis. Nature communications. 2015;6:10063.

[38] Kloos WE, Bannerman TL. Update on clinical significance of coagulase-negative staphylococci. Clinical microbiology reviews. 1994;7(1):117-40.

[39] Schlegelová J, Dendis M, Benedík J, Babák V, Ryšánek D. Staphylococcus aureus isolates from dairy cows and humans on a farm differ in coagulase genotype. Veterinary Microbiology. 2003;92(4):327-34.

Institute of Physics
Dirac House, Temple Back
Bristol BS1 6BE UK

ISSN: 1755-1307
ISBN 978-1-5108-6394-1

9 781510 863941

CURRAN ASSOCIATES INC.
proceedings
.com

VIII International Scientific Conference "Issues of Physics and Technology in Science, Industry and Medicine"

IOP Conference Series: Materials Science and Engineering
Volume 135

Tomsk, Russia
1 – 3 June 2016

IOP
Institute of Physics